Studies in Fuzziness and Soft Computing

Volume 321

About this Series

The series "Studies in Fuzziness and Soft Computing" contains publications on various topics in the area of soft computing, which include fuzzy sets, rough sets, neural networks, evolutionary computation, probabilistic and evidential reasoning, multi-valued logic, and related fields. The publications within "Studies in Fuzziness and Soft Computing" are primarily monographs and edited volumes. They cover significant recent developments in the field, both of a foundational and applicable character. An important feature of the series is its short publication time and world-wide distribution. This permits a rapid and broad dissemination of research results.

More information about this series at http://www.springer.com/series/2941

Bijan Davvaz · Irina Cristea

Fuzzy Algebraic Hyperstructures

An Introduction

Bijan Davvaz
Department of Mathematics
Yazd University
Yazd
Iran

Irina Cristea
University of Nova Gorica CSIT
Nova Gorica
Slovenia

ISSN 1434-9922 ISSN 1860-0808 (electronic)
Studies in Fuzziness and Soft Computing
ISBN 978-3-319-38653-9 ISBN 978-3-319-14762-8 (eBook)
DOI 10.1007/978-3-319-14762-8

Springer Cham Heidelberg New York Dordrecht London

Softcover reprint of the hardcover 1st edition 2015

Printed on acid-free paper

Springer International Publishing AG Switzerland is part of Springer Science+Business Media
(www.springer.com)

Preface

Algebraic hyperstructures represent a natural generalization of classical algebraic structures. In a classical algebraic structure, the composition of two elements is an element, while in an algebraic hyperstructure, the result of this composition is a set. Hyperstructure theory was born in 1934, when Marty, a French mathematician, at the 8th Congress of Scandinavian Mathematicians gave the definition of hypergroup and illustrated some of their applications, with utility in the study of groups, algebraic functions and rational fractions. The first example of hypergroups, which motivated the introduction of these new algebraic structures, was the quotient of a group by any, not necessary normal, subgroup. More exactly, if the subgroup is not normal, then the quotient is not a group, but it is always a hypergroup with respect to a certain hyperoperation.

The fuzzy mathematics forms a branch of mathematics related to fuzzy set theory and fuzzy logic. It started in 1965 after the publication of Zadeh's seminal work [202], entitled none other than: Fuzzy Sets. In the classical set theory, the membership of an element to a set is assessed in binary terms according to a bivalent condition: an element either belongs or not to the set. By contrast, fuzzy set theory permits the gradual assessment of the membership of elements in a set; this is described with the help of a membership function valued in the real interval $[0,1]$. In 1971, Rosenfeld [160] introduced the fuzzy sets in the context of group theory and formulated the concept of a fuzzy subgroup of a group, concept redefined eight years later by Anthony and Sherwood [6]. Since then, many researchers are engaged in extending the concepts of abstract algebra to the framework of the fuzzy setting. There is a considerable amount of work on the association between fuzzy sets and hyperstructures. This work can be classified into three groups. A first group of works studies crisp hyperoperations defined through fuzzy sets. This study was initiated by Corsini, Leoreanu, Davvaz, Cristea, and others. A second group of works concerns the fuzzy hyperalgebras. This is a direct extension of the concept of fuzzy algebras (fuzzy (sub)groups, fuzzy rings, and so on). This approach can be extended to fuzzy polygroups, hypergroups and hyperrings. For example, given a crisp hypergroup $(H,\star)$ and a fuzzy subset μ, then we say that μ is a fuzzy subhypergroup of $(H,\star)$ if every t-cut of μ, denoted μ_t, is a (crisp) subhypergroup of $(H,\star)$. This idea was proposed by Zahedi and his group and later on by Davvaz. A third group deals also with fuzzy hyperstructures, but with a completely different approach. This was studied by Corsini, Kehagias, Tofan, Zahedi and others. The basic idea is the following one: a crisp hyperoperation assigns to every pair of elements a crisp set; a fuzzy hyperoperation assigns to every pair of elements a fuzzy set. This line of research was continuated by Sen, Leoreanu-Fotea, Davvaz, Zhan, etc.

The reader of this book is assumed to have the knowledge of algebraic structures and hyperstructures, presented in the previous five books written on this topic of higher algebra. The first one, "Prolegomena of hypergroup theory" [27], appeared in 1993, where Corsini included the basic notions and results related to hypergroups and obtained in the first five-six decades of this relatively new theory. As its title indicates, this text should be the first book that one reads to become familiar with the theory and to acquire the terminology, notation, and the fundamentals of algebraic hyperstructures. The second book [185], written by Vougiouklis and published in 1994, is dedicated to the study of the representations of the hyperstructures, when the author introduced the H_v-structures as an extension of the classical hyperstructures, obtained by substituting the associativity, commutativity or the distributivity by the corresponding weak axioms (in which the inclusion is considered instead of the equality). The third book [47] contains several applications of hyperstructure theory, where the authors, Corsini and Leoreanu, presented connections of hyperstructures with binary relations, geometry, graphs and hypergraphs, fuzzy sets, rough sets, automata theory, probabilities, etc. The theory of hyperrings is included in the fourth book [83], edited in 2007, and proposed by Davvaz and Leoreanu-Fotea. The last published book [66], appeared in 2013 and written by B. Davvaz, is dedicated to the theory of polygroups and their applications. The present volume is the sixth one in the series of those that deal with hyperstructures, but the first one to be entirely devoted to fuzzy algebraic hyperstructures, despite the fact that fuzzy algebraic hyperstructures have been studied for over twenty years.

One reason we wanted to write this book is that there are a number of topics on fuzzy hyperstructures we consider it is useful to have in a single reference source, most of them being part of our past and actual research. We can divide these topics into three main groups. The first one is represented by the basic fuzzy hyperstructures, as fuzzy polygroups, fuzzy hypergroups and fuzzy hyperrings. The second group contains fuzzy H_v-substructures of H_v-structures, in particular fuzzy H_v-subgroups of H_v-groups and fuzzy H_v-ideals of H_v-rings. In the third group of topics we include some of the latest connections between hypergroups and fuzzy sets, insistead on those that lead to the constructions of new join spaces or L-fuzzy join spaces.

The present book is composed by five chapters. For the sake of completeness, the first chapter contains a review of basic notions of algebraic structures and hyperstructures, that we will use in the others chapters. Many readers, that are already familiar with these theories, may wish to skip them or to regard them as a survey. The second chapter regards fuzzy sets, fuzzy groups and fuzzy polygroups. The following two chapters deal with fuzzy H_v-structures. In particular, in the third chapter, the concept of fuzzy H_v-subgroup of H_v-groups is presented while in the fourth chapter, the concept of fuzzy H_v-ideal of H_v-rings is studied. We decided to consider H_v-groups and H_v-rings because these hyperstructures are the most general form of hypergroups and hyperrings. In the last chapter, we present several connections between hypergroups and fuzzy sets: first, join spaces obtained from fuzzy sets or L-fuzzy sets, then L-fuzzy join spaces. Next we introduce a sequence of join spaces and fuzzy sets associated with a hypergroup and study its properties for different types of hypergroups, like the complete hypergroups and i.p.s. hypergroups. More over we extend this construction to the case of hypergraphs. We conclude Chapter 5 with the study of hypergroupoids connected with fuzzy sets endowed with two membership functions. This book is meant to be the basis in a course of Hyperstructure Theory for graduated students, together with the previous five texts written on this topic.

Many sincere and warm thanks go to our families, for their constant support, while we were engaged on this project.

Bijan Davvaz,
Department of Mathematics,
Yazd University, Yazd, Iran

Irina Cristea,
Center for Systems and Information Technologies,
University of Nova Gorica, Slovenia

Contents

1 **Fundamentals of Algebraic (Hyper)Structures** 1
1.1 Groups 1
1.2 Rings 8
1.3 Polygroups 16
1.4 Hypergroups 22
1.5 Hyperrings 28
1.6 H_v-Structures 34

2 **Fuzzy Polygroups** 39
2.1 What Fuzzy Sets Are? 39
2.2 Groups and Fuzzy Subgroups 42
2.3 Polygroups and Fuzzy Subpolygroups 47
2.4 Generalized Fuzzy Subpolygroups 61
2.5 Roughness in Polygroups 74
2.6 F-Polygroups 79

3 **Fuzzy H_v-Subgroups of H_v-Groups** 91
3.1 Fuzzy H_v-Subgroups and Fuzzy Homomorphisms 91
3.2 Generalized Fuzzy H_v-Subgroups 101
3.3 Intuitionistic Fuzzy H_v-Subgroups 106
3.4 T-Product of Fuzzy H_v-Subgroups 113
3.5 Probabilistic Fuzzy Semihypergroups 121
3.6 F-Hypergroups 123

4 **H_v-Rings (Hyperrings) and H_v-Ideals** 127
4.1 Fuzzy H_v-Ideals 127
4.2 Generalized Fuzzy H_v-Ideals 135
4.3 Interval-Valued Fuzzy H_v-Ideals 143
4.4 Fuzzy Hyperrings 146

5 **Connections between Hypergroups and Fuzzy Sets** 159
5.1 Join Spaces and Fuzzy Sets 159
5.1.1 Constructions Using Fuzzy Sets 159
5.1.2 Generalizations to L-Fuzzy Sets 161
5.2 L-Fuzzy Join Spaces 167
5.3 Fuzzy Grade of a Hypergroupoid 175

5.3.1 The Main Construction 175
5.3.2 Intuitionistic Fuzzy Grade of a Hypergroupoid 189
5.3.3 Some Examples: Complete Hypergroups and i.p.s. Hypergroups 208
5.3.4 Extensions to Hypergraphs 218
5.4 Hypergroupoids and Fuzzy Sets Endowed with Two Membership Functions 222

Bibliography **229**

Index **239**

Chapter 1
Fundamentals of Algebraic (Hyper)Structures

1.1 Groups

If G is a non-empty set, a *binary operation* on G is a function from $G \times G$ to G. There are several commonly used notations for the image of (a, b) under a binary operation: ab (multiplicative notation), $a + b$ (additive notation), $a \cdot b$, $a * b$, etc.

Definition 1.1.1. A *semigroup* is a non-empty set G together with a binary operation on G which is

(1) associative $a(bc) = (ab)c$, for all $a, b, c \in G$.

A *monoid* is a semigroup G which contains a

(2) (two-sided) identity (neutral) element $e \in G$ such that $ae = ea = a$, for all $a \in G$.

A *group* is a monoid G such that

(3) for every $a \in G$ there exists a (two-sided) inverse element a^{-1} such that $a^{-1}a = aa^{-1} = e$.

A group G is said to be *abelian* or *commutative* if its binary operation is commutative.

The order of a group G is the number of elements in the group. It is denoted by $|G|$. If $|G|$ is finite, then the group is called a *finite group*. Otherwise, the group is called an *infinite group*.

EXAMPLE 1 (1) The set of integers $\mathbb{Z}$, the set of rational numbers $\mathbb{Q}$ and the set of real numbers $\mathbb{R}$ are all groups under ordinary addition.

(2) The set $\mathbb{Z}_n = \{0, 1, \ldots, n-1\}$ for $n \geq 1$ is a group under addition modulo n. For any i in $\mathbb{Z}_n$, the inverse of i is $n - i$. This group usually referred to as the *group of integers modulo n*.

(3) Let U_n denote the set of units in $\mathbb{Z}_n$; i.e.,

$$U_n = \{a \in \mathbb{Z}_n \mid \exists b \in \mathbb{Z}_n \text{ such that } ab = 1\}.$$

Then, U_n is an abelian group under multiplication.

B. Davvaz and I. Cristea, *Fuzzy Algebraic Hyperstructures*,
Studies in Fuzziness and Soft Computing 321, DOI: 10.1007/978-3-319-14762-8_1

(4) The subset $\{1, -1, i, -i\}$ of the complex numbers is a group under complex multiplication.
(5) Let $G = \{e, a, b, c\}$ with multiplication as defined by the table below:

$\cdot$	e	a	b	c
e	e	a	b	c
a	a	e	c	b
b	b	c	e	a
c	c	b	a	e

Then, G is an abelian group. This group is often referred to as the *Klein 4-group.*
(6) For a positive integer n, consider the set $C_n = \{a^0, a^1, \ldots, a^{n-1}\}$. On C_n define a binary operation as follows:

$$a^l a^m = \begin{cases} a^{l+m} & \text{if } l + m < n \\ a^{(l+m)-n} & \text{if } l + m \geq n. \end{cases}$$

For every positive integer n, C_n is an abelian group. The group C_n is called the *cyclic group* of order n.
(7) Let X be a non-empty set and S_X the set of all bijections $X \to X$. Under the operation of composition of functions, S_X is a group. The elements of S_X are called *permutations* and S_X is called the *group of permutations* on the set X. If $X = \{1, 2, \ldots, n\}$, then S_X is called the *symmetric group* on n letters and denoted by S_n. Since an element α of S_n is a function on the finite set $X = \{1, 2, \ldots, n\}$, it can be described by listing the elements of X on a line and the image of each element under α directly below it as follows:

$$\begin{pmatrix} 1 & 2 & 3 & \ldots & n \\ i_1 & i_2 & i_3 & \ldots & i_n \end{pmatrix}.$$

(8) The set of all $n \times n$ non-singular (determinant is non-zero) matrices with real entries and with operation matrix multiplication is a group. This group is called the *general linear group* and is denoted by $GL(n, \mathbb{R})$.
(9) The *dihedral group* D_n $(n \geq 3)$ is the group of symmetries of a regular n-sided polygon.
(10) The *quaternion group* is a non-abelian group of order 8. It is often denoted by Q or Q_8 and written in multiplicative form, with the following 8 elements

$$Q = \{1, -1, i, -i, j, -j, k, -k\}.$$

Here 1 is the identity element, $(-1)^2 = 1$ and $(-1)a = a(-1) = -a$ for all a in Q. The remaining multiplication rules can be obtained from the following relation:

$$i^2 = j^2 = k^2 = ijk = -1.$$

(11) Let G_1, G_2 be two groups. The *Cartesian product* of G_1, G_2 is the set $G_1 \times G_2 = \{(x_1, x_2) \mid x_1 \in G_1, x_2 \in G_2\}$. We define a binary operation on $G_1 \times G_2$ by $(x_1, x_2)(y_1, y_2) = (x_1y_1, x_2y_2)$. This is called *multiplication by components*. Then, $G_1 \times G_2$ is a group under the binary operation multiplication by components. This group is called the *direct product* of $G_1 \times G_2$.

Theorem 1.1.2. *Let G be a group.*

(1) *If $xa = xb$ or $ax = bx$, then $a = b$.*
(2) *The identity element e is unique.*
(3) *For all $x \in G$, the inverse element x^{-1} is unique.*
(4) *For all $x \in G$, we have $(x^{-1})^{-1} = x$.*
(5) *For all $a, b \in G$, we have $(ab)^{-1} = b^{-1}a^{-1}$.*

Proof. It is straightforward. ■

Theorem 1.1.3. *Let G be a semigroup with an element e such that*

(1) *$ea = a$, for all $a \in G$.*
(2) *For each $a \in G$ there is an element $a' \in G$ with $a'a = e$.*

Then, G is a group.

Proof. We claim that if $xx = x$, then $x = e$. There is an element $x' \in G$ with $x'x = e$ and $x'(xx) = x'x = e$. On the other hand, $x'(xx) = (x'x)x = ex = x$. Therefore, $x = e$.

If $a'a = e$, we show that $aa' = e$. Now, $(aa')(aa') = a((a'a)a') = a(ea') = aa'$ and so our claim gives $aa' = e$.

If $a \in G$, we must show that $ae = a$. Choose $a' \in G$ with $a'a = e = aa'$. Then, $ae = a(a'a) = (aa')a = ea = a$, as desired. ■

Theorem 1.1.4. *A non-empty set G together with a binary operation is a group if and only if*

(1) *$a(bc) = (ab)c$, for all $a, b, c \in G$.*
(2) *For any $a, b \in G$, the equations $ax = b$ and $ya = b$ have solutions in G.*

Proof. Suppose that G is a group. By the definition of a group, (1) holds in G. Since $a(a^{-1}b) = (aa^{-1})b = eb = b$, $x = a^{-1}b$ is a solution of the equation $ax = b$ in G. Similarly, $y = ba^{-1}$ is a solution of the equation $ya = b$ in G.

Conversely, let G be a non-empty set with a binary operation satisfying (1) and (2). In order to show that G is a group, we only need to show that G satisfies the conditions of Theorem 1.1.3. Consider $a_0 \in G$. Then, by (2), the equation $xa_0 = a_0$ has a solution in G. Thus, there is an element $e \in G$ such that $ea_0 = a_0$. For every $a \in G$, by (2), there is $y \in G$ such that $a = a_0y$. Then,

$$ea = e(a_0y) = (ea_0)y = a_0y = a.$$

So, the first condition of Theorem 1.1.3 holds. Now, for $a \in G$, by (2), there is $a' \in G$ such that $a'a = e$. Hence, the second condition of Theorem 1.1.3 holds, too. ■

Definition 1.1.5. A non-empty subset A of a group G is said to be a *subgroup* of G if, under the operation in G, A itself forms a group.

Lemma 1.1.6. *A non-empty subset A of the group G is a subgroup if and only if*

(1) *$a, b \in A$ implies that $ab \in A$.*
(2) *$a \in A$ implies that $a^{-1} \in A$.*

Proof. It is straightforward. ■

Lemma 1.1.7. *A non-empty subset A of the group G is a subgroup if and only if $a, b \in A$ implies $ab^{-1} \in A$.*

Proof. Suppose that A is a subgroup of G and $a, b \in A$. By Lemma 1.1.6 (2), $b^{-1} \in A$. Consequently, by the closure property of A, $ab^{-1} \in A$.

Conversely, let A be a non-empty subset of G such that for all $a, b \in A$, $ab^{-1} \in A$. Since A is non-empty, there exists $x \in A$. Thus, $xx^{-1} \in A$, i.e., $e \in A$. Let $b \in A$ be an arbitrary element. As $e \in A$, $eb^{-1} = b^{-1} \in A$. Now, let $a, b \in A$ be arbitrary elements. Then, $a, b^{-1} \in A$. Hence, $a(b^{-1})^{-1} \in A$. This implies that $ab \in A$. So, the conditions of Lemma 1.1.6 hold. ■

Corollary 1.1.8. *If G is a group and $\{H_i \mid i \in I\}$ be a non-empty family of subgroups of G, then $\bigcap_{i \in I} H_i$ is a subgroup of G.*

EXAMPLE 2 (1) It is obvious that $\{e\}$ and G are always subgroups of a group G. Group G is called *improper subgroup* of G.
(2) Let G be the group of all real numbers under addition and let H be the set of all integers. Then, H is a subgroup of G.
(3) Let G be the group of all non-zero real numbers under multiplication and let H be the subset of positive rational numbers. Then, H is a subgroup of G.
(4) Let G be the group of all non-zero complex numbers under multiplication and let $H = \{a + bi \mid a^2 + b^2 = 1\}$. Then, H is a subgroup of G.
(5) In $\mathbb{Z}_6$, both $\{0, 3\}$ and $\{0, 2, 4\}$ are subgroups under addition. If p is prime, $\mathbb{Z}_p$ has no proper subgroup.
(6) Consider the general linear group of $n \times n$ matrices over R under multiplication. The subset of matrices that have determinant 1 forms a subgroup.
(7) Let G be a group and $\{H_i \mid i \in I\}$ is a non-empty family of subgroups of G which contain X. Then, $\bigcap_{i \in I} H_i$ is called the subgroup of G *generated by the set* X and denoted by $\langle X \rangle$.
(8) The *center* of a group G, denoted $Z(G)$, is the set of elements that commute with every element of G. The center is a subgroup of G.

Definition 1.1.9. Let A be a subgroup of a group G and $a, b \in G$. Then, a is *right congruence to b modulo A*, if $ab^{-1} \in A$; a is *left congruence to b modulo A*, if $a^{-1}b \in A$.

It is easy to see that the right (respectively, left) congruence modulo A is an equivalence relation on G. The equivalence class of $x \in G$ under the right (respectively, left) equivalence modulo A is the set $Ax = \{ax \mid a \in A\}$ (respectively, $xA = \{xa \mid a \in A\}$. The set Ax is called a *right coset* of A in G and xA is called a *left coset* of A in G.

Theorem 1.1.10. *If N is a subgroup of G, then the following conditions are equivalent:*

(1) *Left and right congruence modulo N define the same equivalence relation on G.*
(2) *Every left coset of N in G is a right coset of N in G.*
(3) *$gN = Ng$, for all $g \in G$.*
(4) *For all $g \in G$, $gNg^{-1} \subseteq N$, where $gNg^{-1} = \{gng^{-1} \mid n \in N\}$.*
(5) *For all $g \in G$, $gNg^{-1} = N$.*

Proof. We prove (4⇒5). We have $gNg^{-1} \subseteq N$. Since (4) also holds for $g^{-1} \in G$, $g^{-1}N(g^{-1})^{-1} \subseteq N$ and so $g^{-1}Ng \subseteq N$. Therefore, for every $n \in N$, we have $n = g(g^{-1}ng)g^{-1} \in gNg^{-1}$ and $N \subseteq gNg^{-1}$. ■

Definition 1.1.11. A subgroup N of G is said to be a *normal subgroup* of G if for every $g \in G$ and $n \in N$, $gng^{-1} \subseteq N$.

Example 3 (1) Every subgroup of an abelian group is normal.
(2) The center $Z(G)$ of a group G is always normal.
(3) The subgroup of rotations in D_n is normal in D_n.
(4) All subgroups of the quaternion group are normal.

Let G be a group and N be a subgroup of G. Denote by G/N the set of distinct (left) cosets with respect to N. In other words, we list all the cosets of the form gN (with $g \in G$) without repetitions and consider each coset as a single element of the newly formed set G/N. Now, we would like to define a binary operation $\star$ on G/N such that $(G/N, \star)$ is a group. It is natural to try to define the operation $\star$ by the formula

$$gN \star kN = gkN, \text{ for all } g, k \in G. \tag{1.1}$$

Before checking the group axioms, we need to find out whether $\star$ is at least well defined. Our first result shows that $\star$ is well defined whenever N is a normal subgroup.

Theorem 1.1.12. *Let G be a group and N be a normal subgroup of G. Then, the operation $\star$ given by (1.1) is well defined.*

Proof. Suppose that $g_1, g_2, k_1, k_2 \in G$ such that $g_1N = g_2N$ and $k_1N = k_2N$. We show that $g_1k_1N = g_2k_2N$. Thus, we need to show that the following implication: $g_1^{-1}g_2 \in N$ and $k_1^{-1}k_2 \in N$ imply that $(g_1k_1)^{-1}g_2k_2 \in N$.

Since $g_1^{-1}g_2 \in N$ and $k_1^{-1}k_2 \in N$, there exist $n, n' \in N$ such that $g_1^{-1}g_2 = n$ and $k_1^{-1}k_2 = n'$. Then, $k_2 = k_1n'$. Now, we obtain

$$(g_1k_1)^{-1}g_2k_2 = k_1^{-1}g_1^{-1}g_2k_2 = k_1^{-1}nk_1n' = (k_1^{-1}nk_1)n'.$$

Since N is a normal subgroup, $k_1^{-1}nk_1 \in N$ and so $(k_1^{-1}nk_1)n' \in N$. Therefore, $(g_1k_1)^{-1}g_2k_2 \in N$ and this completes the proof. ■

The converse of Theorem 1.1.12 is also true, i.e., if the operation $\star$ on G/N is well defined, then N must be normal.

Theorem 1.1.13. *Let G be a group and N be a normal subgroup of G. Then, the quotient set G/N is a group with respect to the operation $\star$ defined by (1.1).*

Proof. Clearly, G/N is closed under $\star$ by the definition of cosets. Associativity of $\star$ follows from the associativity of operation of G. Indeed, for any $g, h, k \in G$, we have

$$\begin{aligned} gN \star (hN \star kN) &= gN \star hkN = (g(hk))N = ((gh)k)N \\ &= ghN \star kN = (gN \star hN) \star kN. \end{aligned}$$

The identity element of G/N is the special coset $N = eN$. Finally, the inverse of a coset gN is the coset $g^{-1}N$. This is because $gN \star g^{-1}N = (gg^{-1})N = eN = N$ and similarly $g^{-1}N \star gN = N$. ■

G/N is called the *quotient group* or *factor group* of G by N. Usually, we write the product in G/N as $gN \cdot kN$ (or even $gNkN$), instead of $gN \star kN$.

Now, we turn to the discussion of functions from one group to another, which "respect" the group structure. These functions are called homomorphisms.

Definition 1.1.14. Let G and G' be groups and $f : G \to G'$ be a function. Then, f is called a *homomorphism* if

$$f(xy) = f(x)f(y), \tag{1.2}$$

for all $x, y \in G$.

Note that in the left side of (1.2), the product xy is taken in G, whereas in the right side the product $f(x)f(y)$ is taken in G'.

An *isomorphism* is a homomorphism that is also a bijection. We say that G is isomorphic to G', denoted by $G \cong G'$, if there exists an isomorphism $f : G \to G'$. An *endomorphism* of a group G is a homomorphism $f : G \to G$.

EXAMPLE 4 (1) Consider the groups $\mathbb{C}^*$ and $\mathbb{R}^+$ under multiplication and the map $f : \mathbb{C}^* \to \mathbb{R}^+$ given by $f(z) = |z|$. Since $|z_1z_2| = |z_1||z_2|$, the equation (1.2) is satisfied and f is a homomorphism.

(2) Consider $\mathbb{R}$ under addition and the group U of complex numbers z with $|z| = 1$ under multiplication. Let $f : \mathbb{R} \to U$ be the map $f(z) = e^{i2\pi x}$. Since

$$f(x+y) = e^{i2\pi(x+y)} = e^{i2\pi x}e^{i2\pi y} = f(x)f(y),$$

f is a homomorphism.

(3) The determinant mapping $det : GL(n, \mathbb{R}) \to \mathbb{R}\backslash\{0\}$ is a homomorphism from $GL(n, \mathbb{R})$ into the group of non-zero real numbers under multiplication, that is for $detAB = (detA)(detB)$, for all $A, B \in GL(n, \mathbb{R})$.

Theorem 1.1.15. *Let G and G' be groups, and let $f : G \to G'$ be a homomorphism. Then,*

(1) *$f(e) = e'$, where e' is the identity in G'.*
(2) *If $a \in G$, then $f(a^{-1}) = f(a)^{-1}$.*
(3) *If $a \in G$ and $n \in \mathbb{Z}$, then $f(a^n) = f(a)^n$.*
(4) *Let A be a subgroup of G. The image $f(A) = \{f(a) \mid a \in A\}$ is a subgroup of G'. For a subgroup B of G', the inverse image*

$$f^{-1}(B) = \{x \in G \mid f(x) \in B\}$$

is a subgroup of G.

Proof. The proofs of these properties of homomorphisms are straightforward and are left as exercises. ■

Definition 1.1.16. Let f be a homomorphism of a group G into a group G' and e' be the identity in G'. The set

$$f^{-1}(e') = \{x \in G \mid f(x) = e'\}$$

is called the *kernel* of f, denoted by $kerf$.

Lemma 1.1.17. *If f is a homomorphism of a group G into a group G', then $kerf$ is a normal subgroup of G.*

Proof. The proof is straightforward. ■

Corollary 1.1.18. *A homomorphism f of G into G' is one to one if and only if $kerf = \{e\}$.*

Now, suppose that two groups G, G' and a homomorphism $f : G \to G'$ are given. By Lemma 1.1.17, $kerf$ is a normal subgroup of G, and thus we can consider the quotient group $G/kerf$. The next theorem, called the *fundamental theorem of homomorphisms* asserts that $G/kerf$ is always isomorphic to the range group $f(G)$.

Theorem 1.1.19. *Let G, G' be groups and $f : G \to G'$ be a homomorphism. Then, $G/kerf \cong f(G)$.*

Proof. Suppose that $K = kerf$ and define the map $\phi : G/K \to f(G)$ by $\phi(gK) = f(g)$, for all $g \in G$. We claim that ϕ is well defined mapping and that ϕ is an isomorphism.

Suppose that $g_1K = g_2K$ for some $g_1, g_2 \in G$. Then, $g_1^{-1}g_2 \in K$. So, $f(g_1^{-1}g_2) = e'$ which implies that $f(g_1^{-1})f(g_2) = e'$. Thus, $f(g_1) = f(g_2)$ and so $\phi(g_1K) = \phi(g_2K)$. This shows that ϕ is well defined.

For any $g_1, g_2 \in G$, we have

$$\phi(g_1K \cdot g_2K) = \phi(g_1g_2K) = f(g_1g_2) = f(g_1)f(g_2) = \phi(g_1K)\phi(g_2K).$$

Thus, ϕ is a homomorphism. Now, if the coset gK lies in the kernel of ϕ, then $e' = \phi(gK) = f(g)$, so $g \in K$. Therefore, we constructed an isomorphism $\phi : G/K \to f(G)$, and thus $G/kerf$ is isomorphic to $f(G)$. ■

Theorem 1.1.20. *Let G be a group, K be a subgroup of G and N be a normal subgroup of G. Then, $K/(K \cap N) \cong KN/N$.*

Proof. It is easy to see that $K \cap N$ is a normal subgroup of K and N is a normal subgroup of KN. We define the function $f : K \to KN/N$ by $f(k) = kN$, for all $k \in K$. We have $f(k_1k_2) = k_1k_2N = k_1N \cdot k_2N = f(k_1)f(k_2)$, for all $k_1, k_2 \in K$. So, f is a homomorphism. Also, we obtain

$$\begin{aligned} kerf &= \{k \in K \mid f(k) = N\} = \{k \in K \mid kN = N\} \\ &= \{k \in K \mid k \in N\} = K \cap N. \end{aligned}$$

Now, the result follows from fundamental theorem of homomorphism. ■

Theorem 1.1.21. *Let G be a group, K, N be normal subgroups of G and $N \subseteq K$. Then, $(G/N)/(K/N) \cong G/K$.*

Proof. Again we let the fundamental theorem of homomorphism do the main work. We define $f : G/N \to G/K$ by $f(gN) = gK$. The reader may check easily that f is an onto homomorphism with kernel K/N. ■

Theorems 1.1.19, 1.1.20 and 1.1.21 are called the *first*, *second* and *third isomorphism theorems*, respectively.

1.2 Rings

In this section, the definition of a ring and some of fundamental properties are presented. A ring is an algebraic system with two binary operations and these operations are usually called *addition* and *multiplication.*

Definition 1.2.1. A non-empty set R is said to be a *ring* if in R there are defined two binary operations, denoted by $+$ and $\cdot$ respectively, such that for all a, b, c in R:

(1) $a + b = b + a$,
(2) $(a + b) + c = a + (b + c)$,
(3) there is an element 0 in R such that $a + 0 = a$,
(4) there exists an element $-a$ in R such that $a + (-a) = 0$,
(5) $(a \cdot b) \cdot c = a \cdot (b \cdot c)$,
(6) $\cdot$ is distributive with respect to $+$, i.e., $a \cdot (b + c) = a \cdot b + a \cdot c$ and $(a + b) \cdot c = a \cdot c + b \cdot c$.

Axioms 1 through 4 merely state that R is an *abelian group* under the operation $+$. The additive identity of a ring is called the *zero element.* If $a \in R$ and $n \in \mathbb{Z}$, then na has its usual meaning for additive groups.

If in addition:

(7) $a \cdot b = b \cdot a$, for all a, b in R,

then R is said to be a *commutative ring.* If R contains an element 1 such that

(8) $1 \cdot a = a \cdot 1 = a$, for all a in R,

then R is said to be a *ring with unit element.*

If R is a system with unit satisfying all the axioms of a ring except possibly $a + b = b + a$ for all $a, b \in R$, then one can show that R is a ring.

For any two elements a, b of a ring R, we shall denote $a + (-b)$ by $a - b$ and for convenience sake we shall usually write ab instead of $a \cdot b$.

EXAMPLE 5 (1) Each of the number sets $\mathbb{Z}$, $\mathbb{Q}$, $\mathbb{R}$ and $\mathbb{C}$ forms a ring with respect to the addition and multiplication.

(2) For every $m \in \mathbb{Z}$, $\{ma \mid a \in \mathbb{Z}\}$ forms a ring with respect to the ordinary addition and multiplication.

(3) The set $\mathbb{Z}_n$ is a ring with respect to the addition and multiplication modulo n.

(4) Let $\mathbb{Z}[i]$ denote the set of all complex numbers of the form $a + bi$, where a and b are integers. Under the usual addition and multiplication of complex numbers, $\mathbb{Z}[i]$ forms a ring called the *ring of Gaussian integers.*

(5) The set of all continuous real-valued functions defined on the interval $[a, b]$ forms a ring, the operations are addition and multiplication of functions.

(6) One of the smallest non-commutative rings is the *Klein* 4-*ring* $(R, +, \cdot)$, where $(R, +)$ is the Klein 4-group $\{0, a, b, c\}$ with 0 the neutral element and the binary operation $\cdot$ given by the following table:

$\cdot$	0	a	b	c
0	0	0	0	0
a	0	a	0	a
b	0	b	0	b
c	0	c	0	c

(7) The set $M_n(\mathbb{R})$ of all $n \times n$ matrices with entries from $\mathbb{R}$ forms a ring with respect to the usual addition and multiplication of matrices. In fact, given an arbitrary ring R, one can consider the ring $M_n(R)$ of $n \times n$ matrices with entries from R.

(8) If G is an abelian group, then $End(G)$, the set of endomorphisms of G, forms a ring, the operations in this ring are the addition and composition of endomorphisms.

(9) This last example is often called the *ring of real quaternions.* Let Q be the set of all symbols $a_0 + a_1 i + a_2 j + a_3 k$, where all the numbers a_0, a_1, a_2

and a_3 are real numbers. Define the equality between two elements of Q as follows: $a_0 + a_1i + a_2j + a_3k = b_0 + b_1i + b_2j + b_3k$ if and only if $a_0 = b_0$, $a_1 = b_1$, $a_2 = b_2$ and $a_3 = b_3$. We define the addition and multiplication on Q by

$$\begin{aligned}
&(a_0 + a_1i + a_2j + a_3k) + (b_0 + b_1i + b_2j + b_3k) \\
&\quad = (a_0 + b_0) + (a_1 + b_1)i + (a_2 + b_2)j + (a_3 + b_3)k, \\
&(a_0 + a_1i + a_2j + a_3k) \cdot (b_0 + b_1i + b_2j + b_3k) \\
&\quad = (a_0b_0 - a_1b_1 - a_2b_2 - a_3b_3) + (a_0b_1 + a_1b_0 + a_2b_3 - a_3b_2)i \\
&\quad +(a_0b_2 + a_2b_0 + a_3b_1 - a_1b_3)j + (a_0b_3 + a_3b_0 + a_1b_2 - a_2b_1)k.
\end{aligned}$$

It is easy to see that Q is a noncommutative ring in which $0 = 0 + 0i + 0j + 0k$ and $1 = 1 + 0i + 0j + 0k$ are the zero and unit elements respectively. Note that the set $\{1, -1, i, -i, j, -j, k, -k\}$ forms a non-abelian group of order 8 under this product.

Lemma 1.2.2. *Let R be a ring.*

(1) *Since a ring is an abelian group under $+$, there are certain things we know from the group theory background, for instance, $-(-a) = a$ and $-(a + b) = -a - b$, for all a, b in R and so on.*
(2) $0a = a0 = 0$, *for all a in R.*
(3) $(-a)b = a(-b) = -(ab)$, *for all a, b in R.*
(4) $(-a)(-b) = ab$, *for all a, b in R.*
(5) $(na)b = a(nb) = n(ab)$, *for all $n \in \mathbb{Z}$ and a, b in R.*
(6) $\left(\sum\limits_{i=1}^{n} a_i\right)\left(\sum\limits_{j=1}^{m} b_j\right) = \sum\limits_{i=1}^{n}\sum\limits_{j=1}^{m} a_ib_j$, *for all a_i, b_j in R.*

Moreover, if R has a unit element 1, then

(7) $(-1)a = -a$, *for all $a \in R$.*
(8) $(-1)(-1) = 1$.

Definition 1.2.3. A non-zero element a is called a *zero-divisor* if there exists a non-zero element $b \in R$ such that either $ab = 0$ or $ba = 0$. A commutative ring is called an *integral domain* if it has no zero-divisors.

The ring of integers is an example of an integral domain. It is easy to verify that a ring R has no zero-divisors if and only if the right and left cancelation laws hold in R.

Definition 1.2.4. If the non-zero elements of a ring R form a multiplicative group, i.e., R has unit element and every element except the zero element has an inverse, then we shall call the ring a *division ring*. A *field* is a commutative division ring.

The inverse of an element a under multiplication will be denoted by a^{-1}.

Example 6 (1) If p is prime, then $\mathbb{Z}_p$ is a field.
(2) $\mathbb{Q}$, $\mathbb{R}$ and $\mathbb{C}$ are examples of fields, whereas $\mathbb{Z}$ is not.
(3) Consider the set $\{a + bx \mid a, b \in \mathbb{Z}_2\}$ with x an "indeterminate". We use the arithmetic addition modulo 2 and multiplication using the "rule" $x^2 = x + 1$. Then, we obtain a field with 4 elements: $\{0, 1, x, 1 + x\}$.

Clearly, every field is an integral domain, but, in general, an integral domain is not a field. For example, the ring of integers is an integral domain, but not all non-zero elements have inverse under multiplication. However, for finite domains of integrity, we have the following theorem.

Theorem 1.2.5. *Any finite ring without zero-divisors is a division ring.*

Proof. Suppose that $R = \{x_1, \dots, x_n\}$ is a finite ring without zero-divisors and $a(\neq 0) \in R$. Then, $ax_1, ax_2, \dots, ax_n$ are all n distinct elements lying in R, as cancellation laws hold in R. Since $a \in R$, there exists $x_i \in R$ such that $a = ax_i$. Then, we have $a(x_i a - a) = a^2 - a^2 = 0$, and so $x_i a = a$. Now, for every $b \in R$ we have $ab = (ax_i)b = a(x_i b)$, hence $b = x_i b$ and further $ba = b(x_i a) = (bx_i)a$ which implies that $b = bx_i$. Hence, x_i is the unit element for R and we denote it by 1. Now, $1 \in R$, so there exists $c \in R$ such that $1 = ac$. Also, $a(ca - 1) = (ac)a - a = a - a = 0$, and so $ca = 1$. Consequently, R is a division ring. ■

Corollary 1.2.6. *A finite integral domain is a field.*

In the study of groups, subgroups play a crucial role. Subrings, the analogous notion in ring theory, play a much less important role than their counterparts in group theory. Nevertheless, subrings are important.

Definition 1.2.7. Let R be a ring and S be a non-empty subset of R, which is closed under the addition and multiplication in R. If S is itself a ring under these operations, then S is called a *subring* of R; more formally, S is a subring of R if the following conditions hold:

$$a, b \in S \text{ implies that } a - b \in S \text{ and } a \cdot b \in S.$$

Example 7 (1) For each positive integer n, the set $n\mathbb{Z} = \{0, \pm n, \pm 2n, \pm 3n, \dots\}$ is a subring of $\mathbb{Z}$.
(2) The set A of all 2×2 matrices of the type $\begin{bmatrix} a & 0 \\ b & c \end{bmatrix}$, where a, b and c are integers, is a subring of the ring $M_2(\mathbb{Z})$.
(3) If R is any ring, then the *center* of R is the set $Z(R) = \{x \in R \mid xy = yx, \forall y \in R\}$. Clearly, the center of R is a subring of R.

In group theory, normal subgroups play a special role, they permit us to construct quotient groups. Now, we introduce the analogous concept for rings.

Definition 1.2.8. A non-empty subset I of a ring R is said to be an *ideal* of R if

(1) I is a subring of R under addition.
(2) For every $a \in I$ and $r \in R$, both ar and ra are in I.

Clearly, each ideal is a subring. For any ring R, $\{0\}$ and R are ideals of R. The ideal $\{0\}$ is called the *trivial ideal.* An ideal I of R such that $I \neq 0$ and $I \neq R$ is called a *proper ideal.* Observe that if R has a unit element and I is an ideal of R, then $I = R$ if and only if $1 \in I$. Consequently, a non-zero ideal I of R is proper if and only if I contains no invertible elements of R. It is easy to see that the intersection of any family of ideals of R is also an ideal.

EXAMPLE 8 (1) For any positive integer n, the set $n\mathbb{Z}$ is an ideal of $\mathbb{Z}$. In fact, every ideal of $\mathbb{Z}$ has this form, for suitable n.
(2) Let R be the ring of all real-valued functions of a real variable. The subset S of all differentiable functions is a subring of R but not an ideal of R.
(3) If

$$R = \left\{ \begin{bmatrix} a & b & c \\ d & e & f \\ 0 & 0 & g \end{bmatrix} \mid a, b, c, d, e, f, g \in \mathbb{Z} \right\}$$

then R is a ring under matrix addition and multiplication. The set

$$R = \left\{ \begin{bmatrix} 0 & 0 & x \\ 0 & 0 & y \\ 0 & 0 & 0 \end{bmatrix} \mid x, y \in \mathbb{Z} \right\}$$

is an ideal of R.
(4) Let R be a ring and let $M_n(R)$ be the ring of matrices over R. If I is an ideal of R, then the set $M_n(I)$ of all matrices with entries in I is an ideal of $M_n(R)$. Conversely, every ideal of $M_n(R)$ is of this type.
(5) Let m be a positive integer such that m is not a square in $\mathbb{Z}$. If $R = \{a + \sqrt{m}b \mid a, b \in \mathbb{Z}\}$, then R is a ring under the operations of sum and product of real numbers. Let p be an odd prime number and consider the set $I_p = \{a + \sqrt{m}b \mid p|a \text{ and } p|b\}$, where $a + \sqrt{m}b \in R$. Then, I_p is an ideal of R.
(6) For ideals I_1, I_2 of a ring R define $I_1 + I_2$ to be the set $\{a + b \mid a \in I_1, b \in I_2\}$ and $I_1 I_2$ to be the set $\left\{ \sum_{i=1}^{n} a_i b_i \mid n \in \mathbb{Z}^+, a_i \in I_1, b_i \in I_2 \right\}$. Then, $I_1 + I_2$ and $I_1 I_2$ are ideals of R.
(7) Let X be a subset of a ring R. Let $\{A_i\}_{i \in \Lambda}$ be the family of all ideals in R which contain X. Then, $\bigcap_{i \in \Lambda} A_i$ is called the *ideal generated by* X. This ideal is denoted by $\langle X \rangle$.

Lemma 1.2.9. *Let R be a commutative ring with unit element whose only ideals are the trivial ideal and R. Then, R is a field.*

Proof. In order to prove this lemma, for any non-zero element $a \in R$ we must find an element $b \in R$ such that $ab = 1$. The set $Ra = \{xa \mid x \in R\}$ is an ideal of R. By our assumptions on R, $Ra = \{0\}$ or $Ra = R$. Since $0 \neq a = 1 \cdot a \in Ra$, $Ra \neq \{0\}$, and so $Ra = R$. Since $1 \in R$, there exists $b \in R$ such that $1 = ba$. ■

Definition 1.2.10. (Quotient ring). Let R be a ring and let I be an ideal of R. In order to define the quotient ring, we consider firstly an equivalence relation on R. We say that the elements $a, b \in R$ are equivalent, and we write $a \sim b$, if and only if $a - b \in I$. If a is an element of R, we denote the corresponding equivalence class by $[a]$. The *quotient ring* of modulo I is the set $R/I = \{[a] \mid a \in R\}$, with a ring structure defined as follows. If $[a], [b]$ are equivalence classes in R/I, then

$$[a] + [b] = [a + b] \text{ and } [a] \cdot [b] = [ab].$$

Since I is closed under addition and multiplication, it follows that the ring structure in R/I is well defined. Clearly, $a + I = [a]$.

Definition 1.2.11. A mapping f from the ring R into the ring R' is said to be a *(ring) homomorphism* if

(1) $f(a + b) = f(a) + f(b)$,
(2) $f(ab) = f(a)f(b)$,

for all $a, b \in R$.

If f is a ring homomorphism from R to R', then $f(0) = 0$ and $f(-a) = -f(a)$, for all $a \in R$. A ring homomorphism $f : R \to R'$ is called an *epimorphism* if f is onto. It is called a *monomorphism* if it is one to one, and an *isomorphism* if it is both one to one and onto. The rings R and R' are said to be *isomorphic* if there exists an isomorphism between them, in this case, we write $R \cong R'$.

Example 9 (1) For any positive integer n, the mapping $k \to k \bmod n$ is a ring homomorphism from $\mathbb{Z}$ onto $\mathbb{Z}_n$.
(2) Let I be an ideal of a ring R. We define $f : R \to R/I$ by $f(a) = a + I$ for all $a \in R$. Then, f is an epimorphism. This map is called a *natural homomorphism.*

Proposition 1.2.12. *Let f be a homomorphism from the ring R to the ring R'. Let S be a subring of R, I be an ideal of R and J be an ideal of R'.*

(1) $f(S) = \{f(a) \mid a \in S\}$ *is a subring of* R'.
(2) *If f is onto, then $f(I)$ is an ideal of R'.*
(3) $f^{-1}(J) = \{r \in R \mid f(r) \in J\}$ *is an ideal of* R.
(4) *If R is commutative, then $f(R)$ is commutative.*
(5) *If R has a unit element 1 and f is onto, then $f(1)$ is the unit element of R'.*

(6) *If f is an isomorphism from R to R', then f^{-1} is an isomorphism from R' to R.*

Proof. It is straightforward. ■

Definition 1.2.13. If f is a ring homomorphism of R into R', then the *kernel* of f is defined by $kerf = \{x \in R \mid f(x) = 0\}$.

Corollary 1.2.14. *If f is a ring homomorphism from R to R', then $kerf$ is an ideal of R.*

Corollary 1.2.15. *A ring homomorphism f from R to R' is one to one if and only if $kerf = \{0\}$.*

We are in a position to establish an important connection between homomorphisms and quotient rings. Many authors prefer to call the next theorem the *fundamental theorem of ring homomorphism.*

Theorem 1.2.16. *Let $f : R \to R'$ be a homomorphism from R to R'. Then, $R/kerf \cong f(R)$.*

Theorem 1.2.17. *Let I and J be two ideals of a ring R. Then, $(I + J)/I \cong J/(I \cap J)$.*

Theorem 1.2.18. *Let I and J be two ideals of a ring R such that $J \subseteq I$. Then, $R/I \cong (R/J)/(I/J)$.*

EXAMPLE 10 (1) $\mathbb{Z}/n\mathbb{Z} \cong \mathbb{Z}_n$.

(2) Let $R = \left\{ \begin{bmatrix} a & b \\ -b & a \end{bmatrix} \mid a, b \in \mathbb{R} \right\}$. We define $\psi : R \to \mathbb{C}$ by $\psi\left(\begin{bmatrix} a & b \\ -b & a \end{bmatrix}\right) = a + bi$. Then, ψ is an isomorphism and so R is isomorphic to the field of complex numbers.

(3) Let R be the ring of all real valued continuous functions defined on the closed unit interval. Then, $I = \{f \in R \mid f(\frac{1}{2}) = 0\}$ is an ideal of R. One can shows that R/I is isomorphic to the real field.

Corollary 1.2.19. *Let R be a ring with unit element 1. The mapping $f : \mathbb{Z} \to R$ given by $f(n) = n1$ is a ring homomorphism.*

Definition 1.2.20. A proper ideal M of R is said to be a *maximal ideal* of R if whenever U is an ideal of R and $M \subseteq U \subseteq R$, then $U = M$ or $U = R$.

EXAMPLE 11 (1) In a division ring, $\langle 0 \rangle$ is a maximal ideal.

(2) In the ring of even integers, $\langle 4 \rangle$ is a maximal ideal.

(3) In the ring of integers, an ideal $n\mathbb{Z}$ is maximal if and only if n is a prime number.

(4) The ideal $\langle x^2 + 1 \rangle$ is maximal in $\mathbb{R}[x]$, the ring of polynomials with coefficient in $\mathbb{R}$.

Theorem 1.2.21. *If R is a commutative ring with unit element and M is an ideal of R, then M is a maximal ideal of R if and only if R/M is a field.*

Proof. Suppose that M is a maximal ideal and let $a \in R$ but $a \notin M$. It suffices to show that $a + M$ has a multiplicative inverse. Consider

$$U = \{ar + b \mid r \in R, b \in M\}.$$

This is an ideal of R that contains M properly. Since M is maximal, we have $U = R$. Thus, $1 \in U$, so there exist $c \in R$ and $d \in M$ such that $1 = ac + d$. Then, $1 + M = ac + d + M = ac + M = (a + M)(c + M)$.

Now, suppose that R/M is a field and U is an ideal of R that contains M properly. Let $a \in U$ but $a \notin M$. Then, $a+M$ is a non-zero element of R/M and so there there exists an element $b+M$ such that $(a+M)(b+M) = 1+M$. Since $a \in U$, we have $ab \in U$. Also, we have $1-ab \in M \subseteq U$. So $1 = (1-ab)+ab \in U$ which implies that $U = R$. ■

Definition 1.2.22. An ideal P in a ring R is said to be *prime* if $P \neq R$ and for any ideals A, B in R

$$AB \subseteq P \;\Rightarrow\; A \subseteq P \text{ or } B \subseteq P.$$

Corollary 1.2.23. *Let R be a commutative ring. An ideal P of R is prime if $P \neq R$ and for any $a, b \in R$*

$$ab \in P \;\Rightarrow\; a \in P \text{ or } b \in P.$$

EXAMPLE 12 (1) A positive integer n is a prime number if and only if the ideal $n\mathbb{Z}$ is a prime ideal in $\mathbb{Z}$.

(2) The prime ideals of $\mathbb{Z} \times \mathbb{Z}$ are $\{0\} \times \mathbb{Z}$, $\mathbb{Z} \times \{0\}$, $p\mathbb{Z} \times \mathbb{Z}$, $\mathbb{Z} \times q\mathbb{Z}$, where p and q are primes.

Theorem 1.2.24. *If R is a commutative ring with unit element and P is an ideal of R, then P is a prime ideal of R if and only if R/P is an integral domain.*

Proof. Suppose that R/P is an integral domain and $ab \in P$. Then, $(a+P)(b+P) = ab + P = P$. So either $a + P = P$ or $b + P = P$; that is either $a \in P$ or $b \in P$. Hence, P is prime.

Now, suppose that P is prime and $(a + P)(b + P) = 0 + P = P$. Then, $ab \in P$ and therefore $a \in P$ or $b \in P$. Thus, one of $a + P$ or $b + P$ is zero. ■

Theorem 1.2.25. *Let R be a commutative ring with unit element. Each maximal ideal of R is a prime ideal.*

Proof. Suppose that M is maximal in R but not prime, so there exist $a, b \in R$ such that $a \notin M$, $b \notin M$ but $ab \in M$. Then, each of the ideals $M + \langle a\rangle$ and $M + \langle b\rangle$ contains M properly. By maximality we obtain $M + \langle a\rangle = R = M + \langle b\rangle$. Therefore, $R^2 = (M + \langle a\rangle)(M + \langle b\rangle) \subseteq M^2 + \langle a\rangle M + M\langle b\rangle + \langle a\rangle\langle b\rangle \subseteq M \subsetneq R$. This is a contradiction. ■

Definition 1.2.26. The *radical* of an ideal I in a commutative ring R, denoted by $Rad(I)$ is defined as

$$Rad(I) = \{r \in R \mid r^n \in I \text{ for some positive integer } n\}.$$

The above definition is equivalent to: The radical of an ideal I in a commutative ring R is

$$Rad(I) = \bigcap_{\substack{P \in Spec(R) \\ I \subseteq P}} P,$$

where $Spec(R)$ is the set of all prime ideals of R.

Proposition 1.2.27. *If $J, I_1, \ldots, I_n$ are ideals in a commutative ring R, then*

(1) $Rad(Rad(J)) = Rad(J)$.
(2) $Rad(I_1 \ldots I_n) = Rad\left(\bigcap_{i=1}^{n} I_i\right) = \bigcap_{i=1}^{n} Rad(I_i)$.

EXAMPLE 13 In the ring of integers

(1) $Rad(12\mathbb{Z}) = 2\mathbb{Z} \cap 3\mathbb{Z} = 6\mathbb{Z}$.
(2) Let $n = p_1^{k_1} \ldots p_r^{k_r}$, where p_i's are distinct prime numbers. Then, we have $Rad(n\mathbb{Z}) = \langle p_1, \ldots, p_r \rangle$.

Definition 1.2.28. Let R be a commutative ring. We define the *Jacobson radical* of R, denoted by $Jac(R)$, as the intersection of all the maximal ideals of R.

1.3 Polygroups

The hyperstructures were introduced by Marty [140] when he first defined a hypergroup as a set equipped with an associative and reproductive hyperoperation. The motivating example was the quotient of a group by any, not necessary normal, subgroup. Algebraic hyperstructures represent a natural extension of classical algebraic structures. In a classical algebraic structure, the composition of two elements is an element, while in an algebraic hyperstructure, the composition of two elements is a set. This section deals with a certain algebraic system called a polygroup. Application of hypergroups have mainly appeared in special subclasses. For example, polygroups which are certain subclasses of hypergroups are studied in [116] by Ioulidis and are used to study color algebra [23, 24]. Quasi-canonical hypergroups (called polygroups by Comer) were introduced in [17], as a generalization of canonical hypergroups, introduced in [147]. Recently, Davvaz [66] published a monograph to the study of polygroup theory. It begins with some basic results concerning group theory and algebraic hyperstructures, which represent the most general algebraic context, in which reality can be modeled.

Definition 1.3.1. A map $\cdot : S \times S \to \mathcal{P}^*(S)$ is called *hyperoperation* on the set S, where S is a non-empty set and $\mathcal{P}^*(S)$ denotes the set of all non-empty subsets of S.

A *hypergroupoid* is a non-empty set S endowed with a hyperoperation $\cdot$ and it is denoted by $(S, \cdot)$.

Definition 1.3.2. A hypergroupoid $(S, \cdot)$ is called a *semihypergroup* if for all $x, y, z \in S$, $(x \cdot y) \cdot z = x \cdot (y \cdot z)$, which means that

$$\bigcup_{u \in x \cdot y} u \cdot z = \bigcup_{v \in y \cdot z} x \cdot v.$$

Definition 1.3.3. A non-empty subset A of a semihypergroup $(S, \cdot)$ is called a *subsemihypergroup* if it is a semihypergroup. In other words, a non-empty subset A of a semihypergroup $(S, \cdot)$ is a subsemihypergroup if $A \cdot A \subseteq A$ and S is called in this case *supersemihypergroup* of A.

If $x \in S$ and A, B are non-empty subsets of S, then

$$A \cdot B = \bigcup_{\substack{a \in A \\ b \in B}} a \cdot b, A \cdot x = A \cdot \{x\}, \text{ and } x \cdot B = \{x\} \cdot B.$$

EXAMPLE 14 (1) Let $(S, \cdot)$ be a semigroup and K be any subsemigroup of S. Then, $S/K = \{xK \mid x \in S\}$ becomes a semihypergroup, where the hyperoperation is defined as follows: $\overline{x} \star \overline{y} = \{\overline{z} \mid z \in \overline{x} \cdot \overline{y}\}$, where $\overline{x} = xK$.

(2) Consider $S = \{a, b, c, d\}$ with the following hyperoperation:

$\cdot$	a	b	c	d
a	a	b	c	d
b	b	$\{a, c\}$	$\{b, c\}$	d
c	c	$\{b, c\}$	$\{a, b\}$	d
d	d	d	d	S

Then, $(S, \cdot)$ is a semihypergroup.

The associativity for semihypergroups can be applied for subsets, i.e., if $(S, \cdot)$ is a semihypergroup, then for all non-empty subsets A, B, C of S, we have $A \cdot (B \cdot C) = (A \cdot B) \cdot C$.

Definition 1.3.4. A *polygroup* is a system $\wp = \langle P, \cdot, e, ^{-1} \rangle$, where $e \in P$, $^{-1}$ is a unitary operation on P, $\cdot$ maps $P \times P$ into the non-empty subsets of P, and the following axioms hold for all x, y, z in P:

(P_1) $(x \cdot y) \cdot z = x \cdot (y \cdot z)$.
(P_2) $e \cdot x = x \cdot e = x$.
(P_3) $x \in y \cdot z$ implies $y \in x \cdot z^{-1}$ and $z \in y^{-1} \cdot x$.

The following elementary facts about polygroups follow easily from the axioms: $e \in x \cdot x^{-1} \cap x^{-1} \cdot x$, $e^{-1} = e$, $(x^{-1})^{-1} = x$, and $(x \cdot y)^{-1} = y^{-1} \cdot x^{-1}$, where $A^{-1} = \{a^{-1} \mid a \in A\}$.

EXAMPLE 15 (1) *Double coset algebra.* Suppose that H is a subgroup of a group G. Define a system $G//H =< \{HgH \mid g \in G\}, *, H, ^{-I} >$, where $(HgH)^{-I} = Hg^{-1}H$ and

$$(Hg_1H) * (Hg_2H) = \{Hg_1hg_2H \mid h \in H\}.$$

The algebra of double cosets $G//H$ is a polygroup introduced in (Dresher and Ore [101]).

(2) *Prenowitz algebras.* Suppose G is a projective geometry with a set P of points and suppose, for $p \neq q$, $\overline{pq}$ denoted the set of all points on the unique line through p and q. Choose an object $I \notin P$ and form the system

$$P_G =< P \cup \{I\}, \cdot, I, ^{-1} >$$

where $x^{-1} = x$ and $I \cdot x = x \cdot I = x$ for all $x \in P \cup \{I\}$ and for $p, q \in P$,

$$p \cdot q = \begin{cases} \overline{pq} - \{p, q\} & \text{if } p \neq q \\ \{p, I\} & \text{if } p = q. \end{cases}$$

P_G is a polygroup (Prenowitz [156]).

(3) Let $< P_1, \cdot, e_1, ^{-1} >$ and $< P_2, *, e_2, ^{-I} >$ be two polygroups. Then, on $P_1 \times P_2$ we can define a hyperproduct as follows: $(x_1, y_1) \circ (x_2, y_2) = \{(x, y) \mid x \in x_1 \cdot x_2,\ y \in y_1 * y_2\}$. We call this the *direct hyperproduct* of P_1 and P_2. Clearly, $P_1 \times P_2$ equipped with the usual direct hyperproduct becomes a polygroup.

(4) [24] Suppose that $\mathcal{A} =< A, \cdot, e, ^{-1} >$ and $\mathcal{B} =< B, \cdot, e, ^{-1} >$ are two polygroups whose elements have been renamed so that $A \cap B = \{e\}$. A new system $\mathcal{A}[\mathcal{B}] =< M, *, e, ^{I} >$ called the *extension* of $\mathcal{A}$ by $\mathcal{B}$ is formed in the following way: Set $M = A \cup B$ and let $e^I = e$, $x^I = x^{-1}$, $e * x = x * e = x$ for all $x \in M$, and for all $x, y \in M \backslash \{e\}$

$$x * y = \begin{cases} x \cdot y & \text{if } x, y \in A \\ x & \text{if } x \in B,\ y \in A \\ y & \text{if } x \in A,\ y \in B \\ x \cdot y & \text{if } x, y \in B,\ y \neq x^{-1} \\ x \cdot y \cup A & \text{if } x, y \in B,\ y = x^{-1}. \end{cases}$$

In this case $\mathcal{A}[\mathcal{B}]$ is a polygroup which is called the *extension* of $\mathcal{A}$ by $\mathcal{B}$.

Definition 1.3.5. A non-empty subset A of a polygroup P is a *subpolygroup* of P if

(1) $a, b \in A$ implies $a \cdot b \subseteq A$.
(2) $a \in A$ implies $a^{-1} \in A$.

Definition 1.3.6. A subpolygroup N of a polygroup P is *normal* in P if $a^{-1} \cdot N \cdot a \subseteq N$, for all $a \in P$.

The following corollaries are direct consequences of the above definitions.

Corollary 1.3.7. *Let N be a normal subpolygroup of a polygroup P. Then,*

(1) $N \cdot a = a \cdot N$, *for all* $a \in P$.
(2) $(N \cdot a) \cdot (N \cdot b) = N \cdot a \cdot b$, *for all* $a, b \in P$.
(3) $N \cdot a = N \cdot b$, *for all* $b \in N \cdot a$.

Corollary 1.3.8. *Let K and N be subpolygroups of a polygroup P with N normal in P. Then,*

(1) $N \cap K$ *is a normal subpolygroup of* K.
(2) $N \cdot K = K \cdot N$ *is a subpolygroup of* P.
(3) N *is a normal subpolygroup of* $N \cdot K$.

Definition 1.3.9. If N is a normal subpolygroup of P, then we define the relation $x \equiv y(modN)$ if and only if $x \cdot y^{-1} \cap N \neq \emptyset$. This relation is denoted by xN_Py.

Lemma 1.3.10. *The relation N_P is an equivalence relation.*

Proof. It is straightforward. ∎

Let $N_P(x)$ be the equivalence class of the element $x \in P$. Suppose that $[P : N] = \{N_P(x) \mid x \in P\}$. On $[P : N]$ we consider the hyperoperation $\odot$ defined as follows: $N_P(x) \odot N_P(y) = \{N_P(z) \mid z \in N_P(x)N_P(y)\}$. For a subpolygroup K of P and $x \in P$, denote the right coset of K by $K \cdot x$ and let P/K is the set of all right cosets of K in P.

Lemma 1.3.11. *If N is a normal subpolygroup of P, then $N \cdot x = N_P(x)$.*

Proof. It is straightforward. ∎

Lemma 1.3.12. *Let N be a normal subpolygroup of P. Then, for all $x, y \in P$, $N \cdot x \cdot y = N \cdot z$* for all $z \in x \cdot y$.

Proof. Suppose that $z \in x \cdot y$. Then, it is clear that $N \cdot z \subseteq N \cdot x \cdot y$. Now, let $a \in N \cdot x \cdot y$. Then, we obtain $y \in (N \cdot x)^{-1} \cdot a$ or $y \in x^{-1} \cdot N \cdot a$, and so $x \cdot y \subseteq x \cdot x^{-1} \cdot N \cdot a$. Since N is a normal subpolygroup, we obtain $x \cdot y \subseteq x \cdot N \cdot x^{-1} \cdot a \subseteq N \cdot a$. Therefore, for every $z \in x \cdot y$, we have $z \in N \cdot a$ which implies that $a \in N \cdot z$. This completes the proof. ∎

Definition 1.3.13. An equivalence relation ρ on a polygroup P is called a *full conjugation* on P if

(1) $x\rho y$ implies $x^{-1}\rho y^{-1}$.
(2) $z \in x \cdot y$ and $z'\rho z$ implies $z' \in x' \cdot y'$ for some $x'\rho x$ and $y'\rho y$.

Lemma 1.3.14. *ρ is a full conjugation of P if and only if*

(1) $\rho(x)^{-1} = \{y^{-1} \mid y \in \rho(x)\} = \rho(x^{-1})$.
(2) $\rho(\rho(x)y) = \rho(x)\rho(y)$.

Proof. Assume that ρ is a full conjugation. Condition (1) of Definition 1.3.13 easily implies (1) of lemma. Condition (2) of definition of conjugation implies that $(\rho(x))(\rho(y))$ is a union of ρ-classes so $\rho((\rho(x))y) \subseteq (\rho(x))(\rho(y))$. Now, assume that $z \in x' \cdot y'$, $x'\rho x$, and $y'\rho y$. By (P_3), $y' \in {x'}^{-1} \cdot z$ so $y \in x'' \cdot z'$, where $x''\rho {x'}^{-1}$ and $z'\rho z$ by (2) of conjugation definition. By (P_3) again and (1),

$$z\rho z' \in (\rho(x)) \cdot y$$

which shows equality in (2). A similar argument establishes that a relation ρ with properties (1) and (2) is a full conjugation. ∎

Corollary 1.3.15. *The equivalence relation N_P is a full conjugation on P.*

Proposition 1.3.16. $< [P : N], \odot, N_P(e), ^{-I} >$ *is a polygroup, where we have* $N_P(a)^{-I} = N_P(a^{-1})$.

Proof. For all $a, b, c \in P$, we have

$$\begin{aligned}
(N_P(a) \odot N_P(b)) \odot N_P(c) &= \{N_P(x) \mid x \in N_P(a)N_P(b)\} \odot N_P(c) \\
&= \{N_P(y) \mid y \in N_P(x)N_P(c),\ x \in N_P(a)N_P(b)\} \\
&= \{N_P(y) \mid y \in N_P(N_P(a)N_P(b))N_P(c)\} \\
&= \{N_P(y) \mid y \in (N_P(a)N_P(b))N_P(c)\}, \\
N_P(a) \odot (N_P(b) \odot N_P(c)) &= N_P(a) \odot \{N_P(x) \mid x \in N_P(b)N_P(c)\} \\
&= \{N_P(y) \mid y \in N_P(a)N_P(x),\ x \in N_P(b)N_P(c)\} \\
&= \{N_P(y) \mid y \in N_P(a)N_P(N_P(b)N_P(c))\} \\
&= \{N_P(y) \mid y \in N_P(a)(N_P(b)N_P(c))\}.
\end{aligned}$$

Since $(N_P(a)N_P(b))N_P(c) = N_P(a)(N_P(b)N_P(c))$, we get $(N_P(a) \odot N_P(b)) \odot N_P(c) = N_P(a) \odot (N_P(b) \odot N_P(c))$. Therefore, $\odot$ is associative. It is easy to see that $N_P(e)$ is the unit element in $[P : N]$, and $N_P(x^{-1})$ is the inverse of the element $N_P(x)$. Now, we show that $N_P(c) \in N_P(a) \odot N_P(b)$ implies $N_P(a) \in N_P(c) \odot N_P(b^{-1})$ and $N_P(b) \in N_P(a^{-1}) \odot N_P(c)$.

We have $N_P(c) \in N_P(a) \odot N_P(b)$, and hence $N_P(c) = N_P(x)$ for some $x \in N_P(a)N_P(b)$. Therefore, there exist $y \in N_P(a)$ and $z \in N_P(b)$ such that $x \in yz$, so $y \in xz^{-1}$. This implies that $N_P(y) \in N_P(x) \odot N_P(z^{-1})$, and so $N_P(a) \in N_P(c) \odot N_P(b^{-1})$. Similarly, we obtain $N_P(b) \in N_P(a^{-1}) \odot N_P(c)$. Therefore, $[P : N]$ is a polygroup. ∎

Corollary 1.3.17. *If N is a normal subpolygroup of P, then* $< P/N, \odot, N, ^{-I} >$ *is a polygroup, where* $N \cdot x \odot N \cdot y = \{N \cdot z \mid z \in x \cdot y\}$ and $(N \cdot x)^{-I} = N \cdot x^{-1}$.

Definition 1.3.18. Let $< P, \cdot, e, ^{-1} >$ and $< P', *, e', ^{-I} >$ be two polygroups. Let f be a mapping from P into P' such that $f(e) = e'$. Then, f is called

(1) an *inclusion homomorphism* if

$$f(x \cdot y) \subseteq f(x) * f(y), \text{ for all } x, y \in P;$$

(2) a *strong homomorphism* or a *good homomorphism* if

$$f(x \cdot y) = f(x) * f(y), \text{ for all } x, y \in P,$$

where for non-empty subset A of P, by $f^{-1}f(A)$ we mean

$$f^{-1}f(A) = \bigcup_{x \in A} f^{-1}f(x);$$

(3) a *homomorphism of type 2* if

$$f^{-1}(f(x) * f(y)) = f^{-1}f(x \cdot y), \text{ for all } x, y \in P_1;$$

(4) a *homomorphism of type 3* if

$$f^{-1}(f(x) * f(y)) = (f^{-1}f(x)) \cdot (f^{-1}f(y)), \text{ for all } x, y \in P_1;$$

(5) a *homomorphism of type 4* if

$$f^{-1}(f(x)*f(y)) = f^{-1}f(x \cdot y) = (f^{-1}f(x)) \cdot (f^{-1}f(y)), \text{ for all } x, y \in P_1.$$

Clearly, a strong homomorphism f is an *isomorphism* if f is one to one and onto.

Lemma 1.3.19. *Let* $< P, \cdot, e, ^{-1} >$ *and* $< P', *, e', ^{-I} >$ *be two polygroups and* f *be a strong homomorphism from* P *into* P'. *Then,* $f(a^{-1}) = f(a)^{-1}$, *for all* $a \in P$.

Proof. Since P is a polygroup, $e \in a \cdot a^{-1}$ for all $a \in P$. Then, we have $f(e) \in f(a) * f(a^{-1})$ or $e' \in f(a) * f(a^{-1})$ which implies $f(a^{-1}) \in f(a)^{-1} * e'$. Thus, $f(a^{-1}) = f(a)^{-1}$, for all $a \in P$. ■

Definition 1.3.20. Let $< P, \cdot, e, ^{-1} >$ and $< P', *, e', ^{-I} >$ be two polygroups and f be a strong homomorphism from P into P'. Then, the *kernel* of f is the set $ker f = \{x \in P \mid f(x) = e'\}$.

It is trivial that $ker f$ is a subpolygroup of P but in general is not normal in P.

Lemma 1.3.21. *Let* f *be a strong homomorphism from* P *into* P'. *Then,* f *is injective if and only if* $ker f = \{e\}$.

Proof. Let $y, z \in P$ be such that $f(y) = f(z)$ Then, $f(y) * f(y^{-1}) = f(z) * f(y^{-1})$. It follows that $f(e) \in f(y \cdot y^{-1}) = f(z \cdot y^{-1})$, and so there exists $x \in z \cdot y^{-1}$ such that $e' = f(e) = f(x)$. Thus, if $ker f = \{e\}$, $x = e$, whence $y = z$. Now, let $x \in ker f$. Then, $f(x) = e' = f(e)$. Thus, if f is injective, we conclude that $x = e$. ■

The proofs of the following theorems are similar to the proofs of Theorems 1.1.19, 1.1.20 and 1.1.21.

Theorem 1.3.22. *Let f be a strong homomorphism from P into P' with kernel K such that K is a normal subpolygroup of P. Then, $P/K \cong f(P)$.*

Theorem 1.3.23. *If K and N are subpolygroups of a polygroup P, with N normal in P, then $K/N \cap K \cong N \cdot K/N$.*

Theorem 1.3.24. *If K and N are normal subpolygroups of a polygroup P such that $N \subseteq K$, then K/N is a normal subpolygroup of P/N and $(P/N)/(K/N) \cong P/K$.*

Corollary 1.3.25. *If N_1, N_2 are normal subpolygroups of P_1, P_2 respectively, then $N_1 \times N_2$ is a normal subpolygroup of $P_1 \times P_2$ and $(P_1 \times P_2)/(N_1 \times N_2) \cong P_1/N_1 \times P_2/N_2$.*

1.4 Hypergroups

The hypergroup theory both extends some well-known group results and introduces new topics, thus leading to a wide variety of applications, as well as to a broadening of the investigation fields. A comprehensive review of the theory of hyperstructures appears in [27, 47].

Definition 1.4.1. A hypergroupoid $(H, \cdot)$ is called a *quasihypergroup* if for all a of H we have $a \cdot H = H \cdot a = H$. This condition is also called the *reproduction axiom.*

Definition 1.4.2. A hypergroupoid $(H, \cdot)$ which is both a semihypergroup and a quasihypergroup is called a *hypergroup.*

EXAMPLE 16 (1) If H is a non-empty set and for all x, y of H, we define $x \cdot y = H$, then $(H, \cdot)$ is a hypergroup, called the *total hypergroup.*

(2) Let $(S, \cdot)$ be a semigroup and let P be a non-empty subset of S. For all x, y of S, we define $x \star y = xPy$. Then, $(S, \star)$ is a semihypergroup. If $(S, \cdot)$ is a group, then $(S, \star)$ is a hypergroup, called a *P-hypergroup.*

(3) If G is a group and for all x, y of G, $\langle x, y\rangle$ denotes the subgroup generated by x and y, then we define $x \star y = \langle x, y\rangle$. We obtain that $(G, \star)$ is a hypergroup.

(4) Let $(G, \cdot)$ be a group and H be a normal subgroup of G. For all x, y of G, we define $x \star y = xyH$. Then, $(G, \star)$ is a hypergroup.

(5) Let $(G, \cdot)$ be a group and let H be a non-normal subgroup of it. If we denote $G/H = \{xH \mid x \in G\}$, then $(G/H, \star)$ is a hypergroup, where for all xH, yH of G/H, we have $xH \star yH = \{zH \mid z \in xHy\}$.

A hypergroup for which the hyperproduct of any two elements has exactly one element is a group. Indeed, let $(H, \cdot)$ be a hypergroup, such that for all x, y of H, we have $|x \cdot y| = 1$. Then, $(H, \cdot)$ is a semigroup, such that for all a, b in H, there exist x and y for which we have $a = b \cdot x$ and $a = y \cdot b$. It follows that $(H, \cdot)$ is a group.

Definition 1.4.3. A non-empty subset K of a hypergroup $(H, \cdot)$ is called a *subhypergroup* if it is a hypergroup.

Hence, a non-empty subset K of a hypergroup $(H, \cdot)$ is a subhypergroup if for all a of K we have $a \cdot K = K \cdot a = K$.

There are several kinds of subhypergroups. In what follows, we introduce closed, invertible, ultraclosed and conjugable subhypergroups.

Definition 1.4.4. Let $(H, \cdot)$ be a hypergroup and K be a subhypergroup of it. We say that K is:

- *closed on the left* (*on the right*) if for all k_1, k_2 of K and x of H, from $k_1 \in x \cdot k_2$ ($k_1 \in k_2 \cdot x$, respectively), it follows that $x \in K$;
- *invertible on the left* (*on the right*) if for all x, y of H, from $x \in K \cdot y$ ($x \in y \cdot K$), it follows that $y \in K \cdot x$ ($y \in x \cdot K$, respectively);
- *ultraclosed on the left* (*on the right*) if for all x of H, we have $K \cdot x \cap (H \setminus K) \cdot x = \emptyset$ ($x \cdot K \cap x \cdot (H \setminus K) = \emptyset$);
- *conjugable on the right* if it is closed on the right and for all $x \in H$, there exists $x' \in H$ such that $x' \cdot x \subseteq K$.

We say that K is *closed* (*invertible, ultraclosed, conjugable*) if it is closed (invertible, ultraclosed, conjugable respectively) on the left and on the right.

Example 17 (1) Let $(A, \cdot)$ be a hypergroup, $H = A \cup T$, where T is a set with at least three elements and $A \cap T = \emptyset$. We define the hyperoperation $\otimes$ on H, as follows:
- if $(x, y) \in A^2$, then $x \otimes y = x \cdot y$;
- if $(x, t) \in A \times T$, then $x \otimes t = t \otimes x = t$;
- if $(t_1, t_2) \in T \times T$, then $t_1 \otimes t_2 = t_2 \otimes t_1 = A \cup (T \setminus \{t_1, t_2\})$.

Then, $(H, \otimes)$ is a hypergroup and $(A, \otimes)$ is a ultraclosed, non-conjugable subhypergroup of H.

(2) Let $(A, \cdot)$ be a total hypergroup, with at least two elements and let $T = \{t_i\}_{i \in \mathbb{N}}$ such that $A \cap T = \emptyset$ and $t_i \neq t_j$ for $i \neq j$. We define the hyperoperation $\otimes$ on $H = A \cup T$ as follows:
- if $(x, y) \in A^2$, then $x \otimes y = A$;
- if $(x, t) \in A \times T$, then $x \otimes t = t \otimes x = (A \setminus \{x\}) \cup T$;
- if $(t_i, t_j) \in T \times T$, then $t_i \otimes t_j = t_j \otimes t_i = A \cup \{t_{i+j}\}$.

Then, $(H, \otimes)$ is a hypergroup and $(A, \otimes)$ is a non-closed subhypergroup of H.

(3) Let us consider the group $(\mathbb{Z}, +)$ and the subgroups $S_i = 2^i\mathbb{Z}$, where $i \in \mathbb{N}$. For any $x \in \mathbb{Z} \setminus \{0\}$, there exists a unique integer $n(x)$, such that $x \in S_{n(x)} \setminus S_{n(x)+1}$. Define the following commutative hyperoperation on $\mathbb{Z} \setminus \{0\}$:
- if $n(x) < n(y)$, then $x \cdot y = x + S_{n(y)}$;
- if $n(x) = n(y)$, then $x \cdot y = S_{n(x)} \setminus \{0\}$;
- if $n(x) > n(y)$, then $x \cdot y = y + S_{n(x)}$.

Notice that if $n(x) < n(y)$, then $n(x+y) = n(x)$. Then, $(\mathbb{Z} \setminus \{0\}, \cdot)$ is a hypergroup and for all $i \in \mathbb{N}$, $(S_i \setminus \{0\}, \cdot)$ is an invertible subhypergroup of $\mathbb{Z} \setminus \{0\}$.

Lemma 1.4.5. *A subhypergroup K is invertible on the right if and only if $\{x \cdot K\}_{x \in H}$ is a partition of H.*

Proof. If K is invertible on the right and $z \in x \cdot K \cap y \cdot K$, then $x, y \in z \cdot K$, whence $x \cdot K \subseteq z \cdot K$ and $y \cdot K \subseteq z \cdot K$. It follows that $x \cdot K = z \cdot K = y \cdot K$. Conversely, if $\{x \cdot K\}_{x \in H}$ is a partition of H and $x \in y \cdot K$, then $x \cdot K \subseteq y \cdot K$, whence $x \cdot K = y \cdot K$ and so we have $x \in y \cdot K = x \cdot K$. Hence, for all x of H we have $x \in x \cdot K$. From here, we obtain that $y \in y \cdot K = x \cdot K$. ■

If A and B are subsets of H such that we have $H = A \cup B$ and $A \cap B = \emptyset$, then we denote $H = A \oplus B$.

Theorem 1.4.6. *If a subhypergroup K of a hypergroup $(H, \cdot)$ is ultraclosed, then it is closed and invertible.*

Proof. First we check that K is closed. For $x \in K$, we have $K \cap x \cdot (H \setminus K) = \emptyset$ and from $H = x \cdot K \cup x \cdot (H \setminus K)$, we obtain $x \cdot (H \setminus K) = H \setminus K$, which means that $K \cdot (H \setminus K) = H \setminus K$. Similarly, we obtain $(H \setminus K) \cdot K = H \setminus K$, hence K is closed. Now,we show that $\{x \cdot K\}_{x \in H}$ is a partition of H. Let $y \in x \cdot K \cap z \cdot K$. It follows that $y \cdot K \subseteq x \cdot K$ and $y \cdot (H \setminus K) \subseteq x \cdot K \cdot (H \setminus K) = x \cdot (H \setminus K)$. From $H = x \cdot K \oplus x \cdot (H \setminus K) = y \cdot K \oplus y \cdot (H \setminus K)$, we obtain $x \cdot K = y \cdot K$. Similarly, we have $z \cdot K = y \cdot K$. Hence, $\{x \cdot K\}_{x \in H}$ is a partition of H, and according to the above lemma, it follows that K is invertible on the right. Similarly, we can show that K is invertible on the left. ■

Theorem 1.4.7. *If a subhypergroup K of a hypergroup $(H, \cdot)$ is invertible, then it is closed.*

Proof. Let $k_1, k_2 \in K$. If $k_1 \in x \cdot k_2 \subseteq x \cdot K$, then $x \in k_1 \cdot K \subseteq K$. Similarly, from $k_1 \in k_2 \cdot x$, we obtain $x \in K$. ■

We denote the set $\{e \in H \mid \exists x \in H,$ such that $x \in x \cdot e \cup e \cdot x\}$ by I_p and we call it the *set of partial identities* of H.

Theorem 1.4.8. *A subhypergroup K of a hypergroup $(H, \cdot)$ is ultraclosed if and only if K is closed and $I_p \subseteq K$.*

Proof. Suppose that K is closed and $I_p \subseteq K$. First, we show that K is invertible on the left. Suppose there are x, y of H such that $x \in K \cdot y$ and $y \notin K \cdot x$. Hence, $y \in (H \setminus K) \cdot x$, whence $x \in K \cdot (H \setminus K) \cdot x = (H \setminus K) \cdot x$, since K is closed. We obtain that $I_p \cap (H \setminus K) \neq \emptyset$, which is a contradiction. Hence, K is invertible on the left. Now, we check that K is ultraclosed on the left. Suppose there are a and x in H such that $a \in K \cdot x \cap (H \setminus K) \cdot x$. It follows that $x \in K \cdot a$, since K is invertible on the left. We obtain $a \in (H \setminus K) \cdot x \subseteq (H \setminus K) \cdot K \cdot a = (H \setminus K) \cdot a$, since K is closed. This means that $I_p \cap (H \setminus K) \neq \emptyset$, which is a contradiction. Therefore, K is ultraclosed on the left and similarly it is ultraclosed on the right.

Conversely, suppose that K is ultraclosed. According to Theorems 1.4.6 and 1.4.7, K is closed. Now, suppose that $I_p \cap (H \setminus K) \neq \emptyset$, which means that there is $e \in H \setminus K$ and there is $x \in H$, such that $x \in e \cdot x$, for instance. We obtain $x \in (H \setminus K) \cdot x$, whence $K \cdot x \subseteq (H \setminus K) \cdot x$, which contradicts that K is ultraclosed. Hence, $I_p \subseteq K$. ■

Theorem 1.4.9. *If a subhypergroup K of a hypergroup $(H, \cdot)$ is conjugable, then it is ultraclosed.*

Proof. Let $x \in H$. Denote $B = x \cdot K \cap x \cdot (H \setminus K)$. Since K is conjugable it follows that K is closed and there exists $x' \in H$, such that $x' \cdot x \subseteq K$. We obtain

$$\begin{aligned} x' \cdot B &= x' \cdot (x \cdot K \cap x \cdot (H \setminus K)) \\ &\subseteq K \cap x' \cdot x \cdot (H \setminus K) \\ &\subseteq K \cap K \cdot (H \setminus K) \\ &\subseteq K \cap (H \setminus K) = \emptyset. \end{aligned}$$

Hence, $B = \emptyset$, which means that K is ultraclosed on the right. Similarly, we check that K is ultraclosed on the left. ■

Definition 1.4.10. Let $(H_1, \cdot)$ and $(H_2, *)$ be two hypergroups. A map $f : H_1 \to H_2$, is called

(1) a *homomorphism* or *inclusion homomorphism* if for all x, y of H, we have $f(x \cdot y) \subseteq f(x) * f(y)$;
(2) a *good homomorphism* if for all x, y of H, we have $f(x \cdot y) = f(x) * f(y)$;
(3) a *very good homomorphism* if it is good and for all x, y of H, we have $f(x/y) = f(x)/f(y)$ and $f(x \backslash y) = f(x) \backslash f(y)$, where $x/y = \{z \in H \mid x \in z \cdot y\}$ and $x \backslash y = \{u \in H \mid y \in x \cdot u\}$;
(4) an *isomorphism* if it is a good homomorphism, and its inverse f^{-1} is a homomorphism, too.

Definition 1.4.11. Let $(H, \cdot)$ be a semihypergroup and R be an equivalence relation on H. If A and B are non-empty subsets of H, then we define

$$\begin{aligned} &A\overline{R}B \text{ means that } \forall a \in A, \exists b \in B \text{ such that } aRb \text{ and} \\ &\qquad\qquad\qquad \forall b' \in B, \exists a' \in A \text{ such that } a'Rb'; \\ &A\overline{\overline{R}}B \text{ means that } \forall a \in A, \forall b \in B, \text{ we have } aRb. \end{aligned}$$

Definition 1.4.12. The equivalence relation R is called

(1) *regular on the right* (*on the left*) if for all x of H, from aRb, it follows that $(a \cdot x)\overline{R}(b \cdot x)$ ($(x \cdot a)\overline{R}(x \cdot b)$ respectively);
(2) *strongly regular on the right* (*on the left*) if for all x of H, from aRb, it follows that $(a \cdot x)\overline{\overline{R}}(b \cdot x)$ ($(x \cdot a)\overline{\overline{R}}(x \cdot b)$ respectively);
(3) R is called *regular* (*strongly regular*) if it is regular (strongly regular) on the right and on the left.

Theorem 1.4.13. *Let $(H, \cdot)$ be a semihypergroup and R be an equivalence relation on H.*

(1) *If R is regular, then H/R is a semihypergroup, with respect to the following hyperoperation: $\overline{x} \otimes \overline{y} = \{\overline{z} \mid z \in x \cdot y\}$.*
(2) *If the above hyperoperation is well defined on H/R, then R is regular.*

Proof. (1) First, we check that the hyperoperation $\otimes$ is well defined on H/R. Consider $\overline{x} = \overline{x_1}$ and $\overline{y} = \overline{y_1}$. We check that $\overline{x} \otimes \overline{y} = \overline{x_1} \otimes \overline{y_1}$. We have xRx_1 and yRy_1. Since R is regular, it follows that $(x \cdot y)\overline{\overline{R}}(x_1 \cdot y)$, $(x_1 \cdot y)\overline{\overline{R}}(x_1 \cdot y_1)$ whence $(x \cdot y)\overline{\overline{R}}(x_1 \cdot y_1)$. Hence, for all $z \in x \cdot y$, there exists $z_1 \in x_1 \cdot y_1$ such that zRz_1, which means that $\overline{z} = \overline{z_1}$. It follows that $\overline{x} \otimes \overline{y} \subseteq \overline{x_1} \otimes \overline{y_1}$ and similarly we obtain the converse inclusion. Now, we check the associativity of $\otimes$. Let $\overline{x}, \overline{y}, \overline{z}$ be arbitrary elements in H/R and $\overline{u} \in (\overline{x} \otimes \overline{y}) \otimes \overline{z}$. This means that there exists $\overline{v} \in \overline{x} \otimes \overline{y}$ such that $\overline{u} \in \overline{v} \otimes \overline{z}$. In other words, there exist $v_1 \in x \cdot y$ and $u_1 \in v \cdot z$, such that vRv_1 and uRu_1. Since R is regular, it follows that there exists $u_3 \in v_1 \cdot z \subseteq x \cdot (y \cdot z)$ such that $u_1 R u_3$. From here, we obtain that there exists $u_4 \in y \cdot z$ such that $u_3 \in x \cdot u_4$. We have $\overline{u} = \overline{u_1} = \overline{u_3} \in \overline{x} \otimes \overline{u_4} \subseteq \overline{x} \otimes (\overline{y} \otimes \overline{z})$. It follows that $(\overline{x} \otimes \overline{y}) \otimes \overline{z} \subseteq \overline{x} \otimes (\overline{y} \otimes \overline{z})$. Similarly, we obtain the converse inclusion.

(2) Let aRb and x be an arbitrary element of H. If $u \in a \cdot x$, then $\overline{u} \in \overline{a} \otimes \overline{x} = \overline{b} \otimes \overline{x} = \{\overline{v} \mid v \in b \cdot x\}$. Hence, there exists $v \in b \cdot x$ such that uRv, whence $(a \cdot x)\overline{R}(b \cdot x)$. Similarly we obtain that R is regular on the left. ■

Corollary 1.4.14. *If $(H, \cdot)$ is a hypergroup and R is an equivalence relation on H, then R is regular if and only if $(H/R, \otimes)$ is a hypergroup.*

Proof. If H is a hypergroup, then for all x of H we have $H \cdot x = x \cdot H = H$, whence we obtain $H/R \otimes \overline{x} = \overline{x} \otimes H/R = H/R$. According to the above theorem, it follows that $(H/R, \otimes)$ is a hypergroup. ■

Notice that if R is regular on a (semi)hypergroup H, then the canonical projection $\pi : H \to H/R$ is a good epimorphism. Indeed, for all x, y of H and $\overline{z} \in \pi(x \cdot y)$, there exists $z' \in x \cdot y$ such that $\overline{z} = \overline{z'}$. We have $\overline{z} = \overline{z'} \in \overline{x} \otimes \overline{y} = \pi(x) \otimes \pi(y)$. Conversely, if $\overline{z} \in \pi(x) \otimes \pi(y) = \overline{x} \otimes \overline{y}$, then there exists $z_1 \in x \cdot y$ such that $\overline{z} = \overline{z_1} \in \pi(x \cdot y)$.

Theorem 1.4.15. *If $(H, \cdot)$ and $(K, *)$ are semihypergroups and $f : H \to K$ is a good homomorphism, then the equivalence ρ^f associated with f, i.e., $x\rho^f y \Rightarrow f(x) = f(y)$, is regular and $\varphi : f(H) \to H/\rho^f$, defined by $\varphi(f(x)) = \overline{x}$, is an isomorphism.*

Proof. Let $h_1 \rho^f h_2$ and a be an arbitrary element of H. If $u \in h_1 \cdot a$, then

$$f(u) \in f(h_1 \cdot a) = f(h_1) * f(a) = f(h_2) * f(a) = f(h_2 \cdot a).$$

Then, there exists $v \in h_2 \cdot a$ such that $f(u) = f(v)$, which means that $u\rho^f v$. Hence, ρ^f is regular on the right. Similarly, it can be shown that ρ^f is regular on the left. On the other hand, for all $f(x), f(y)$ of $f(H)$, we have

$$\varphi(f(x) * f(y)) = \varphi(f(x \cdot y)) = \{\overline{z} \mid z \in x \cdot y\} = \overline{x} \otimes \overline{y} = \varphi(f(x)) \otimes \varphi(f(y)).$$

Moreover, if $\varphi(f(x)) = \varphi(f(y))$, then $x\rho^f y$, so φ is injective and clearly, it is also surjective. Finally, for all $\overline{x}, \overline{y}$ of H/ρ^f we have

$$\begin{aligned}\varphi^{-1}(\overline{x} \otimes \overline{y}) &= \varphi^{-1}(\{\overline{z} \mid z \in x \cdot y\}) = \{f(z) \mid z \in x \cdot y\} \\ &= f(x \cdot y) = f(x) * f(y) = \varphi^{-1}(\overline{x}) * \varphi^{-1}(\overline{y}).\end{aligned}$$

Therefore, φ is an isomorphism. ■

Theorem 1.4.16. *Let $(H, \cdot)$ be a semihypergroup and R be an equivalence relation on H.*

(1) *If R is strongly regular, then H/R is a semigroup, with respect to the following operation: $\overline{x} \otimes \overline{y} = \{\overline{z} \mid z \in x \cdot y\}$.*
(2) *If the above operation is well defined on H/R, then R is strongly regular.*

Proof. (1) For all x, y of H, we have $(x \cdot y)\overline{\overline{R}}(x \cdot y)$. Hence, $\overline{x} \otimes \overline{y} = \{\overline{z} \mid z \in x \cdot y\} = \{\overline{z}\}$, which means that $\overline{x} \otimes \overline{y}$ has exactly an element. Therefore, $(H/R, \otimes)$ is a semigroup.

(2) If aRb and x is an arbitrary element of H, we check that $(a \cdot x)\overline{\overline{R}}(b \cdot x)$. Indeed, for all $u \in a \cdot x$ and all $v \in b \cdot x$ we have $\overline{u} = \overline{a} \otimes \overline{x} = \overline{b} \otimes \overline{x} = \overline{v}$, which means that uRv. Hence, R is strongly regular on the right and similarly, it can be shown that it is strongly regular on the left. ■

Corollary 1.4.17. *If $(H, \cdot)$ is a hypergroup and R is an equivalence relation on H, then R is strongly regular if and only if $(H/R, \otimes)$ is a group.*

Proof. It is obvious. ■

Theorem 1.4.18. *If $(H, \cdot)$ is a semihypergroup, $(S, *)$ is a semigroup and $f : H \to S$ is a homomorphism, then the equivalence ρ^f associated with f is strongly regular.*

Proof. Let $a\rho^f b$, $x \in H$ and $u \in a \cdot x$. It follows that

$$f(u) = f(a) * f(x) = f(b) * f(x) = f(b \cdot x).$$

Hence, for all $v \in b \cdot x$, we have $f(u) = f(v)$, which means that $u\rho^f v$. Hence, ρ^f is strongly regular on the right and similarly, it is strongly regular on the left. ■

Definition 1.4.19. Let $(H, \cdot)$ be a semihypergroup and $n > 1$ be a natural number. We say that

$$x\beta_n y \text{ if there exists } a_1, \ldots, a_n \text{ in } H, \text{ such that } \{x, y\} \subseteq \prod_{i=1}^{n} a_i.$$

Let $\beta = \bigcup_{n \geq 1} \beta_n$, where $\beta_1 = \{(x, x) \mid x \in H\}$ is the diagonal relation on H. Clearly, the relation β is reflexive and symmetric. Denote by β^* the transitive closure of β.

Theorem 1.4.20. *β^* is the smallest strongly regular relation on H.*

Proof. We show that

(1) β^* is a strongly regular relation on H.
(2) If R is a strongly regular relation on H, then $\beta^* \subseteq R$.

(1) Suppose that $a\beta^*b$ and x is an arbitrary element of H. It follows that there exist $x_0 = a, x_1, \ldots, x_n = b$ such that for all $i \in \{0, 1, \ldots, n-1\}$ we have $x_i\beta x_{i+1}$. Let $u_1 \in a \cdot x$ and $u_2 \in b \cdot x$. We check that $u_1\beta^*u_2$. From $x_i\beta x_{i+1}$ it follows that there exists a hyperproduct P_i, such that $\{x_i, x_{i+1}\} \subseteq P_i$ and so $x_i \cdot x \subseteq P_i \cdot x$ and $x_{i+1} \cdot x \subseteq P_i \cdot x$, which means that $x_i \cdot x\overline{\overline{\beta}}x_{i+1} \cdot x$. Hence, for all $i \in \{0, 1, \ldots, n-1\}$ and for all $s_i \in x_i \cdot x$ we have $s_i\beta s_{i+1}$. If we consider $s_0 = u_1$ and $s_n = u_2$, then we obtain $u_1\beta^*u_2$. Then, β^* is strongly regular on the right and similarly, it is strongly regular on the left.

(2) We have $\beta_1 = \{(x, x) \mid x \in H\} \subseteq R$, since R is reflexive. Suppose that $\beta_{n-1} \subseteq R$ and show that $\beta_n \subseteq R$. If $a\beta_n b$, then there exist $x_1, \ldots, x_n$ in H, such that $\{a, b\} \subseteq \prod_{i=1}^{n} x_i$. Hence, there exists u, v in $\prod_{i=1}^{n-1} x_i$, such that $a \in u \cdot x_n$ and $b \in v \cdot x_n$. We have $u\beta_{n-1}v$ and according to the hypothesis, we obtain uRv. Since R is strongly regular, it follows that aRb. Hence, $\beta_n \subseteq R$. By induction, it follows that $\beta \subseteq R$, whence $\beta^* \subseteq R$. ■

Hence, the relation β^* is the smallest equivalence relation on H, such that the quotient H/β^* is a group.

Definition 1.4.21. β^* is called the *fundamental relation* on H and H/β^* is called the *fundamental group.*

The kernel of the canonical map $\varphi : H \to H/\beta^*$ is called the *core* or *heart* of H and is denoted by ω_H. Here, we also denote by ω_H the unit of H/β^*.

1.5 Hyperrings

The more general structure that satisfies the ring-like axioms is the hyperring in the general sense: $(R, +, \cdot)$ is a hyperring if $+$ and $\cdot$ are two hyperoperations such that $(R, +)$ is a hypergroup and $\cdot$ is an associative hyperoperation, which is distributive with respect to $+$. There are different types of hyperrings. If only the addition $+$ is a hyperoperation and the multiplication $\cdot$ is a usual operation, then we say that R is an additive hyperring. A special case of this type is the hyperring introduced by Krasner [130].

Definition 1.5.1. A *Krasner hyperring* is an algebraic structure $(R, +, \cdot)$ which satisfies the following axioms:

(1) $(R, +)$ is a *canonical hypergroup*, i.e.,
   (a) for every $x, y, z \in R$, $x + (y + z) = (x + y) + z$,
   (b) for every $x, y \in R$, $x + y = y + x$,

(c) there exists $0 \in R$ such that $0 + x = \{x\}$ for every $x \in R$,
(d) for every $x \in R$ there exists a unique element $x' \in R$ such that $0 \in x + x'$;
(We shall write $-x$ for x' and we call it the *opposite* of x.)
(e) $z \in x + y$ implies $y \in -x + z$ and $x \in z - y$.

(2) $(R, \cdot)$ is a semigroup having zero as a bilateral absorbing element, i.e., $x \cdot 0 = 0 \cdot x = 0$.
(3) The multiplication is distributive with respect to the hyperoperation $+$.

The following elementary facts follow easily from the axioms: $-(-x) = x$ and $-(x + y) = -x - y$, where $-A = \{-a \mid a \in A\}$. Also, for all $a, b, c, d \in R$ we have $(a + b) \cdot (c + d) \subseteq a \cdot c + b \cdot c + a \cdot d + b \cdot d$. In Definition 1.5.1, for simplicity of notations we write sometimes xy instead of $x \cdot y$ and in (c), $0 + x = x$ instead of $0 + x = \{x\}$.

A Krasner hyperring $(R, +, \cdot)$ is called *commutative* (*with unit element*) if $(R, \cdot)$ is a commutative semigroup (with unit element).

EXAMPLE 18 (1) Let $R = \{0, 1, 2\}$ be a set with the hyperoperation $+$ and the binary operation $\cdot$ defined as follow:

$+$	0	1	2
0	0	1	2
1	1	1	R
2	2	R	2

and

$\cdot$	0	1	2
0	0	0	0
1	0	1	2
2	0	1	2

Then, $(R, +, \cdot)$ is a Krasner hyperring.

(2) The first construction of a hyperring appeared in Krasner's paper [130] and it is the following one: Consider $(F, +, \cdot)$ a field, G a subgroup of $(F^*, \cdot)$ and take $F/G = \{aG \mid a \in F\}$ with the hyperaddition and the multiplication given by

$$aG \oplus bG = \{cG \mid c \in aG + bG\},$$
$$aG \odot bG = abG.$$

Then, $(F/G, \oplus, \odot)$ is a hyperring. If $(F, +, \cdot)$ is a unitary ring and G is a subgroup of the monoid $(F^*, \cdot)$ such that $xG = Gx$, for all $x \in F$, then $(F/G, \oplus, \odot)$ is a Krasner hyperring with identity.

(3) Let $(A, +, \cdot)$ be a ring and N a normal subgroup of its multiplicative semigroup. Then, the multiplicative classes $\overline{x} = xN$ $(x \in A)$ form a partition of R, and let $\overline{A} = A/N$ be the set of these classes. If for all $\overline{x}, \overline{y} \in \overline{A}$, we define

$$\overline{x} \oplus \overline{y} = \{\overline{z} \mid z \in \overline{x} + \overline{y}\}, \text{ and } \overline{x} * \overline{y} = \overline{x \cdot y},$$

then the obtained structure is a Krasner hyperring.

(4) Let R be a commutative ring with identity. We set $\overline{R} = \{\overline{x} = \{x, -x\} \mid x \in R\}$. Then, $\overline{R}$ becomes a Krasner hyperring with respect to the hyperoperation $\overline{x} \oplus \overline{y} = \{\overline{x + y}, \overline{x - y}\}$ and multiplication $\overline{x} \otimes \overline{y} = \overline{x \cdot y}$.

Definition 1.5.2. Let $(R, +, \cdot)$ be a Krasner hyperring and A be a non-empty subset of R. Then, A is said to be a *subhyperring* of R if $(A, +, \cdot)$ is itself a Krasner hyperring.

The subhyperring A of R is *normal* in R if and only if $x + A - x \subseteq A$ for all $x \in R$.

Definition 1.5.3. A subhyperring A of a Krasner hyperring R is a *left (right) hyperideal* of R if $r \cdot a \in A \quad (a \cdot r \in A)$ for all $r \in R,\ a \in A$. A is called a *hyperideal* if A is both a left and a right hyperideal.

Lemma 1.5.4. *A non-empty subset A of a Krasner hyperring R is a left (right) hyperideal if and only if*

(1) $a, b \in A$ *implies* $a - b \subseteq A$.
(2) $a \in A,\ r \in R$ *imply* $r \cdot a \in A \quad (a \cdot r \in A)$.

Proof. It is straightforward. ■

Definition 1.5.5. Let A and B be non-empty subsets of a Krasner hyperring R.

- The sum $A + B$ is defined by

$$A + B = \{x \mid x \in a + b \text{ for some } a \in A,\ b \in B\}.$$

- The product AB is defined by

$$AB = \left\{x \mid x \in \sum_{i=1}^{n} a_i b_i, a_i \in A, b_i \in B, n \in \mathbb{Z}^+\right\}.$$

If A and B are hyperideals of R, then $A + B$ and AB are also hyperideals of R.

The second type of a hyperring was introduced by Rota [158] in 1982. The multiplication is a hyperoperation, while the addition is an operation, that is why she called it a multiplicative hyperring.

Definition 1.5.6. A triple $(R, +, \cdot)$ is called a *multiplicative hyperring* if

(1) $(R, +)$ is an abelian group.
(2) $(R, \cdot)$ is a semihypergroup.
(3) For all $a, b, c \in R$, we have $a \cdot (b + c) \subseteq a \cdot b + a \cdot c$ and $(b + c) \cdot a \subseteq b \cdot a + c \cdot a$.
(4) For all $a, b \in R$, we have $a \cdot (-b) = (-a) \cdot b = -(a \cdot b)$.

If in (3) we have equalities instead of inclusions, then we say that the multiplicative hyperring is *strongly distributive*.

An element e in R, such that for all $a \in R$, we have $a \in a \cdot e \cap e \cdot a$, is called a *weak identity* of R.

EXAMPLE 19 (1) Let $(R, +, \cdot)$ be a ring and I be an ideal of it. We define the following hyperoperation on R: For all $a, b \in R$, $a * b = a \cdot b + I$. Then, $(R, +, *)$ is a strongly distributive hyperring. Indeed, first of all, $(R, +)$ is an abelian group. Then, for all $a, b, c \in R$, we have
$a*(b*c) = a*(b \cdot c+I) = \bigcup_{h \in I} a*(b \cdot c+h) = \bigcup_{h \in I} a \cdot (b \cdot c+h)+I = a \cdot b \cdot c+I$
and similarly, we have $(a*b)*c = a \cdot b \cdot c + I$. Moreover, for all $a, b, c \in R$, we have $a * (b + c) = a \cdot (b + c) + I = a \cdot b + a \cdot c + I = a * b + a * c$ and similarly, we have $(b + c) * a = b * a + c * a$. Finally, for all $a, b \in R$, we have $a * (-b) = a \cdot (-b) + I = (-a) \cdot b + I = (-a) * b$ and $-(a * b) = (-a \cdot b) + I = (-a) \cdot b + I = a * (-b)$.

(2) Let $(R, +, \cdot)$ be a non-zero ring. For all $a, b \in R$ we define the hyperoperation $a * b = \{a \cdot b, 2a \cdot b, 3a \cdot b, \ldots\}$. Then, $(R, +, *)$ is a multiplicative hyperring, which is not strongly distributive. Notice that for all $a \in R$, we have $a * 0 = 0 * a = \{0\}$.

Proposition 1.5.7. *If $(R, +, \cdot)$ is a multiplicative hyperring, then for all $a, b, c \in R$,*

$$a \cdot (b - c) \subseteq a \cdot b - a \cdot c \quad \textit{and} \quad (b - c) \cdot a \subseteq b \cdot a - c \cdot a.$$

If $(R, +, \cdot)$ is a strongly distributive, then for all $a, b, c \in R$,

$$a \cdot (b - c) = a \cdot b - a \cdot c \quad \textit{and} \quad (b - c) \cdot a = b \cdot a - c \cdot a.$$

Proof. The statement follows from the conditions (3) and (4) of Definition 1.5.6 ■

Proposition 1.5.8. *In a strong distributive hyperring $(R, +, \cdot)$, we have $0 \in a \cdot 0$ and $0 \in 0 \cdot a$, for all $a \in R$.*

Proof. The statement follows from the above proposition, by considering b=c. ■

Theorem 1.5.9. *For a strongly distributive hyperring $(R, +, \cdot)$, the following statements are equivalent:*

(1) *There exists $a \in R$ such that $|0 \cdot a| = 1$.*
(2) *There exists $a \in R$ such that $|a \cdot 0| = 1$.*
(3) $|0 \cdot 0| = 1$.
(4) $|a \cdot b| = 1$, *for all $a, b \in R$.*
(5) $(R, +, \cdot)$ *is a ring.*

Proof. (2⇒3): Suppose $a \neq 0$. For all $a \in R$ we have $0 \cdot 0 = (a - a) \cdot 0 = a \cdot 0 - a \cdot 0$ and so by (2), it follows that $0 \cdot 0 = \{0\}$, whence we obtain (3).

(3⇒4): For all $a \in R$, we have $0 \cdot 0 = a \cdot 0 - a \cdot 0$ and so by (3) it follows that $|a \cdot 0| = 1$, otherwise if we suppose that there exist $x \neq y$ elements of $a \cdot 0$, then $0 \cdot 0$ would contain $x - y \neq 0$ and 0, a contradiction. On the other hand, for all $a, b \in R$ we have $a \cdot 0 = a \cdot (b - b) = a \cdot b - a \cdot b$, whence it follows that $a \cdot b$ contains only an element. The other implications (4⇒5) and (5⇒2) are immediate. Similarly, the condition (1) is equivalent to (3), (4) and (5). ■

Corollary 1.5.10. *A strongly distributive hyperring $(R, +, \cdot)$ is a ring if and only if there exist $a_0, b_0 \in R$ such that $|a_0 \cdot b_0| = 1$.*

Proof. According to the above theorem, it is sufficient to check that $|a_0 \cdot 0|=1$. We have $a_0 \cdot 0 = a_0 \cdot (b_0 - b_0) = a_0 \cdot b_0 - a_0 \cdot b_0$, whence we obtain that $a_0 \cdot 0$ contains only 0. ■

Notice that there exist multiplicative hyperrings, which are not strongly distributive and for which we have $a * 0 = \{0\}$ for all $a \in R$.

Definition 1.5.11. A hyperring $(R, +, \cdot)$ is called *unitary* if it contains an element u, such that $a \cdot u = u \cdot a = \{a\}$ for all $a \in R$.

We obtain the following result.

Theorem 1.5.12. *Every unitary strongly distributive hyperring $(R, +, \cdot)$ is a ring.*

Proof. If u is the unit element, then we have $u \cdot u = \{u\}$ and according to the above corollary, it follows that R is a ring. ■

Theorem 1.5.13. *In any multiplicative hyperring $(R, +, \cdot)$, if there are $a, b \in R$ such that $|a \cdot b| = 1$, then $0 \cdot 0 = \{0\}$.*

Proof. We have $a \cdot 0 = a \cdot (b - b) \subseteq a \cdot b - a \cdot b = \{0\}$. On the other hand, $0 \cdot 0 = (a - a) \cdot 0 \subseteq a \cdot 0 - a \cdot 0$. But this must also be $\{0\}$, since $a \cdot 0$ is a singleton. ■

Corollary 1.5.14. *In any unitary multiplicative hyperring $(R, +, \cdot)$, we have $0 \cdot 0 = \{0\}$.*

Definition 1.5.15. Let $(R, +, \cdot)$ be a multiplicative hyperring and H be a non-empty subset of R. We say that H is a *subhyperring* of $(R, +, \cdot)$ if $(H, +, \cdot)$ is a multiplicative hyperring.

In other words, H is a subhyperring of $(R, +, \cdot)$ if $H - H \subseteq H$ and for all $x, y \in H$, $x \cdot y \subseteq H$.

Definition 1.5.16. We say that H is a *hyperideal* of $(R, +, \cdot)$ if $H - H \subseteq H$ and for all $x, y \in H$, $r \in R$, $x \cdot r \cup r \cdot x \subseteq H$.

The intersection of two subhyperrings of a multiplicative hyperring $(R, +, \cdot)$ is a subhyperring of R. The intersection of two hyperideals of a multiplicative hyperring $(R, +, \cdot)$ is a hyperideal of R. Moreover, any intersection of subhyperrings of a multiplicative hyperring is a subhyperring, while any intersection of hyperideals of a multiplicative hyperring is a hyperideal.

In this manner, we can consider the *hyperideal generated by any subset S* of $(R, +, \cdot)$, which is the intersection of all hyperideals of R, which contain S.

For each multiplicative hyperring $(R, +, \cdot)$, the *zero hyperideal* is the hyperideal generated by the additive identity 0. Contrary to what happens in ring theory, the zero hyperideal can contain other elements than 0. If we denote the zero hyperideal of R by $< 0 >$, then we have

$$\langle 0 \rangle = \Big\{ \sum_i x_i + \sum_j y_j + \sum_k z_k \mid \text{each sum is finite and for each } i, j, k \text{ there exist } r_i, s_j, t_k, u_k \in R \text{ such that } x_i \in r_i \cdot 0,\ y_j \in 0 \cdot s_j,\ z_k \in t_k \cdot 0 \cdot u_k \Big\}.$$

Denote by $H \oplus K$ the hyperideal generated by $H \cup K$, where H and K are hyperideals of $(R, +, \cdot)$.

Theorem 1.5.17. *If H and K are hyperideals of R, then*

$$H \oplus K = \{h + k \mid h \in H, k \in K\}.$$

Proof. Denote the set $\{h + k \mid h \in H, k \in K\}$ by I. Then, I is a hyperideal of R, which contains H and K.

Moreover, if J is a hyperideal of R, containing H and K, then $I \subseteq J$. Hence, we have $I = H \oplus K$. ■

Notice that the above theorem can be extended to an whichever family of hyperideals.

If we denote by $\mathcal{I}$ the set of all hyperideals of a multiplicative hyperring $(R, +, \cdot)$, then $(\mathcal{I}, \subseteq)$ is a complete lattice. The infimum of any family of hyperideals is their intersection, while the supremum is the hyperideal generated by their union.

Definition 1.5.18. A *homomorphism (good homomorphism)* between two multiplicative hyperrings $(R, +, \circ)$ and $(R', +', \circ')$ is a map $f : R \to R'$ such that for all x, y of R, we have $f(x+y) = f(x) +' f(y)$ and $f(x \circ y) \subseteq f(x) \circ' f(y)$ ($f(x \circ y) = f(x) \circ' f(y)$ respectively).

The following definition introduces a hyperring in general form. Both addition and multiplication are hyperoperations, that satisfy a set of conditions.

Definition 1.5.19. A hyperringoid $(H, \oplus, \odot)$ is called a *hyperring* if the following conditions are satisfied.

(1) $(H, \oplus)$ is a commutative hypergroup.
(2) $(H, \odot)$ is a semihypergroup.
(3) For all $x, y, z \in H$, $(x \oplus y) \odot z = (x \odot z) \oplus (y \odot z)$, $z \odot (x \oplus y) = (z \odot x) \oplus (z \odot y)$.
(4) For all $x \in H$ and all $u \in \omega_{(H,\oplus)}$, $x \odot u \subseteq \omega_{(H,\oplus)} \supseteq u \odot x$.

Several authors considered $(H, \oplus)$ as a canonical hypergroup in Definition 1.5.19.

1.6 H_v-Structures

H_v-structures were first introduced by Vougiouklis in the fourth AHA congress (1990)[187]. The concept of H_v-structures constitutes a generalization of the well-known algebraic hyperstructures (hypergroup, hyperring, and so on). Actually some axioms concerning the above hyperstructures such as the associative law, the distributive law and so on are replaced by their corresponding weak axioms. Since the quotients of the H_v-structures with respect to the fundamental equivalence relations (β^*, γ^*, etc.) are always ordinary structures, we can say that they are by virtue structures and this is why they are called H_v-structures. The reader will find in [69, 83, 185] some basic definitions and theorems about the H_v-structures.

The study of H_v-structure theory has been pursued in many directions by Vougiouklis, Davvaz, Spartalis and many others.

Definition 1.6.1. The hyperstructure $(H, \cdot)$ is called an *H_v-group* if

(1) $x \cdot (y \cdot z) \cap (x \cdot y) \cdot z \neq \emptyset$, for all $x, y, z \in H$.
(2) $a \cdot H = H \cdot a = H$, for all $a \in H$.

A motivation to obtain the above structures is explained by the following example.

EXAMPLE 20 [184] Let $(G, \cdot)$ be a group and R an equivalence relation on G. In G/R consider the hyperoperation $\odot$ such that $\overline{x} \odot \overline{y} = \{\overline{z} \mid z \in \overline{x} \cdot \overline{y}\}$, where $\overline{x}$ denotes the class of the element x. Then, $(G, \odot)$ is an H_v-group which is not always a hypergroup.

EXAMPLE 21 [184] On the set $\mathbb{Z}_{mn}$ consider the hyperoperation $\oplus$ defined by setting $\overline{0} \oplus m = \{\overline{0}, m\}$ and $x \oplus y = x + y$ for all $(x, y) \in \mathbb{Z}^2_{mn} - \{(\overline{0}, m)\}$. Then, $(\mathbb{Z}_{mn}, \oplus)$ becomes an H_v-group. $\oplus$ is weak associative but not associative.

EXAMPLE 22 [181] Consider the group $(\mathbb{Z}^n, +)$ and take $m_1, \ldots, m_n \in \mathbb{N}$. We define a hyperoperation $\oplus$ in $\mathbb{Z}^n$ as follows:

$$\begin{array}{l}(m_1, 0, \ldots, 0) \oplus (0, 0, \ldots, 0) = \{(m_1, 0, \ldots, 0), (0, 0, \ldots, 0)\},\\ (0, m_1, \ldots, 0) \oplus (0, 0, \ldots, 0) = \{(0, m_1, \ldots, 0), (0, 0, \ldots, 0)\},\\ (0, 0, \ldots, m_n) \oplus (0, 0, \ldots, 0) = \{(0, 0, \ldots, m_n), (0, 0, \ldots, 0)\},\end{array}$$

and $+ = \oplus$ in the remaining cases. Then, $(\mathbb{Z}^n, \oplus)$ is an H_v-group.

Definition 1.6.2. Let $(H_1, \cdot)$, $(H_2, *)$ be two H_v-groups. A map $f : H_1 \to H_2$ is called an *H_v-homomorphism* or *weak homomorphism* if

$$f(x \cdot y) \cap f(x) * f(y) \neq \emptyset, \text{ for all } x, y \in H_1.$$

f is called an *inclusion homomorphism* if

$$f(x \cdot y) \subseteq f(x) * f(y), \text{ for all } x, y \in H_1.$$

Finally, f is called a *strong homomorphism* if

$$f(x \cdot y) = f(x) * f(y), \text{ for all } x, y \in H_1.$$

If f is onto, one to one and strong homomorphism, then it is called an *isomorphism*, if moreover f is defined on the same H_v-group then it is called an *automorphism*. It is an easy verification that the set of all automorphisms in H, written $AutH$, is a group.

On a set H several H_v-structures can be defined. A partial order on those hyperstructures is introduced as follows.

Definition 1.6.3. [187] Let $(H, \cdot)$, $(H, *)$ be two H_v-groups defined on the same set H. We call $(\cdot)$ *less than or equal to* $(*)$, and write $\cdot \leq *$, if there is $f \in Aut(H, *)$ such that $x \cdot y \subseteq f(x * y)$ for all $x, y \in H$.

In [184], it is proved that all the quasihypergroups with two elements are H_v-groups. It is also proved that up to the isomorphism there are exactly 18 different H_v-groups.

If a hyperoperation is weak associative then every greater hyperoperation, defined on the same set is also weak associative. In [182], using this property, the set of all H_v-groups with a scalar unit defined on a set with three elements is determined, see [153, 195].

Let $(H, \cdot)$ be an H_v-group. The relation β^* is the smallest equivalence relation on H such that the quotient H/β^*, the set of all equivalence classes, is a group. β^* is called the *fundamental equivalence relation* on H.

According to [185] if $\mathcal{U}$ denotes the set of all the finite products of elements of H, then a relation β can be defined on H whose transitive closure is the fundamental relation β^*. The relation β is as follows: for x and y in H we write $x\beta y$ if and only if $\{x, y\} \subseteq u$ for some $u \in \mathcal{U}$. We can rewrite the definition of β^* on H as follows:
$a\beta^* b$ if and only if there exist $z_1, \ldots, z_{n+1} \in H$ with $z_1 = a,\ z_{n+1} = b$ and $u_1, \ldots, u_n \in \mathcal{U}$ such that

$$\{z_i, z_{i+1}\} \subseteq u_i \quad (i = 1, \ldots, n).$$

Suppose that $\beta^*(a)$ is the equivalence class containing $a \in H$. Then the product $\odot$ on H/β^* is defined as follows:

$$\beta^*(a) \odot \beta^*(b) = \{\beta^*(c) |\ c \in \beta^*(a) \cdot \beta^*(b)\} \text{ for all } a, b \in H.$$

It is proved in [185] that $\beta^*(a) \odot \beta^*(b)$ is the singleton $\{\beta^*(c)\}$ for all $c \in \beta^*(a) \cdot \beta^*(b)$. In this way H/β^* becomes a hypergroup. If we put $\beta^*(a) \odot \beta^*(b) = \beta^*(c)$, then H/β^* becomes a group.

Definition 1.6.4. A multivalued system $(R, +, \cdot)$ is an H_v-*ring* if

(1) $(R, +)$ is an H_v-group.
(2) $(R, \cdot)$ is an H_v-semigroup.
(3) $\cdot$ is weak distributive with respect to $+$, i.e., for all x, y, z in R we have

$$x \cdot (y + z) \cap x \cdot y + x \cdot z \neq \emptyset \text{ and } (x + y) \cdot z \cap x \cdot z + y \cdot z \neq \emptyset.$$

An H_v-ring may be commutative with respect either to $+$ or $\cdot$. If H is commutative with respect to both $+$ and $\cdot$, then we call it a *commutative H_v-ring.* If there exists $u \in R$ such that $x \cdot u = u \cdot x = \{x\}$ for all $x \in R$, then u is called the *scalar unit* of R and is denoted by 1.

Definition 1.6.5. An H_v-ring $(R, +, \cdot)$ is called a *dual H_v-ring* if $(R, \cdot, +)$ is an H_v-ring. If both $(+)$, $(\cdot)$ are weak commutative then R is called a *weak commutative dual H_v-ring.*

EXAMPLE 23 [191] Let $(H, *)$ be an H_v-group, then for every hyperoperation $(\circ)$ such that $\{x, y\} \subseteq x \circ y$ for all $x, y \in H$, the hyperstructure $(H, *, \circ)$ is a dual H_v-ring.

EXAMPLE 24 [100] In the set $\mathbb{R}^n$ ($\mathbb{R}$ is the set of real numbers) we define three hyperoperations as follows:

$$\begin{aligned} x \uplus y &= \{r(x+y) \mid r \in [0,1]\ \}, \\ x \otimes y &= \{x + r(y-x) \mid r \in [0,1]\ \}, \\ x \Box y &= \{x + ry \mid r \in [0,1]\ \}. \end{aligned}$$

Then, the hyperstructure $(\mathbb{R}^n, *, \circ)$ where $*, \circ \in \{\uplus, \otimes, \}$ is a weak commutative dual H_v-ring.

Definition 1.6.6. Let R be an H_v-ring. A non-empty subset I of R is called an *H_v-ideal* if the following axioms hold.

(1) $(I, +)$ is an H_v-subgroup of $(R, +)$.
(2) $I \cdot R \subseteq I$ and $R \cdot I \subseteq I$.

Definition 1.6.7. Let R_1, R_2 be two H_v-rings, the map $f : R_1 \to R_2$ is called an *H_v-homomorphism* or *weak homomorphism* if, for all $x, y \in R_1$, the following conditions hold: $f(x+y) \cap f(x) + f(y) \neq \emptyset$ and $f(x \cdot y) \cap f(x) \cdot f(y) \neq \emptyset$.

f is called an *inclusion homomorphism* if, for all $x, y \in R_1$, the following conditions hold: $f(x + y) \subseteq f(x) + f(y)$ and $f(x \cdot y) \subseteq f(x) \cdot f(y)$.

Finally, f is called a *strong homomorphism* if for all $x, y \in R_1$ we have $f(x + y) = f(x) + f(y)$ and $f(x \cdot y) = f(x) \cdot f(y)$.

If R_1 and R_2 are H_v-rings and there exists a strong one to one homomorphism from R_1 onto R_2, then R_1 and R_2 are called *isomorphic.*

Let $(R, +, \cdot)$ be an H_v-ring. We define the relation γ^* as the smallest equivalence relation on R such that the quotient R/γ^*, the set of all equivalence classes, is a ring [170]. In this case, γ^* is called the *fundamental equivalence relation* on R and R/γ^* is called the *fundamental ring.*

Let $\mathcal{U}$ be the set of all finite sums of products of elements of R. We define the relation γ on R as follows:

$$a \gamma b \text{ if and only if } \{a, b\} \subseteq u \text{ for some } u \in \mathcal{U}.$$

Let us denote $\widehat{\gamma}$ the transitive closure of γ. Then we can rewrite the definition of $\widehat{\gamma}$ on R as follows:

$a \widehat{\gamma} b$ if and only if there exist $z_1, z_2, \ldots, z_{n+1} \in R$ with $z_1 = a, z_{n+1} = b$ and $u_1, \ldots, u_n \in \mathcal{U}$ such that $\{z_i, z_{i+1}\} \subseteq u_i \quad (i = 1, \ldots, n)$.

Theorem 1.6.8. [170, 185] *The fundamental relation γ^* is the transitive closure of the relation γ.*

Now, we consider any ring $(R, +, \cdot)$ and let P_1, P_2 be non-empty subsets of R. We shall make use of the following right P-hyperoperations:

$$xP_1^*y = x + y + P_1, \; xP_2^*y = xyP_2, \text{ for all } x, y \in R.$$

We denote the center of the semigroup $(R, \cdot)$ by $Z(R)$.

Theorem 1.6.9. [171] *Let $(R, +, \cdot)$ be a ring and P_1, P_2 be non-empty subsets of R. If $0 \in P_1$ and $P_2 \cap Z(R) \neq \emptyset$, then (R, P_1^*, P_2^*) is an H_v-ring called P-H_v-ring.*

Definition 1.6.10. Let J be an H_v-ideal of (R, P_1^*, P_2^*). Since $0 \in RP_2^*J \cap JP_2^*R \subseteq J$ and $P_1 = 0P_1^*0 \subseteq J$, we have $JP_1^*x = J + x = xP_1^*J$, for all $x \in R$. Moreover, the addition $\oplus$ and the multiplication $\otimes$ between classes are defined in a usual manner:

$$(JP_1^*x) \oplus (JP_1^*y) = \{JP_1^*z |\; z \in (JP_1^*x)P_1^*(JP_1^*y)\} = \{J + x + y\},$$

$$(JP_1^*x) \otimes (JP_1^*y) = \{J + w |\; w \in (JP_1^*x)P_2^*(JP_1^*y)\} = \{J + w |\; w \in xyP_2\}.$$

Theorem 1.6.11. [171] *If (R, P_1^*, P_2^*) is a P-H_v-ring and J is an H_v-ideal, then $(R/J, \oplus, \otimes)$ is a multiplicative H_v-ring.*

Theorem 1.6.12. [171] *Let (R, P_1^*, P_2^*) be a P-H_v-ring and J is an H_v-ideal of R. If H is an H_v-subring of R containing P_1, then $HP_1^*J/J \cong H/H \cap J$.*

Theorem 1.6.13. [171] *Let J, K be two H_v-ideals of the P-H_v-ring (R, P_1^*, P_2^*). If $J \subseteq K$, then $(R/J)/(K/J) \cong R/K$.*

Chapter 2
Fuzzy Polygroups

2.1 What Fuzzy Sets Are?

Fuzzy sets are sets whose elements have degrees of membership. Fuzzy sets have been introduced by L. A. Zadeh (1965) as an extension of the classical notion of set [202]. In classical set theory, the membership of elements in a set is assessed in binary terms according to a bivalent condition: an element either belongs or does not belong to the set. By contrast, fuzzy set theory permits the gradual assessment of the membership of elements in a set; this is described with the aid of a membership function valued in the real unit interval $[0,1]$. Fuzzy sets generalize classical sets, since the characteristic functions of classical sets are special cases of the membership functions of fuzzy sets, if the latter only take values 0 or 1.

In our daily life , we usually want to seek opinions from professional persons with the best qualifications, for examples, the best medical doctors can provide the best diagnostics, the best pilots can provide the best navigation suggestions for airplanes etc. It is therefore desirable to incorporate the knowledge of these experts into some automatic systems so that it would become helpful for other people to make appropriate decisions which are (almost) as good as the decisions made by the top experts. With this aim in mind, our task is to design a system that would provide the best advice from the best experts in the field. However, one of the main hurdles of this incorporation is that the experts are usually unable to describe their knowledge by using precise and exact terms. For example, in order to describe the size of certain type of a tumor, a medical doctor would rarely use the exact numbers. Instead he would say something like "the size is between 1.4 and 1.6 cm ". Also, an expert would usually use some words from a natural language, e.g., "the size of the tumor is approximately 1.5 cm, with an error of about 0.1 cm". Thus, under such circumstances, the way to formalize the statements given by an expert is one of the main objectives of fuzzy logic.

Definition 2.1.1. Let X be a set. A *fuzzy subset* A of X is characterized by a membership function $\mu_A : X \to [0,1]$ which associates with each point $x \in X$ its *grade* or *degree of membership* $\mu_A(x) \in [0,1]$.

B. Davvaz and I. Cristea, *Fuzzy Algebraic Hyperstructures*,
Studies in Fuzziness and Soft Computing 321, DOI: 10.1007/978-3-319-14762-8_2

EXAMPLE 25 We can define a possible membership function for the fuzzy subsets of real numbers close to zero as follows:

$$\mu_A(x) = \frac{1}{1+10x^2}.$$

Using this function, we can determine the grade of each real number in this fuzzy subset, which signifies the degree to which that number is close to 0. For instance, the number 3 is assigned a grade of 0.01, the number 1 a grade of 0.09, the number 0.25 a grade of 0.62, and the number 0 a grade of 1. We might intuitively expect that by performing some operation on the function corresponding to the set of numbers close to 0, we could obtain a function representing the set of numbers very close to 0. One possible way of accomplishing this is to square the function, i.e.,

$$\mu_A(x) = \left(\frac{1}{1+10x^2}\right)^2.$$

We could also generalize this function to a family of functions representing the set of real numbers close to any given number a as follows:

$$\mu_A(x) = \frac{1}{1+10(x-a)^2}.$$

Definition 2.1.2. Let A and B be fuzzy subsets of X.

- $A \subseteq B$ if and only if $\mu_A(x) \leq \mu_B(x)$, for all $x \in X$.
- $A = B$ if and only if $\mu_A(x) = \mu_B(x)$, for all $x \in X$.
- $C = A \cup B$ if and only if $\mu_C(x) = \max\{\mu_A(x), \mu_B(x)\}$, for all $x \in X$.
- $D = A \cap B$ if and only if $\mu_D(x) = \min\{\mu_A(x), \mu_B(x)\}$, for all $x \in X$.

The *complement* of A, denoted by A^c, is defined by

$$\mu_{A^c}(x) = 1 - \mu_A(x), \text{ for all } x \in X.$$

Notice that when the range of membership functions is restricted to the set $\{0,1\}$, these functions perform precisely as the corresponding operators for crisp subsets.

For the sake of simplicity, we will represent every fuzzy subset by its membership function.

REMARK 1 In the definition of intersection and union of two fuzzy subsets, there are the traditional employed operations of the minimum, maximum, and the traditionally defined negation, and we use them in principle in the entire book. However, sometimes we use in the book later extensions of these operations, notably t-norms/conorms, wherever possible in this particular book. An extension of our results to the general case of t-norms/conorms, and other types of negation, as well as in a more general sense to later generalizations of many other concepts, operations, etc. of fuzzy sets theory and fuzzy logic, would probably in principle be possible but it may be non-trivial and we leave this for our future works.

Definition 2.1.3. Let f be a mapping from a set X to a set Y. Let μ be a fuzzy subset of X and λ be a fuzzy subset of Y. Then, the *inverse image* $f^{-1}(\lambda)$ of λ is the fuzzy subset of X defined by $f^{-1}(\lambda)(x) = \lambda(f(x))$, for all $x \in X$. The *image* $f(\mu)$ of μ is the fuzzy subset of Y defined by

$$f(\mu)(y) = \begin{cases} \sup\{\mu(t) \mid t \in f^{-1}(y)\} & \text{if } f^{-1}(y) \neq \emptyset \\ 0 & \text{otherwise,} \end{cases}$$

for all $y \in Y$.

It is not difficult to see that the following assertions hold.

(1) If $\{\lambda_i\}_{i \in I}$ is a family of fuzzy subsets of Y, then

$$f^{-1}\left(\bigcup_{i \in I} \lambda_i\right) = \bigcup_{i \in I} f^{-1}(\lambda_i) \text{ and } f^{-1}\left(\bigcap_{i \in I} \lambda_i\right) = \bigcap_{i \in I} f^{-1}(\lambda_i).$$

(2) If μ is a fuzzy subset of X, then $\mu \subseteq f^{-1}(f(\mu))$. Moreover, if f is one to one, then $f^{-1}(f(\mu)) = \mu$.
(3) If λ is a fuzzy subset of Y, then $f(f^{-1}(\lambda)) \subseteq \lambda$. Moreover, if f is onto, then $f\left(f^{-1}(\lambda)\right) = \lambda$.

Let f be a mapping from a set X to a set Y and μ be a fuzzy subset of X. Then, μ is called *f-invariant* if $f(x) = f(y)$ implies that $\mu(x) = \mu(y)$, for all $x, y \in X$.

Although the range of values between 0 and 1, inclusive, is the one most commonly used for representing membership grades, any arbitrary set with some natural full or partial ordering can in fact be used. Elements of this set are not required to be numbers as long as the ordering among them can be interpreted as representing various strengths of membership degree. This generalized membership function has the form

$$\mu_A : X \to L,$$

where L denotes any set that is at least partially ordered. Since L is most frequently a lattice, fuzzy subsets defined by this generalized membership grade function are called *L-fuzzy subsets*, where L is intended as an abbreviation for lattice.

Let $(X, \leq, \vee, \wedge)$ be a lattice with the minimum element 0 and the maximum 1. Given an L-fuzzy subset $\mu : X \to X$, the *α-cut* of μ is defined by $\mu_\alpha = \{x \in X \mid \mu(x) \geq \alpha\}$, with $\alpha \in X$.

We recall here some properties of the α-cuts, that can be find in [150].

Proposition 2.1.4. *For any L-fuzzy subset μ on X, with α-cuts $\{\mu_\alpha\}_{\alpha \in X}$, the following statements are valid.*

(1) *For all $\alpha, \beta \in X$, if $\alpha \leq \beta$, then $\mu_\beta \subseteq \mu_\alpha$.*
(2) *For all $I \subseteq X$, we have: $\bigcap_{\alpha \in I} \mu_\alpha = \mu_{\vee I}$.*
(3) $\mu_0 = X$.

Proposition 2.1.5. *Consider a family of sets* $\{\widetilde{\mu}_\alpha\}_{\alpha \in X}$ *which satisfy the following properties.*

(1) *For all* $\alpha, \beta \in X$, *if* $\alpha \leq \beta$, *then* $\widetilde{\mu}_\beta \subseteq \widetilde{\mu}_\alpha$.
(2) *For all* $I \subseteq X$, *we have:* $\bigcap\limits_{\alpha \in I} \widetilde{\mu}_\alpha = \widetilde{\mu}_{\vee I}$.
(3) $\widetilde{\mu}_0 = X$.

Define the L-fuzzy subset μ *as follows: for all* $x \in X, \mu(x) = \bigvee\{\alpha \in X \mid x \in \widetilde{\mu}_\alpha\}$. *Then, for all* $\alpha \in X$, *we have* $\mu_\alpha = \widetilde{\mu}_\alpha$.

Proposition 2.1.6. *For any L-fuzzy subsets* μ *and* λ *on* X, *we have* $\mu = \lambda$ *if and only if, for any* $\alpha \in X$, *we have* $\mu_\alpha = \lambda_\alpha$.

2.2 Groups and Fuzzy Subgroups

After the introduction of fuzzy sets by Zadeh, reconsideration of the concept of classical mathematics began. On the other hand, because of the importance of group theory in mathematics, as well as its many areas of application, the notion of fuzzy subgroups was defined by Rosenfeld in 1971 [160] and its structure was investigated. Das characterized fuzzy subgroups by their level subgroups in [57], since then many notions of fuzzy group theory can be equivalently characterized with the help of notion of level subgroups. Algebraic structures play a prominent role in mathematics with wide ranging applications in many disciplines such as theoretical physics, computer sciences, control engineering, information sciences, coding theory, topological spaces and so on. This provides sufficient motivations to researchers to review various concepts and results from the realm of abstract algebra in the broader framework of fuzzy setting.

In 1979, Anthony and Sherwood [6] redefined the fuzzy subgroup using the statistical triangular norm. This notion was introduced by Schweizer and Sklar [165] in order to generalize the ordinary triangle inequality in a metric space to the more general probabilistic metric space.

Definition 2.2.1. Let G be a group. A fuzzy subset μ of G is called a *fuzzy subgroup* if

(1) $\min\{\mu(x), \mu(y)\} \leq \mu(xy)$, for all $x, y \in G$,
(2) $\mu(x) \leq \mu(x^{-1})$, for all $x \in G$.

The following elementary facts about fuzzy subgroups follow easily from the axioms: $\mu(x) = \mu(x^{-1})$ and $\mu(x) \leq \mu(e)$, for all $x \in G$. Also, μ satisfies conditions (1) and (2) of Definition 2.2.1 if and only if $\min\{\mu(x), \mu(y)\} \leq \mu(xy^{-1})$, for all $x, y \in G$.

EXAMPLE 26 (1) Let $G = S_3$ be the symmetric group of degree 3 and $r_1, r_2, r_3 \in [0,1]$ such that $r_3 \le r_2 \le r_1$. We define μ by

$$\mu(x) = \begin{cases} r_1 \text{ if } x = e \\ r_2 \text{ if } x = (1\ 2) \\ r_3 \text{ otherwise.} \end{cases}$$

Clearly, μ is a fuzzy subgroup of G.

(2) Consider the infinite group $\mathbb{Z}$ with respect to the usual addition. We define μ by

$$\mu(x) = \begin{cases} 0.9 \text{ if } x \in 2\mathbb{Z} \\ 0.8 \text{ if } x \in 2\mathbb{Z}+1. \end{cases}$$

Then, μ is a fuzzy subgroup of $\mathbb{Z}$.

Definition 2.2.2. Let μ be a fuzzy subset of a set X. For $t \in [0,1]$, we define μ_t as follows:

$$\mu_t = \{x \in X \mid \mu(x) \ge t\}.$$

μ_t is called the *level set* of μ.

Theorem 2.2.3. *Let G be a group and μ be a fuzzy subset of G. Then, μ is a fuzzy subgroup of G if and only if μ_t ($\neq \emptyset$) is a subgroup of G for every $t \in [0,1]$.*

Proof. Suppose that μ is a fuzzy subgroup of G and let $t \in [0,1]$. Clearly, $e \in \mu_t$. Thus, μ_t is a non-empty set. Let $x, y \in \mu_t$. Then, $\mu(x) \ge t$ and $\mu(y) \ge t$. Since μ is a fuzzy subgroup, we have $\mu(xy^{-1}) \ge \min\{\mu(x), \mu(y)\} \ge t$. Hence, $xy^{-1} \in \mu_t$ and so μ_t is a subgroup of G.

Conversely, suppose that μ_t is a subgroup of G, for all $t \in [0,1]$. Let $x, y \in G$, and let $\mu(x) = t_1$ and $\mu(y) = t_2$. Let $t = \min\{t_1, t_2\}$. Then, $x, y \in \mu_t$ and $\mu(e) \ge t$. By hypothesis, μ_t is a subgroup of G and so $xy^{-1} \in \mu_t$. Hence, $\mu(xy^{-1}) \ge t = \min\{t_1, t_2\} = \min\{\mu(x), \mu(y)\}$. Then, μ is a fuzzy subgroup of G. ■

Theorem 2.2.4. *Let G, G' be two groups and let f be a homomorphism of G into G'.*

(1) *If μ is a fuzzy subgroup of G, then $f(\mu)$ is a fuzzy subgroup of G'.*
(2) *If λ is a fuzzy subgroup of G', then $f^{-1}(\lambda)$ is a fuzzy subgroup of G.*

Proof. (1) Let $a, b \in G'$. Suppose either $a \notin f(G)$ or $b \notin f(G)$. Then, $\min\{f(\mu)(a), f(\mu)(b)\} = 0 \le f(\mu)(ab)$. Assume that $a \notin f(G)$. Then, $a^{-1} \notin f(G)$. Thus, $f(\mu)(a) = 0 = f(\mu)(a^{-1})$. Now, suppose that $a = f(x)$ and $b = f(y)$ for some $x, y \in G$. Then,

$$\begin{aligned} f(\mu)(ab) &= \sup\{\mu(z) \mid z \in G, f(z) = ab\} \\ &\ge \sup\{\mu(xy) \mid x, y \in G, f(x) = a, f(y) = b\} \\ &\ge \sup\{\min\{\mu(x), \mu(y)\} \mid x, y \in G, f(x) = a, f(y) = b\} \\ &= \min\{\sup\{\mu(x) \mid x \in G, f(x) = a\}, \sup\{\mu(y) \mid y \in G, f(y) = b\}\} \\ &= \min\{f(\mu)(a), f(\mu)(b)\}. \end{aligned}$$

Also,

$$\begin{aligned} f(\mu)(a^{-1}) &= \sup\left\{\mu(z) \mid z \in G, f(z) = a^{-1}\right\} \\ &= \sup\left\{\mu(z^{-1}) \mid z \in G, f(z^{-1}) = a\right\} \\ &= f(\mu)(a). \end{aligned}$$

Hence, $f(\mu)$ is a fuzzy subgroup of G'.

(2) Let $x, y \in G$. Then,

$$\begin{aligned} f^{-1}(\lambda)(xy) &= \lambda(f(xy)) = \lambda(f(x)f(y)) \\ &\geq \min\{\lambda(f(x)), \lambda(f(y))\} \\ &= \min\left\{f^{-1}(\lambda)(x), f^{-1}(\lambda)(y)\right\}. \end{aligned}$$

Further,

$$f^{-1}(\lambda)(x^{-1}) = \lambda(f(x^{-1})) = \lambda(f(x)^{-1}) = \lambda(f(x)) = f^{-1}(\lambda)(x).$$

Hence, $f^{-1}(\lambda)$ is a fuzzy subgroup of G. ■

Definition 2.2.5. We define the binary operation "$\circ$" and the unitary operation $^{-1}$ on the set of all fuzzy subsets of G as follows. Let μ, λ be fuzzy subsets of G and $x \in G$ be arbitrary. Then,

$$(\mu \circ \lambda)(x) = \sup\{\min\{\mu(y), \lambda(z)\} \mid y, z \in G, yz = x\} \text{ and } \mu^{-1}(x) = \mu(x^{-1}).$$

We call $\mu \circ \lambda$ the *product* of μ and λ, and μ^{-1} the *inverse* of μ.

It is easy to verify that the binary operation $\circ$ in Definition 2.2.5 is associative.

Theorem 2.2.6. *Let μ be a fuzzy subset of a group G. Then, μ is a fuzzy subgroup if and only if μ satisfies the following conditions.*

(1) $\mu \circ \mu \subseteq \mu$.
(2) $\mu^{-1} \subseteq \mu$.

Proof. It is straightforward. ■

Corollary 2.2.7. *Let μ, λ be fuzzy subgroups of a group G. Then, $\mu \circ \lambda$ is a fuzzy subgroup of G if and only if $\mu \circ \lambda = \lambda \circ \mu$.*

Lemma 2.2.8. *Let $\{\mu_i\}_{i \in I}$ be a family of fuzzy subgroups of a group G. Then, $\cap\{\mu_i \mid i \in I\}$ is a fuzzy subgroup of G.*

Definition 2.2.9. Let μ be a fuzzy subset of a group G. Let

$$< \mu > = \cap\{\lambda \mid \mu \subseteq \lambda,\ \lambda \text{ is a fuzzy subgroup of } G\}.$$

Then, $< \mu >$ is called the fuzzy subgroup of G *generated* by μ.

Clearly, $< \mu >$ is the smallest fuzzy subgroup of G which contains μ.

Theorem 2.2.10. *Let μ be a fuzzy subset of a group G. Then, the following assertions are equivalent:*

(1) $\mu(xy) = \mu(yx)$, *for all $x, y \in G$; in this case, μ is called an abelian fuzzy subset of G.*
(2) $\mu(xyx^{-1}) = \mu(y)$, *for all $x, y \in G$.*
(3) $\mu(xyx^{-1}) \geq \mu(y)$, *for all $x, y \in G$.*
(4) $\mu(xyx^{-1}) \leq \mu(y)$, *for all $x, y \in G$.*
(5) $\mu \circ \lambda = \lambda \circ \mu$, *for every fuzzy subset λ of G.*

Proof. It is straightforward. ∎

Definition 2.2.11. Let μ be a fuzzy subgroup of a group G. Then, μ is called a *fuzzy normal subgroup* of G if it is an abelian fuzzy subset of G.

Theorem 2.2.12. *Let μ be a fuzzy subset of a group G. Then, μ is a fuzzy normal subgroup of G if and only if μ_t $(\neq \emptyset)$ is a normal subgroup of G, for all $t \in [0, 1]$.*

Proof. Suppose that μ is a fuzzy normal subgroup of G. Let $t \in [0, 1]$. Since μ is a fuzzy subgroup of G, μ_t is a subgroup of G. If $x \in G$ and $y \in \mu_t$, then it follows from Theorem 2.2.10 that $\mu(xyx^{-1}) = \mu(y) \geq t$. Thus, $xyx^{-1} \in \mu_t$. Hence, μ_t is a normal subgroup of G.

Conversely, assume that μ_t is a normal subgroup of G, for all $t \in [0, 1]$. Then, we have that μ is a fuzzy subgroup of G. Let $x, y \in G$ and $t = \mu(y)$. Then, $y \in \mu_t$ and so $xyx^{-1} \in \mu_t$. Hence, $\mu(xyx^{-1}) \geq t = \mu(y)$. That is, μ satisfies condition (3) of Theorem 2.2.10. Consequently, μ is a fuzzy normal subgroup of G. ∎

In what follows, we like to define the fuzzy quotients as the set of fuzzy t-cosets and give a structure to it by defining the operation between two fuzzy t-cosets.

Definition 2.2.13. Let μ be a fuzzy subgroup of a group G. For any $a \in G$ and $t \in [0, \mu(e)]$, a fuzzy subset μa of G is called a *fuzzy right t-coset* of μ in G if

$$\mu a(x) = \min\{\mu(xa), t\}, \quad \text{for all } x \in G.$$

Similarly, a fuzzy subset $a\mu$ of G is called a *fuzzy left t-coset* of μ in G if

$$a\mu(x) = \min\{\mu(ax), t\}, \quad \text{for all } x \in G.$$

The following example shows that for a fuzzy subgroup μ of G, fuzzy left t-cosets need not be equal to corresponding fuzzy right t-cosets.

EXAMPLE 27 Consider $G = S_3$, the symmetric group of order 3. Then, G can be expressed as $G = \{e, (1\ 2), (1\ 3), (2\ 3), (1\ 2\ 3), (1\ 3\ 2)\}$. Consider a fuzzy subgroup on G by

$$\mu(x) = \begin{cases} 1 & \text{if } x = e \text{ or } x = (1\ 2) \\ 0.5 & \text{otherwise.} \end{cases}$$

Then, it can be seen that for $t = 1$, $(1\ 2\ 3)\mu \neq \mu(1\ 2\ 3)$.

Proposition 2.2.14. *If μ is a fuzzy normal subgroup of G and a an arbitrary element of G, then the fuzzy right t-coset μa is same as the fuzzy left t-coset $a\mu$.*

Proof. Let μ be a fuzzy normal subgroup of G, $a \in G$, and $t \in [0, \mu(e)]$. Then, for any $x \in G$, $\mu(xa) = \mu(ax)$ and so $\mu a(x) = \min\{\mu(xa), t\} = \min\{\mu(ax), t\} = a\mu(x)$. ■

Consider the set $G/\mu^t = \{\mu a \mid a \in G\}$ of all fuzzy right t-cosets of μ. The following theorem shows that $|G/\mu_t| = |G/\mu^t|$ having the same number of elements.

Theorem 2.2.15. *Let μ be a fuzzy normal subgroup of a group G and $N = \mu_t$ be the level set of μ for some $t \in [0, 1]$. Then, $Na = Nb$ if and only if $\mu a = \mu b$, for all elements $a, b \in G$.*

Proof. Let a, b be arbitrary elements of G and $\mu a = \mu b$. Then,

$$\mu a(b^{-1}) = \mu b(b^{-1}) = \min\{\mu(bb^{-1}), t\} = \min\{1, t\} = t.$$

But, $\mu a(b^{-1}) = \min\{\mu(ab^{-1}), t\}$ which shows that $\mu(ab^{-1}) \geq t$ and hence $ab^{-1} \in N$, so $Na = Nb$.

Conversely, let $Na = Nb$ for any $a, b \in G$ and let x be any element in G. Then, $Nax = Nbx$. Observe that $\mu(ax) < t$ and $\mu(bx) \geq t$ implies that $Nbx = N \neq Nax$ which is not true. Similarly, $\mu(ax) \geq t$ and $\mu(bx) < t$ implies that $Nax = N \neq Nbx$ which is false again. Hence, either both $\mu(ax)$ and $\mu(bx)$ are greater than or equal to t, or both are less than t. In the first case,

$$\mu a(x) = \min\{\mu(ax), t\} = t \text{ and } \mu b(x) = \min\{\mu(bx), t\} = t,$$

where as in the other case,

$$\mu a(x) = \mu(ax) < t \text{ and } \mu b(x) = \mu(bx) < t.$$

In this case, since $Na = Nb$, we have $a = nb$ for some $n \in N$ and so

$$\begin{aligned} \mu b(x) &= \mu(bx) = \mu(n^{-1}ax) \\ &\geq \min\{\mu(n^{-1}), \mu(ax)\} \\ &= \mu(ax) = \mu a(x) = \mu(nbx) \\ &\geq \min\{\mu(n), \mu(bx)\} = \mu(bx) = \mu b(x). \end{aligned}$$

Thus, $\mu a(x) = \mu b(x)$, for all $x \in G$ and the proof is completed. ■

Now, we give a structure on G/μ^t by defining the operation $*$ between two fuzzy right t-cosets as

$$\mu a * \mu b = \mu ab.$$

Proposition 2.2.16. *If μ is a fuzzy normal subgroup of a group G, then the operation $*$ defined on G/μ^t is well defined.*

Proof. It is straightforward. ■

Then, $(G/\mu^t, *)$ becomes a group and is called the *quotient group* of G relative to the fuzzy normal subgroup μ.

2.3 Polygroups and Fuzzy Subpolygroups

The study of fuzzy hyperstructures is an interesting research topic of fuzzy sets. There is a considerable amount of work on the connections between fuzzy sets and hyperstructures. This work can be classified into three groups. A *first* group of papers studies *crisp* hyperoperations defined through fuzzy sets. This was initiated by Corsini in [28] and continued by himself and then by him together with Leoreanu [47]. For example let $\mu : H \to [0,1]$ be a fuzzy subset of a non-empty set H. Corsini in [28] defined on H the following hyperoperation:

$$x \circ y = y \circ x = \{z \in H \mid \mu(x) \leq \mu(z) \leq \mu(y)\}$$

for all $(x, y) \in H^2$ such that $\mu(x) \leq \mu(y)$. Then, $(H, \circ)$ is a join space. This connection will be presented in Section 5.1.1.

A *second* group of papers concerns the *fuzzy hyperalgebras*. This is a direct extension of the concept of fuzzy algebras (fuzzy (sub)groups, fuzzy lattices, fuzzy rings, etc). This approach can be extended to fuzzy hypergroups. For example, given a crisp hypergroup $(H, \cdot)$ and a fuzzy subset μ, then we say that μ is a fuzzy (sub)hypergroups of $(H, \cdot)$ if every level set of μ, say μ_t, is a (crisp) subhypergroup of $(H, \cdot)$. Zahedi, Bolurian and Hasankhani in 1995 [205] introduced the concept of a fuzzy subpolygroup. Davvaz in 1999 [76] introduced the concept of a fuzzy subhypergroup of a hypergroup which is a generalization of the concept of fuzzy subgroup and fuzzy subpolygroup.

A *third* approach involves something also called *fuzzy hypergroup*, but it is completely different from what we described above. It was studied by Corsini, Tofan, Zahedi, Hasankhani and then studied by Kehagias, Konstantinidou, Serafimidis, and many others. The basic idea is the following: a *crisp* hyperoperation assigns to every pair of elements a *crisp* set; a *fuzzy* hyperoperation assigns to every pair of elements a *fuzzy* set. More details about this connection can be found in Section 5.2.

Definition 2.3.1. [205] Let P be a polygroup. A fuzzy subset μ of P is called a *fuzzy subpolygroup* if

(1) $\min\{\mu(x), \mu(y)\} \leq \mu(z)$, for all $x, y \in P$ and for all $z \in x \cdot y$.
(2) $\mu(x) \leq \mu(x^{-1})$, for all $x \in P$.

Notice that Definition 2.3.1 is a generalization of Definition 2.2.1.

The following elementary facts about fuzzy subpolygroups follow easily from the axioms: $\mu(x) = \mu(x^{-1})$ and $\mu(x) \leq \mu(e)$, for all $x \in P$.

A *strong level set* $\mu_t^>$ of a fuzzy subset μ in P is defined by

$$\mu_t^> = \{x \in P \mid \mu(x) > t\}.$$

Theorem 2.3.2. *Let P be a polygroup and μ be a fuzzy subset of P. Then, the following are equivalent:*

(1) *μ is a fuzzy subpolygroup of P.*
(2) *Each non-empty strong level set of μ is a subpolygroup of P.*
(3) *Each non-empty level set of μ is a subpolygroup of P.*

Proof. The proof is similar to the proof of Theorem 2.2.3, by considering the suitable modifications. ■

Definition 2.3.3. Let μ be a fuzzy subpolygroup of P. Then, μ is said to be *normal* if for all $x, y \in P$,

$$\mu(z) = \mu(z'), \text{ for all } z \in x \cdot y \text{ and } z' \in y \cdot x.$$

It is obvious that if μ is a fuzzy normal subpolygroup of P, then for all $x, y \in P$,

$$\mu(z) = \mu(z'), \text{ for all } z, z' \in x \cdot y.$$

Theorem 2.3.4. *Let μ be a fuzzy subpolygroup of P. Then, the following conditions are equivalent:*

(1) *μ is a fuzzy normal subpolygroup of P.*
(2) *$\mu(z) = \mu(y)$, for all $x, y \in P$ and for all $z \in x \cdot y \cdot x^{-1}$.*
(3) *$\mu(z) \geq \mu(y)$, for all $x, y \in P$ and for all $z \in x \cdot y \cdot x^{-1}$.*
(4) *$\mu(z) \geq \mu(y)$, for all $x, y \in P$ and for all $z \in x^{-1} \cdot y^{-1} \cdot x \cdot y$.*

Proof. (1⇒2): Let $x, y \in P$ and $z \in x \cdot y \cdot x^{-1}$. Then, $z \in x \cdot p$, where $p \in y \cdot x^{-1}$. Since $p \in y \cdot x^{-1}$, we obtain $y \in p \cdot x$. Thus, by hypothesis we have $\mu(z) = \mu(y)$.

(2⇒3): It is obvious.

(3⇒1): Let $x, y \in P$ and $z \in x \cdot y$, $z' \in y \cdot x$. Hence, $y \in z' \cdot x^{-1}$ and so $z \in x \cdot y \subseteq x \cdot z' \cdot x^{-1}$. Thus, by hypothesis we obtain $\mu(z) \geq \mu(z')$. Similarly, we have $\mu(z') \geq \mu(z)$. Therefore, $\mu(z') = \mu(z)$.

(3⇒4): Let $x, y \in P$ and $z \in x^{-1} \cdot y^{-1} \cdot x \cdot y$. Then, $z \in q \cdot y$, where $q \in x^{-1} \cdot y^{-1} \cdot x$. Thus, $\mu(q) \geq \mu(y^{-1}) = \mu(y)$. So, $\mu(z) \geq \min\{\mu(q), \mu(y)\} = \mu(y)$.

(4⇒3): Let $x, y \in P$ and $z \in x \cdot y \cdot x^{-1}$. Then, $z \in x \cdot y \cdot x^{-1} \subseteq x \cdot y \cdot x^{-1} \cdot y^{-1} \cdot y$. So, we assume that $z \in q \cdot y$, where $q \in x \cdot y \cdot x^{-1} \cdot y^{-1}$. Thus, by (4) we have $\mu(q) \geq \mu(y^{-1}) = \mu(y)$. Therefore, $\mu(z) \geq \min\{\mu(y), \mu(q)\} = \mu(y)$. ■

Lemma 2.3.5. *Let μ be a fuzzy subpolygroup of P. Then, μ is normal in P if and only if μ_t $(\neq \emptyset)$ is a normal subpolygroup of P, for all $t \in [0,1]$.*

Proof. It is straightforward. ∎

Theorem 2.3.6. *Let $(P, \cdot)$, $(P', *)$ be polygroups, μ be a (normal) fuzzy subpolygroup of P, λ be a (normal) fuzzy subpolygroup of P', and $f : P \to P'$ be a function.*

(1) *If f is a strong homomorphism, then $f(\mu)$ is a fuzzy subpolygroup of P'.*
(2) *If f is a strong homomorphism, then $f^{-1}(\lambda)$ is a (normal) fuzzy subpolygroup of P.*
(3) *If f is an onto homomorphism of type 3 and μ is f-invariant, then $f(\mu)$ is a fuzzy subpolygroup of P'.*
(4) *If f is an onto homomorphism of type 4 and μ is f-invariant, then $f(\mu)$ is a (normal) fuzzy subpolygroup of P'.*

Proof. (1) First, we suppose that μ is a fuzzy subpolygroup of P. In order to prove that $f(\mu)$ is a fuzzy subpolygroup of P', by Theorem 2.3.2, it is sufficient to show that each non-empty strong level set $f(\mu)_t^{>}$ is a subpolygroup of P'. So, suppose that $f(\mu)_t^{>}$ is non-empty for some $t \in [0,1)$. Let $y_1, y_2 \in f(\mu)_t^{>}$. We show that $y_1 * y_2 \subseteq f(\mu)_t^{>}$. We have $f(\mu)(y_1) > t$ and $f(\mu)(y_2) > t$ which implies that

$$\sup_{x \in f^{-1}(y_1)} \{\mu(x)\} > t \text{ and } \sup_{x \in f^{-1}(y_2)} \{\mu(x)\} > t.$$

Therefore, there exist $x_1 \in f^{-1}(y_1)$ and $x_2 \in f^{-1}(y_2)$ such that $\mu(x_1) > t$ and $\mu(x_2) > t$. Since f is a strong homomorphism, we have $y_1 * y_2 = f(x_1) * f(x_2) = f(x_1 \cdot x_2)$. Let $z \in y_1 * y_2$. Then, there exists $x' \in x_1 \cdot x_2$ such that $f(x') = z$. Thus, $f(x) = f(x') = z$. Since μ is a fuzzy subpolygroup of P, we have

$$\mu(x') \geq \min\{\mu(x_1), \mu(x_2)\} > t.$$

Therefore,

$$f(\mu)(z) = \sup_{x \in f^{-1}(z)} \{\mu(x)\} > t.$$

Hence, $z \in f(\mu)_t^{>}$. Thus, $y_1 * y_2 \subseteq f(\mu)_t^{>}$.

Next, for $y \in f(\mu)_t^{>}$, we have $f(\mu)(y) > t$. Thus,

$$\sup_{x \in f^{-1}(y)} \{\mu(x)\} > t.$$

Hence,

$$\sup_{x^{-1} \in f^{-1}(y^{-1})} \{\mu(x^{-1})\} > t.$$

Therefore, $f(\mu)(y^{-1}) > t$ implies that $y^{-1} \in f(\mu)_t^{>}$.

(2) The proof is similar to the proof of Theorem 2.2.4, by considering the suitable modifications.

(3) Since f is an onto homomorphism of type 3, we have

$$f^{-1}(y_1 * y_2) = f^{-1}(y_1) \cdot f^{-1}(y_2), \text{ for all } y_1, y_2 \in P'.$$

Now, let y_1, y_2 be two arbitrary elements of P'. Then, we prove that

$$f(\mu)(z) \geq \min\{f(\mu)(y_1), f(\mu)(y_2)\}, \text{ for all } z \in y_1 * y_2.$$

Let $z \in y_1 * y_2$. Then, since μ is f-invariant, we have $f(\mu)(z) = \mu(x)$, for some $x \in f^{-1}(z)$. Since f is of type 3, we obtain

$$x \in f^{-1}(z) \subseteq f^{-1}(y_1 * y_2) = f^{-1}(y_1) \cdot f^{-1}(y_2).$$

Therefore, there exist $x_1 \in f^{-1}(y_1)$ and $x_2 \in f^{-1}(y_2)$ such that $x \in x_1 \cdot x_2$, and hence we have

$$\begin{aligned} f(\mu)(z) = \mu(x) &\geq \min\{\mu(x_1), \mu(x_2)\}, \text{ where } x_1 \in f^{-1}(y_1),\ x_2 \in f^{-1}(y_2) \\ &= \min\{f(\mu)(y_1), f(\mu)(y_2)\}, \end{aligned}$$

since μ is f-invariant. Now, $x \in f^{-1}(z) \Leftrightarrow x^{-1} \in f^{-1}(z^{-1})$. So, for an arbitrary element $z \in P$ we have

$$\begin{aligned} f(\mu)(z) &= \mu(x), \text{ for some } x \in f^{-1}(z) \\ &= \mu(x^{-1}), \text{ for some } x^{-1} \in f^{-1}(z^{-1}) \\ &= f(\mu)(z^{-1}), \text{ since } \mu \text{ is } f\text{-invariant.} \end{aligned}$$

(4) Suppose that μ is a fuzzy normal subpolygroup of P. By (3), it is enough to show that $f(\mu)$ is normal. In order to show that $f(\mu)$ is a fuzzy normal subpolygroup of P', it is sufficient to show that each non-empty strong level set $f(\mu)_t^>$ is a normal subpolygroup of P'. So, suppose that $f(\mu)_t^>$ is a non-empty strong level set of $f(\mu)$. Now, we show that

$$y * f(\mu)_t^> * y^{-1} \subseteq f(\mu)_t^>, \text{ for all } y \in P'.$$

Let $z \in y * f(\mu)_t^> * y^{-1}$. Then, $z \in y * p_0$, for some $p_0 \in f(\mu)_t^> * y^{-1}$. Therefore, $p_0 \in p_1 * y^{-1}$, for some $p_1 \in f(\mu)_t^>$. Now $p_1 \in f(\mu)_t^>$ implies that $f(\mu)(p_1) > t$. Thus, there exists $x_{p_1} \in f^{-1}(p_1)$ such that $\mu(x_{p_1}) > t$.

Let $f(x) = y$. Thus, $f(x^{-1}) = y^{-1}$ and $f(x_{p_1}) = p_1$. Thus, we have

$$p_0 \in p_1 * y^{-1} = f(x_{p_1}) * f(x^{-1}) = f(x_{p_1} \cdot x^{-1}),$$

since f is a homomorphism of type 4. Therefore, there exists $x_{p_0} \in x_{p_1} \cdot x^{-1}$ such that $f(x_{p_0}) = p_0$. Now, $z \in y * p_0 = f(x) * f(x_{p_0}) = f(x \cdot x_{p_0})$. This implies that there exists $x_z \in x \cdot x_{p_0} \subseteq x \cdot x_{p_1} \cdot x^{-1}$ such that $z = f(x_z)$. Now, since μ is normal, each non-empty strong level set $\mu_t^>$ is a normal subpolygroup of P. Therefore,

$$x_z \in x \cdot x_{p_1} \cdot x^{-1} \subseteq x \cdot \mu_t^> \cdot x^{-1} \subseteq \mu_t^>.$$

So, $\mu(x_z) > t$. Hence, $f(\mu)(z) = \sup\limits_{x \in f^{-1}(z)} \{\mu(x)\} > t$. Thus, $z \in f(\mu)_t^>$. ■

Definition 2.3.7. Let P be a polygroup, μ be a fuzzy subpolygroup of P and $t \in [0, \mu(e)]$. Then, the fuzzy subset $\mu^t a$ of P which is defined by

$$\mu^t a(x) = \min\left\{ \inf_{z \in x \cdot a} \{\mu(z)\}, t \right\}$$

is called the *fuzzy right t-coset* of μ. The *fuzzy left t-coset* $a\mu^t$ of μ is defined similarly.

Notice that the above definition is a generalization of Definition 2.2.13.

If μ is a fuzzy normal subpolygroup of P, then it is easy to see that $a\mu^t = \mu^t a$.

We denote the set of all fuzzy right t-cosets of μ by P/μ^t, i.e., $P/\mu^t = \{\mu^t a \mid a \in P\}$.

Theorem 2.3.8. *Let P be a polygroup, μ be a fuzzy normal subpolygroup of P and $t \in [0, \mu(e)]$. Then, there is a bijection between P/μ^t and P/μ_t.*

Proof. We define $\psi : P/\mu^t \to P/\mu_t$, by $\psi(\mu^t a) = \mu_t a$, for all $a \in P$. First, we show that

$$\mu_t a = \mu_t b \Leftrightarrow \mu^t a = \mu^t b, \text{ for all } a, b \in P.$$

Let $\mu^t a = \mu^t b$. Then, $\mu^t b(b^{-1}) = \min\{\mu(z_0), t\}$, for some $z_0 \in b^{-1} \cdot b$. Since $e \in b^{-1} \cdot b$, so by Definition 2.3.3, we have $\mu(z_0) = \mu(e) \geq t$, for $t \in [0, \mu(e)]$. Therefore, $\mu^t b(b^{-1}) = t$. Thus, $\mu^t a(b^{-1}) = t$. But, we have $t = \mu^t a(b^{-1}) = \min\{\mu(z_0'), t\}$, for some $z_0' \in b^{-1} \cdot a$. Thus, $\mu(z_0') \geq t$, for some $z_0' \in b^{-1} \cdot a$. So, the normality of μ implies that $\mu(z) \geq t$, for all $z \in b^{-1} \cdot a$. That is $b^{-1} \cdot a \subseteq \mu_t$. Now, choose $y_0 \in b^{-1} \cdot a$. Thus, $a \in b \cdot y_0$, and if $x \in \mu_t a$, then we obtain $x \in a\mu_t$. Therefore, $x \in a \cdot n$, for some $n \in \mu_t$, and hence $x \in a \cdot n \subseteq b \cdot y_0 \cdot n \subseteq b\mu_t$. So, $\mu_t a \subseteq b\mu_t = \mu_t b$. Similarly, we can show that $\mu_t b \subseteq \mu_t a$. Hence, $\mu_t b = \mu_t a$.

Conversely, let $\mu_t b = \mu_t a$. Then, $a \cdot x \cdot \mu_t = b \cdot x \cdot \mu_t$, for all $x \in P$. Also, for an arbitrary element $x \in P$, we obtain $\mu^t a(x) = \min\{\mu(z_0), t\}$, for some $z_0 \in a \cdot x$ and $\mu^t b(x) = \min\{\mu(z_0'), t\}$, for some $z_0' \in b \cdot x$. Now, we show that

$$\mu(z_0) \geq t, \ \mu(z_0') \geq t \ \text{ or } \ \mu(z_0) < t, \ \mu(z_0') < t.$$

In order to show this, we suppose on the contrary that $\mu(z_0) < t, \ \mu(z_0') \geq t$. Since for any $w \in b \cdot x$ we have $\mu(w) = \mu(z_0') \geq t$, so $b \cdot x \subseteq \mu_t$, which implies that $b \cdot x \cdot \mu_t = \mu_t$. Since $z_0 \notin \mu_t$, we obtain $a \cdot x \not\subseteq \mu_t$, and by some manipulations we can see that $a \cdot x \cdot \mu_t \neq \mu_t$. Therefore, we have $a \cdot x \cdot \mu_t \neq b \cdot x \cdot \mu_t$, which is a contradiction. Similarly, we can show that it is impossible to have $\mu(z_0) \geq t, \ \mu(z_0') < t$. Now, if $\mu(z_0) \geq t, \ \mu(z_0') \geq t$, then $\mu^t a(x) = \mu^t b(x) = t$, for all $x \in P$. If $\mu(z_0) < t, \ \mu(z_0') < t$, then it is easy to see that

$$\mu^t a(x) \geq \mu^t b(x) \ \text{ and } \ \mu^t b(x) \geq \mu^t a(x), \ \text{for all } x \in P.$$

Thus, $\mu^t a = \mu^t b$. Therefore, ψ is well defined and one to one. It is obvious that ψ is onto. ∎

Corollary 2.3.9. *Let P be a polygroup, μ be a fuzzy normal subpolygroup of P, $a \in P$ and $t \in [0, \mu(e)]$. Then, $\mu^t z = \mu^t a$, for all $z \in \mu_t a$.*

Now, we will induce a polygroup structure on P/μ^t.

Theorem 2.3.10. *Let P be a polygroup, μ be a fuzzy normal subpolygroup of P and $t \in [0, \mu(e)]$. Then, $(P/\mu^t, *)$ is a polygroup (called the polygroup of fuzzy t-cosets induced by μ and t), where the hyperoperation $*$ is defined as follows:*

$$* : P/\mu^t \times P/\mu^t \to \wp^*(P/\mu^t)$$
$$(\mu^t a, \mu^t b) \mapsto \{\mu^t z \mid z \in \mu_t \cdot a \cdot b\},$$

and $^{-1}$ on P/μ^t is defined by $(\mu^t a)^{-1} = \mu^t a^{-1}$.

Proof. First, we show that $*$ is well defined. Let $\mu^t a = \mu^t a_1$, $\mu^t b = \mu^t b_1$. Then, by Theorem 2.3.8, we have $\mu_t \cdot a \cdot b = \mu_t \cdot a_1 \cdot b_1$. If $\mu^t z \in \mu^t a * \mu^t b$, then $\mu^t z = \mu^t w$, for some $w \in \mu_t \cdot a \cdot b = \mu_t \cdot a_1 \cdot b_1$. So, by Theorem 2.3.8, we have $\mu_t \cdot z = \mu_t \cdot w$. Since $z \in \mu_t \cdot z$, we conclude that $z \in \mu_t \cdot w$. Therefore, $z \in \mu_t \cdot a_1 \cdot b_1$. Thus, we obtain $\mu^t z \in \mu^t a_1 * \mu^t b_1$. Hence, $\mu^t a * \mu^t b \subseteq \mu^t a_1 * \mu^t b_1$. Similarly, we have $\mu^t a_1 * \mu^t b_1 \subseteq \mu^t a * \mu^t b$. Now, we prove that the hyperoperation $*$ is associative. Let $a, b, c \in P$. Then,

$$\begin{aligned}(\mu^t a * \mu^t b) * \mu^t c &= \bigcup_{z \in \mu_t \cdot a \cdot b} (\mu^t z * \mu^t c) \\ &= \bigcup_{z \in \mu_t \cdot a \cdot b} \{\mu^t k \mid k \in \mu_t \cdot z \cdot c\} \\ &= \{\mu^t x \mid x \in \mu_t \cdot a \cdot b \cdot c\} \\ &= \bigcup_{s \in \mu_t \cdot b \cdot c} \{\mu^t w \mid w \in \mu_t \cdot a \cdot s\} \\ &= \bigcup_{s \in \mu_t \cdot b \cdot c} (\mu^t a * \mu^t s) \\ &= \mu^t a * (\mu^t b * \mu^t c).\end{aligned}$$

Clearly, $\mu^t e$ is the unit element in P/μ^t and the inverse element of $\mu^t a$ is $\mu^t a^{-1}$. Finally, we have

$$\mu^t z \in \mu^t x * \mu^t y \Rrightarrow \mu^t x \in \mu^t z * \mu^t y^{-1} \text{ and } \mu^t y \in \mu^t x^{-1} * \mu^t z.$$

Therefore, the proof is complete. ■

Now, let $\{P_\alpha \mid \alpha \in I\}$ be a collection of polygroups and $\prod\limits_{\alpha \in I} P_\alpha = \{< x_\alpha > \mid x_\alpha \in P_\alpha\}$ be the cartesian product of P_α $(\alpha \in I)$. Then, on $\prod\limits_{\alpha \in I} P_\alpha$ we can define a hyperoperation as follows: $< x_\alpha > \odot < y_\alpha >= \{< z_\alpha > \mid z_\alpha \in x_\alpha \cdot y_\alpha,\ \alpha \in I\}$. We call this the *direct product* of P_α $(\alpha \in I)$. It is easy to see that $\prod\limits_{\alpha \in I} P_\alpha$ equipped with the direct product becomes a polygroup.

Definition 2.3.11. Let $\{P_\alpha | \ \alpha \in I\}$ be a collection of polygroups and let μ_α be a fuzzy subset of P_α, for all $\alpha \in I$. We define *direct product* $\prod\limits_{\alpha \in I} \mu_\alpha$ by

$$\left(\prod_{\alpha \in I} \mu_\alpha\right)(x) = \inf\{\mu_\alpha(x_\alpha) | \ \alpha \in I\},$$

where $x =< x_\alpha >$ and $< x_\alpha >$ denotes an element of the direct product $\prod\limits_{\alpha \in I} P_\alpha$.

Proposition 2.3.12. *Let $\{P_\alpha \mid \alpha \in I\}$ be a collection of polygroups and let μ_α be a fuzzy subpolygroup of P_α. Then, $\prod\limits_{\alpha \in I} \mu_\alpha$ is a fuzzy subpolygroup of $\prod\limits_{\alpha \in I} P_\alpha$.*

Proof. Suppose that $x =< x_\alpha >$, $y =< y_\alpha > \in \prod\limits_{\alpha \in I} P_\alpha$. Then, for every $z =< z_\alpha > \in x \odot y =< x_\alpha > \odot < y_\alpha >$ we have

$$\begin{aligned}
\left(\prod_{\alpha \in I} \mu_\alpha\right)(z) &= \inf_{\alpha \in I}\{\mu_\alpha(z_\alpha)\} \geq \inf_{\alpha \in I}\{\min\{\mu_\alpha(x_\alpha), \mu_\alpha(y_\alpha)\}\} \\
&= \min\left\{\inf_{\alpha \in I}\{\mu_\alpha(x_\alpha)\}, \inf_{\alpha \in I}\{\mu_\alpha(y_\alpha)\}\right\} \\
&= \min\left\{\left(\prod_{\alpha \in I} \mu_\alpha\right)(x), \left(\prod_{\alpha \in I} \mu_\alpha\right)(y)\right\}.
\end{aligned}$$

Now, suppose that $x =< x_\alpha > \in \prod\limits_{\alpha \in I} P_\alpha$, we have $x^{-1} =< x_\alpha^{-1} >$ and so

$$\left(\prod_{\alpha \in I} \mu_\alpha\right)(x^{-1}) = \inf_{\alpha \in I}\{\mu_\alpha(x_\alpha^{-1})\} \geq \left(\prod_{\alpha \in I} \mu_\alpha\right)(x). \qquad \blacksquare$$

Corollary 2.3.13. *Let P_1, P_2 be two polygroups and μ, λ fuzzy subpolygroups of P_1, P_2 respectively. Then, $(\mu \times \lambda)_t = \mu_t \times \lambda_t$.*

Proposition 2.3.14. *Let P_1, P_2 be two polygroups and μ be a fuzzy subpolygroup of $P_1 \times P_2$. Then, μ_1, μ_2 are fuzzy subpolygroups of P_1, P_2 respectively, where*

$$\begin{aligned}
\mu_1(x) &= \mu(x, e_2), \quad x \in P_1 \\
\mu_2(y) &= \mu(e_1, y), \quad y \in P_2.
\end{aligned}$$

Moreover, $\mu_1 \times \mu_2 \subseteq \mu$.

Proof. We show that μ_1 is a fuzzy subpolygroup of P_1. Suppose that $x, y \in P_1$. Then, for every $z \in x \cdot y$, we have

$$\mu_1(z) = \mu(z, e_2) \geq \min\{\mu(x, e_2), \ \mu(y, e_2)\} = \min\{\mu_1(x), \ \mu_1(y)\}$$

and $\mu_1(x^{-1}) = \mu(x^{-1}, e_2) = \mu(x, e_2) = \mu_1(x)$. Similarly, we can show that μ_2 is a fuzzy subpolygroup of P_2. Now, let $(x, y) \in P_1 \times P_2$. Then,

$$(\mu_1 \times \mu_2)(x, y) = \min\{\mu_1(x), \mu_2(y)\} = \min\{\mu(x, e_2), \mu(e_1, y)\} \leq \mu(x, y). \ \blacksquare$$

Corollary 2.3.15. *Let μ be a fuzzy subset of a polygroup P. Then, $\mu \times \mu$ is a fuzzy subpolygroup of $P \times P$ if and only if μ is a fuzzy subpolygroup of P.*

Proposition 2.3.16. *Let P_1, P_2 be two polygroups and μ, λ be fuzzy subsets of P_1, P_2, respectively. If $\mu(e_1) = \lambda(e_2) = 1$ and $\mu \times \lambda$ is a fuzzy subpolygroup of $P_1 \times P_2$, then μ and λ are fuzzy subpolygroups of P_1, P_2, respectively.*

Proof. Suppose that $\mu_1(x) = (\mu \times \lambda)(x, e_2)$, $\mu_2(y) = (\mu \times \lambda)(e_1, y)$. Then, it is enough to show that $\mu(x) = \mu_1(x)$ and $\lambda(y) = \mu_2(y)$. Let $x \in P_1$. Then, $\mu(x) = \min\{\mu(x), 1\} = \min\{\mu(x), \lambda(e_2)\} = (\mu \times \lambda)(x, e_2) = \mu_1(x)$. Similarly, $\lambda(y) = \mu_2(y)$. Now, the proof follows from Proposition 2.3.14. ■

Definition 2.3.17. Let P be a polygroup, μ be a fuzzy subset of P and β^* the fundamental relation on P. Define the fuzzy subset μ_{β^*} on P/β^* as follows:

$$\begin{aligned} &\mu_{\beta^*} : P/\beta^* \to [0,1] \\ &\mu_{\beta^*}(\beta^*(x)) = \sup_{a \in \beta^*(x)} \{\mu(a)\}. \end{aligned}$$

Theorem 2.3.18. *Let P be a polygroup and μ be a fuzzy subpolygroup of P. Then μ_{β^*} is a fuzzy subgroup of the group P/β^*.*

Proof. Suppose that $\beta^*(x)$ and $\beta^*(y)$ are two elements of P/β^*. We can write:

$$\begin{aligned} \min\left\{\mu_{\beta^*}(\beta^*(x)), \mu_{\beta^*}(\beta^*(y))\right\} &= \min\left\{ \sup_{a \in \beta^*(x)} \{\mu(a)\}, \sup_{b \in \beta^*(y)} \{\mu(b)\} \right\} \\ &= \sup_{\substack{a \in \beta^*(x) \\ b \in \beta^*(y)}} \{\min\{\mu(a), \mu(b)\}\} \\ &\leq \sup_{\substack{a \in \beta^*(x) \\ b \in \beta^*(y)}} \left\{ \inf_{z \in a \cdot b} \{\mu(z)\} \right\} \\ &\leq \sup_{\substack{a \in \beta^*(x) \\ b \in \beta^*(y)}} \left\{ \sup_{z \in a \cdot b} \{\mu(z)\} \right\} \\ &\leq \sup_{\substack{a \in \beta^*(x) \\ b \in \beta^*(y)}} \left\{ \sup_{z \in \beta^*(a \cdot b)} \{\mu(z)\} \right\} \\ &= \sup_{\substack{a \in \beta^*(x) \\ b \in \beta^*(y)}} \{\mu_{\beta^*}(\beta^*(a \cdot b))\} \\ &= \sup_{\substack{a \in \beta^*(x) \\ b \in \beta^*(y)}} \{\mu_{\beta^*}(\beta^*(a) \odot \beta^*(b))\} \\ &= \mu_{\beta^*}(\beta^*(x) \odot \beta^*(y)). \end{aligned}$$

Now, suppose that $\beta^*(x)$ is an arbitrary element of P/β^*. Then,

$$\begin{aligned} \mu_{\beta^*}\left(\beta^*(x)^{-1}\right) &= \mu_{\beta^*}(\beta^*(x^{-1})) \\ &= \sup_{a \in \beta^*(x^{-1})} \{\mu(a)\} = \sup_{a \in \beta^*(x^{-1})} \{\mu(a^{-1})\} \\ &= \sup_{b \in \beta^*(x)} \{\mu(b)\} = \mu_{\beta^*}(\beta^*(x)). \end{aligned}$$

Thus, the proof is complete. ■

Theorem 2.3.19. *Let P_1, P_2 be plygroups and β_1^*, β_2^* and β^* be fundamental equivalence relations on P_1, P_2 and $P_1 \times P_2$, respectively. If μ, λ are fuzzy subpolygroups of P_1, P_2, respectively, then $(\mu \times \lambda)_{\beta^*} = \mu_{\beta_1^*} \times \lambda_{\beta_2^*}$.*

Proof. Since $\mu \times \lambda$ is a fuzzy subpolygroup of $P_1 \times P_2$, then by Theorem 2.3.18, $(\mu \times \lambda)_{\beta^*}$ is a fuzzy subgroup of the group $(P_1 \times P_2)/\beta^*$.

Now, assume that $x \in P_1$ and $y \in P_2$. Then,

$$\begin{aligned}
(\mu \times \lambda)_{\beta^*}(\beta^*(x,y)) &= \sup_{(a,b)\in\beta^*(x,y)} \{(\mu \times \lambda)(a,b)\} \\
&= \sup_{(a,b)\in\beta^*(x,y)} \{\min\{\mu(a), \lambda(b)\}\} \\
&= \sup_{\substack{a\in\beta_1^*(x)\\ b\in\beta_2^*(y)}} \{\min\{\mu(a), \lambda(b)\}\} \\
&= \min\left\{ \sup_{a\in\beta_1^*(x)} \{\mu(a)\},\ \sup_{b\in\beta_2^*(y)} \{\lambda(b)\} \right\} \\
&= \min\left\{\mu_{\beta_1^*}(\beta_1^*(x)), \lambda_{\beta_2^*}(\beta_2^*(y))\right\} \\
&= (\mu_{\beta_1^*} \times \lambda_{\beta_2^*})(\beta_1^*(x), \beta_2^*(y)). \blacksquare
\end{aligned}$$

Definition 2.3.20. A *t-norm* is a mapping $T : [0,1] \times [0,1] \to [0,1]$ satisfying, for all $x, y, z \in [0,1]$,

(1) $T(x,1) = x$.
(2) $T(x,y) = T(y,x)$.
(3) $T(x,T(y,z)) = T(T(x,y),z)$.
(4) $T(x,y) \leq T(x,z)$ whenever $y \leq z$.

Let T be a t-norm on $[0,1]$. We define $T_{i=1}^n x_i$, and D_T as follows:

$$\begin{aligned}
T_{i=1}^n x_i &= T(T_{i=1}^{n-1} x_i, x_n) = T(x_1, x_2, \ldots, x_n), \\
D_T &= \{x \in [0,1] \mid T(x,x) = x\}.
\end{aligned}$$

It is clear that when $T = \wedge$, D_T coincides with $[0,1]$. In particular, for every t-norm T, we have $T(x_1, x_2, \ldots, x_n) \leq x_1 \wedge x_2 \wedge \ldots \wedge x_n$ for all $x_1, x_2, \ldots, x_n \in [0,1]$. There exist uncountably t-norms. The following are the four basic t-norms T_M, T_P, T_L, and T_D given by, respectively:

$$\begin{aligned}
&T_M(x,y) = \min(x,y), && \text{(minimum)} \\
&T_P(x,y) = x \cdot y, && \text{(product)} \\
&T_L(x,y) = \max(x+y-1, 0), && \text{(Lukasiewicz t-norm)} \\
&T_D(x,y) = \begin{cases} 0 & \text{if } (x,y) \in [0,1)^2 \\ \min(x,y) & \text{otherwise.} \end{cases} && \text{(drastic product)}
\end{aligned}$$

Let T_1 and T_2 be two *t*-norms. T_2 is said to *dominate* T_1 and write $T_1 \ll T_2$ if for all $a, b, c, d \in [0,1]$,

$$T_1(T_2(a,c), T_2(b,d)) \leq T_2(T_1(a,b), T_1(c,d))$$

and T_1 is said *weaker than* T_2 or T_2 is *stronger than* T_1 and write $T_1 \leq T_2$ if for all $x, y \in [0,1]$, $T_1(x,y) \leq T_2(x,y)$.

Since a t-norm (triangular norm) T is a generalization of the minimum function, Anthony and Sherwood in [6] replaced the axiom $\min\{\mu(x), \mu(y)\} \leq \mu(xy)$ occurring in the definition of a fuzzy subgroup by the inequality $T(\mu(x), \mu(y)) \leq \mu(xy)$.

Goguen in [109] generalized the fuzzy subsets of X, to L-subsets, as a function from X to a lattice L. From now on, in this section, L is a complete lattice, i.e., there is a partial order $\leq$ on L such that, for any $S \subseteq L$, infimum of S and supremum of S exist and these will be denoted by $\bigwedge_{s \in S} \{s\}$ and $\bigvee_{s \in S} \{s\}$, respectively. In particular, for any elements $a, b \in L$, $\inf\{a, b\}$ and $\sup\{a, b\}$ will be denoted by $a \wedge b$ and $a \vee b$, respectively. Also, L is a distributive lattice with a least element 0 and a greatest element 1. If $a, b \in L$, we write $a \geq b$ if $b \leq a$, and $a > b$ if $a \geq b$ and $a \neq b$.

Now, we adopt the following definition of triangular norm on a lattice. Note that, as a lattice, the real interval $[0,1]$ is a complete lattice. A binary composition T on the lattice $(L, \leq, \vee, \wedge)$ which contains 0 and 1, is a triangular norm on L if the four axioms of the above definition are satisfied for all $x, y, z \in L$. One writes $T(x,y)$ and xTy interchangeable. Let S and T be triangular norms on L. If $S(x,y) \leq T(x,y)$ for all $x, y \in L$, one writes $S \leq T$. The meet $\wedge$ is a triangular norm on L. Now, let $I_T = \{x \in L \mid T(x,x) = x\}$, which is the set of all T-idempotent elements of L. Under the partial ordering induced by the partial ordering $\leq$ of L, I_T is a complete lattice with join $\vee$ and meet T.

We consider L-subsets of X in the sense of Goguen [109]. Accordingly an L-subset of X is a mapping of X into L. If L is the unit interval $[0,1]$ of real numbers, these are the usual fuzzy subsets of X. For a non-empty set X, let $F^L(X) = \{\mu | \ \mu \text{ is an } L\text{-subset of } X\}$. Let $\mu, \lambda, \mu_i (i \in I)$, be in $F^L(X)$. Then, the inclusion $\mu \subseteq \lambda$, the intersection $\mu \cap \lambda$ and the union $\mu \cup \lambda$ are defined in $F^L(X)$ as follows:

$$\begin{aligned}
&\mu \subseteq \lambda \Leftrightarrow \mu(x) \leq \lambda(x), \text{ for all } x \in X, \\
&(\mu \cap \lambda)(x) = \mu(x) \wedge \lambda(x), \\
&(\mu \cup \lambda)(x) = \mu(x) \vee \lambda(x).
\end{aligned}$$

One defines arbitrary intersection and arbitrary union in $F^L(X)$ as follows:

$$\left(\bigcap_{i \in I} \mu_i\right)(x) = \wedge\{\mu_i(x) \mid i \in I\} \text{ and } \left(\bigcup_{i \in I} \mu_i\right)(x) = \vee\{\mu_i(x) | \ i \in I\}.$$

Let f be a mapping from a set X into a set Y, and let $\mu \in F^L(X)$. Then, we define $f(\mu) \in F^L(Y)$, by

$$f(\mu)(y) = \begin{cases} \bigvee\limits_{x \in f^{-1}(y)} \{\mu(x)\} & \text{if } f^{-1}(y) \neq \emptyset \\ 0 & \text{otherwise,} \end{cases}$$

for all $y \in Y$. If $\lambda \in F^L(Y)$, then $f^{-1}(\lambda)(x) = \lambda(f(x))$, for all $x \in X$.

Definition 2.3.21. Let T be a t-norm on the complete lattice $(L, \leq, \vee, \wedge)$. An L-subset $\mu \in F^L(P)$ of the polygroup P is a TL-*subpolygroup* of P if the following axioms hold.

(1) $Im(\mu) \subseteq I_T$.
(2) $T(\mu(x), \mu(y)) \leq \bigwedge_{\alpha \in x \cdot y} \mu(\alpha)$, for all $x, y \in P$.
(3) $\mu(x) \leq \mu(x^{-1})$, for all $x \in P$.

Let μ be a TL-subpolygroup of P. Then, μ is said to be *normal* if for all $x, y \in P$,

$$\mu(y) \leq \mu(z), \text{ for all } z \in x \cdot y \cdot x^{-1}.$$

Proposition 2.3.22. *Let μ be a TL-subpolygroup of P. Then, $\mu(e) \geq \mu(x)$ for all $x \in P$.*

Proof. Since $e \in x \cdot x^{-1}$ for all $x \in P$, we have

$$\mu(e) \geq \bigwedge_{\alpha \in x \cdot x^{-1}} \mu(\alpha) \geq T(\mu(x), \mu(x^{-1})) = T(\mu(x), \mu(x)) = \mu(x),$$

because $\mu(x) \in I_T$. ■

Theorem 2.3.23. *Let T be a t-norm on the complete lattice $(L, \leq, \vee, \wedge)$, μ be an L-subset of P such that $Im(\mu) \subseteq I_T$ and $b = \bigvee Im(\mu)$. Then, the following two statements are equivalent:*

(1) *μ is a TL-subpolygroup of P.*
(2) *$\mu^{-1}[a, b]$ is a subpolygroup of P, whenever $a \in I_T$ and $0 < a \leq b$.*

Proof. (1⇒2): Suppose that $a \in I_T$, and $0 < a \leq b$. Since $b = \mu(e)$, $\mu^{-1}[a, b]$ is non-empty. Now, if $x, y \in \mu^{-1}[a, b]$, then $\bigwedge_{\alpha \in x \cdot y} \mu(\alpha) \geq \mu(x) T \mu(y) \geq aTa = a$ which implies that $x \cdot y \subseteq \mu^{-1}[a, b]$. Since $\mu(x^{-1}) \geq \mu(x) \geq a$, $x^{-1} \in \mu^{-1}[a, b]$, and so $\mu^{-1}[a, b]$ is a subpolygroup of P.

(2⇒1): Suppose that $x \in P$ and $\mu(x) = a$. If $a = 0$, then we obtain $\mu(x) = 0 \leq \mu(x^{-1})$. Otherwise, we have $0 < a < b$ and $a \in Im(\mu) \subseteq I_T$. So, $x \in \mu^{-1}[a, b]$, which is a subpolygroup of P, and hence $x^{-1} \in \mu^{-1}[a, b]$. Therefore, $\mu(x) = a \leq \mu(x^{-1})$. Let $x, y \in P$. Since $Im(\mu) \subseteq I_T$, both $\mu(x)$ and $\mu(y)$ are in I_T. We have

$$\begin{aligned}(\mu(x) T \mu(y)) T (\mu(x) T \mu(y)) &= \mu(x) T (\mu(y) T \mu(x)) T \mu(y) \\ &= \mu(x) T (\mu(x) T \mu(y)) T \mu(y) \\ &= (\mu(x) T \mu(x)) T (\mu(y) T \mu(y)) \\ &= \mu(x) T \mu(y),\end{aligned}$$

and so $\mu(x) T \mu(y) \in I_T$. Assume that $a = \mu(x) T \mu(y)$. If $a = 0$, then $\mu(x) T \mu(y) = 0 \leq \bigwedge_{\alpha \in x \cdot y} \mu(\alpha)$. So, we let $0 < a = \mu(x) T \mu(y) \leq \mu(x) \wedge \mu(y) \leq \mu(x) \leq b$. Hence, $x, y \in \mu^{-1}[a, b]$ which implies that $x \cdot y \subseteq \mu^{-1}[a, b]$. Therefore, $\mu(x) T \mu(y) \leq \bigwedge_{\alpha \in x \cdot y} \mu(\alpha)$. ■

Corollary 2.3.24. *Let $S \subseteq P$. Then, the characteristic function χ_S is a TL-subpolygroup of P if and only if S is a subpolygroup of P.*

Let T be a t-norm on the complete lattice $(L, \leq, \vee, \wedge)$, and let $\{\mu_i\}_{i\in I}$ be a family of TL-subpolygroups of P. Then, $\bigcap_{i\in I} \mu_i$ is a TL-subpolygroup of P.

Corollary 2.3.25. *Let $f : P_1 \to P_2$ be a strong homomorphism and let μ be any TL-subpolygroup in P_1. Then, $f(\mu)$ is a TL-subpolygroup in P_2.*

Definition 2.3.26. Let P_1, P_2 be polygroups and μ, λ be TL-subpolygroups of P_1, P_2, respectively. The *product* of μ, λ is defined to be the L-subset $\mu \times \lambda$ of $P_1 \times P_2$ with

$$(\mu \times \lambda)(x, y) = T(\mu(x), \lambda(y)), \quad \text{for all} \quad (x, y) \in P_1 \times P_2.$$

Corollary 2.3.27. *In the above definition, $\mu \times \lambda$ is a TL-subpolygroup of $P_1 \times P_2$.*

Proof. Suppose that (x_1, x_2), $(y_1, y_2) \in P_1 \times P_2$. For every $(\alpha_1, \alpha_2) \in (x_1, x_2) \odot (y_1, y_2)$ we have

$$\begin{aligned}
(\mu \times \lambda)(\alpha_1, \alpha_2) &= T(\mu(\alpha_1), \lambda(\alpha_2)) \\
&\geq T(T(\mu(x_1), \mu(y_1)), T(\lambda(x_2), \lambda(y_2)) \\
&= T(T(T(\mu(x_1), \mu(y_1)), \lambda(x_2)), \lambda(y_2)) \\
&= T(T(\lambda(x_2), T(\mu(x_1), \mu(y_1)), \lambda(y_2))) \\
&= T(T(T(\lambda(x_2), \mu(x_1)), \mu(y_1), \lambda(y_2))) \\
&= T(\lambda(y_2), T(\mu(y_1), T(\lambda(x_2), \mu(x_1))) \\
&= T(T(\mu(x_1), \lambda(x_2)), T(\mu(y_1), \lambda(y_2)) \\
&= T((\mu \times \lambda)(x_1, x_2), (\mu \times \lambda)(y_1, y_2)).
\end{aligned}$$

Taking the infimum in the complete lattice $(L, \leq, \vee, \wedge)$ over all $(\alpha_1, \alpha_2) \in (x_1, x_2) \odot (y_1, y_2)$ we obtain

$$\bigwedge_{(\alpha_1,\alpha_2)\in(x_1,x_2)\odot(y_1,y_2)} (\mu \times \lambda)(\alpha_1, \alpha_2) \geq T((\mu \times \lambda)(x_1, x_2), (\mu \times \lambda)(y_1, y_2)).$$

Also, we have

$$\begin{aligned}
(\mu \times \lambda)(x_1, x_2) &= T(\mu(x_1), \lambda(x_2)) \\
&\leq T(\mu(x_1^{-1}), \lambda(x_2^{-1})) \\
&= (\mu \times \lambda)(x_1^{-1}, x_2^{-1}) \\
&= (\mu \times \lambda)(x_1, x_2)^{-1}. \ \blacksquare
\end{aligned}$$

Definition 2.3.28. Let μ be a normal TL-subpolygroup of a polygroup P. We define $x \sim y (mod \mu)$ if there exists $a \in x \cdot y^{-1}$ such that $\mu(a) = \mu(e)$.

Lemma 2.3.29. *The relation $\sim$ is an equivalence relation.*

Proof. For every $x, y, z \in P$, we have

(1) $e \in x \cdot x^{-1}$ implies that $x \sim x(mod\mu)$.
(2) $x \sim y(mod\mu)$ then there exists $a \in x \cdot y^{-1}$ such that $\mu(a) = \mu(e)$. Since $\mu(a) = \mu(a^{-1})$ and $a^{-1} \in y \cdot x^{-1}$ we obtain $y \sim x(mod\mu)$.
(3) If $x \sim y(mod\mu)$ and $y \sim z(mod\mu)$, then there exist $a \in x \cdot y^{-1}$ and $b \in y \cdot z^{-1}$ such that $\mu(a) = \mu(b) = \mu(e)$. Therefore, we have $x \in a \cdot y$ and $z^{-1} \in y^{-1} \cdot b$ which imply that $x \cdot z^{-1} \subseteq a \cdot y \cdot y^{-1} \cdot b$. Suppose that $t \in x \cdot z^{-1}$ is arbitrary. Then, $t \in a \cdot y \cdot y^{-1} \cdot b$. Since $t \in y \cdot y^{-1} \cdot a \cdot y \cdot y^{-1} \cdot b \cdot y \cdot y^{-1}$, there exist $c \in y^{-1} \cdot a \cdot y$ and $d \in y^{-1} \cdot b \cdot y$ such that $t \in y \cdot c \cdot d \cdot y^{-1}$. Thus, there exists $q \in c \cdot d$ such that $t \in y \cdot q \cdot y^{-1}$. Since μ is normal and $c \in y^{-1} \cdot a \cdot y$, we obtain $\mu(a) \leq \mu(c)$ and so $\mu(e) \leq \mu(c)$. Hence, $\mu(c) = \mu(e)$. Similarly, since μ is normal and $d \in y^{-1} \cdot b \cdot y$, we obtain $\mu(b) \leq \mu(d)$ and so $\mu(e) \leq \mu(d)$. Hence, $\mu(d) = \mu(e)$. From $q \in c \cdot d$, we conclude that

$$\mu(q) \geq T(\mu(c), \mu(d)) = T(\mu(e), \mu(e)) = \mu(e),$$

and so $\mu(q) = \mu(e)$. Since $t \in y \cdot q \cdot y^{-1}$, $\mu(q) \leq \mu(t)$. Hence, $\mu(e) \leq \mu(t)$ and so $\mu(t) = \mu(e)$. Therefore, $x \sim z(mod\mu)$. ■

Corollary 2.3.30. *If $x \sim y(mod\mu)$, then for all $t \in x \cdot y^{-1}$ we have $\mu(t) = \mu(e)$.*

Corollary 2.3.31. *If $x \sim y(mod\mu)$, then $\mu(x) = \mu(y)$.*

Proof. Since $x \sim y(mod\mu)$, there exists $a \in x \cdot y^{-1}$ such that $\mu(a) = \mu(e)$. Since $a \in x \cdot y^{-1}$, it follows that $x \in a \cdot y$ and so

$$\mu(y) = T(\mu(y), \mu(y)) \leq T(\mu(e), \mu(y)) = T(\mu(a), \mu(y)) \leq \mu(x).$$

Similarly, we have $\mu(x) \leq \mu(y)$. ■

Lemma 2.3.32. *If P is a commutative polygroup, then the equivalence relation $\sim$ is a conjugation on P.*

Proof. Suppose that $x \sim y(mod\mu)$. Then, there exists $a \in x \cdot y^{-1}$ such that $\mu(a) = \mu(e)$. Since $a^{-1} \in y \cdot x^{-1}$, $\mu(a^{-1}) = \mu(a)$ and P is commutative, we obtain $a^{-1} \in x^{-1} \cdot (y^{-1})^{-1}$ and $\mu(a^{-1}) = \mu(e)$. Thus, $x^{-1} \sim y^{-1}(mod\mu)$. To verify the second condition, we have $z \in x \cdot y$ and $z \sim z'(mod\mu)$. Hence, there exists $r \in z^{-1} \cdot z'$ with $\mu(r) = \mu(e)$. We get $z' \in z \cdot r$ and so $z' \in x \cdot y \cdot r$. In this case, there exists $y' \in y \cdot r$. Therefore, we have $y \sim y'(mod\mu)$, $x \sim x(mod\mu)$ and $z' \in x \cdot y'$. ■

Lemma 2.3.33. *If P is a polygroup, then the equivalence relation $\sim$ is a strongly regular relation.*

Proof. Suppose that $x \sim y$ and $a \in P$. We show that $a \cdot x \overline{\overline{\sim}} a \cdot y$. Let $u \in a \cdot x$ and $v \in a \cdot y$ be arbitrary. Then, $u \cdot v^{-1} \subseteq a \cdot x \cdot y^{-1} \cdot a^{-1}$. Hence, for every

$r \in u \cdot v^{-1}$, $r \in a \cdot q \cdot a^{-1}$, where $q \in x \cdot y^{-1}$. Since μ is normal, $\mu(r) \geq \mu(q)$. Since $x \sim y$, $\mu(q) = \mu(e)$. Therefore, $\mu(r) = \mu(e)$, which implies that $u \sim v$. Similarly, we can show that $x \cdot a \overline{\approx} y \cdot a$. ■

Suppose that $\mu[x]$ is the equivalence class containing x, we denote $P/\sim$ the set of all equivalence classes, i.e., $P/\sim = \{\mu[x] \mid x \in P\}$. By Lemma 2.3.33, we can define

$$\mu[x] \odot \mu[y] = \{\mu[z] \mid z \in x \cdot y\},$$

for all $\mu[x],\ \mu[y] \in P/\sim$.

Corollary 2.3.34. *If P is a polygroup, then $< P/\sim, \odot, \mu[e], ^{-I} >$ is a group, where $\mu[x]^{-I} = \mu[x^{-1}]$.*

Definition 2.3.35. Let T be a t-norm on the complete lattice $(L, \leq, \vee, \wedge)$ and let $< P_1, \cdot, e_1, ^{-1} >$ and $< P_2, *, e_2, ^{-1} >$ be polygroups and μ, λ be TL-subpolygroups of P_1, P_2, respectively. Then,

$$(x_1, y_1) \approx (x_2, y_2)(mod(\mu \times \lambda)) \text{ if and only if } x_1 \sim x_2(mod\mu), y_1 \sim y_2(mod\lambda),$$

for all $(x_1, y_1),\ (x_2, y_2) \in P_1 \times P_2$.

Theorem 2.3.36. *Let T be a t-norm on the complete lattice $(L, \leq, \vee, \wedge)$ and let $< P_1, \cdot, e_1, ^{-1} >$ and $< P_2, *, e_2, ^{-1} >$ be polygroups and μ, λ be TL-subpolygroups of P_1, P_2 respectively. Then,*

$$(P_1/\sim) \times (P_2/\sim) \cong (P_1 \times P_2)/\approx.$$

Proof. We define $\varphi : (P_1/\sim) \times (P_2/\sim) \to (P_1 \times P_2)/\approx$ by setting $\varphi((\mu[x], \lambda[y])) = (\mu \times \lambda)[(x, y)]$. We prove that φ is well defined. Suppose that $(\mu[x_1], \lambda[y_1]) = (\mu[x_2], \lambda[y_2])$. Then, there exist $r_1 \in x_1 \cdot x_2^{-1}$, $r_2 \in y_1 * y_2^{-1}$ with $\mu(r_1) = \mu(e_1), \lambda(r_2) = \lambda(e_2)$. So, $(r_1, r_2) \in (x_1, y_1) \odot (x_2, y_2)^{-1}$ and

$$(\mu \times \lambda)(r_1, r_2) = T(\mu(r_1), \lambda(r_2)) = T(\mu(e_1), \lambda(e_2)) = (\mu \times \lambda)(e_1, e_2).$$

Therefore, $(\mu \times \lambda)[(x_1, y_1)] = \mu \times \lambda[(x_2, y_2)]$. Thus, φ is well defined. It is easy to see that φ is an isomorphism of polygroups. ■

Definition 2.3.37. Let P be a polygroup and μ be a L-subset of P. The L-subset μ_{β^*} on P/β^* is defined as follows:

$$\mu_{\beta^*} : P/\beta^* \to L$$

$$\mu_{\beta^*}(\beta^*(x)) = \bigvee_{a \in \beta^*(x)} \{\mu(a)\}.$$

Theorem 2.3.38. *Let P be a polygroup and μ be a TL-subpolygroup of P. Then, μ_{β^*} is a TL-subgroup of P/β^*.*

Proof. We consider P/β^* as a polygroup (since every group is a polygroup). Since the canonical map $\phi : P \to P/\beta^*$ is a strong epimorphism, then using Corollary 2.3.25, the proof is complete. ■

2.4 Generalized Fuzzy Subpolygroups

$(\in, \in \vee q)$-fuzzy subgroup was introduce in an earlier paper of Bhakat and Das [12, 13] by using the combined notions of "belongingness" and "quasicoincidence" of fuzzy points and fuzzy sets. In fact, the $(\in, \in \vee q)$-fuzzy subgroup is an important and useful generalization of Rosenfeld's fuzzy subgroup. Moreover, a generalization of Rosenfeld's fuzzy subgroup, and Bhakat and Das's fuzzy subgroup is given in [201].

Now, in this section, using the notion of "belongingness $(\in)$" and "quasi-coincidence (q)" of fuzzy points with fuzzy sets, the concept of $(\in, \in \vee q)$-fuzzy subpolygroup is introduced. The study of $(\in, \in \vee q)$-fuzzy normal subpolygroups of a polygroup are dealt with. Characterization and some of the fundamental properties of such fuzzy subpolygroups are obtained. $(\in, \in \vee q)$-fuzzy cosets determined by $(\in, \in \vee q)$-fuzzy subpolygroups are discussed. Finally, we give the definition of a fuzzy subpolygroup with thresholds which is a generalization of an ordinary fuzzy subpolygroup and an $(\in, \in \vee q)$-fuzzy subpolygroup and discuss relations between two fuzzy subpolygroups.

A fuzzy subset μ of P of the form

$$\mu(y) = \begin{cases} t(\neq 0) & \text{if } y = x \\ 0 & \text{if } y \neq x \end{cases}$$

is said to be *fuzzy point with support* x and *value* t and is denoted by x_t. A fuzzy point x_t is said to be *belong to* (respectively, *quasi-coincident with*) a fuzzy set μ, written as $x_t \in \mu$ (respectively, $x_t q \mu$) if $\mu(x) \geq t$ (resp. $\mu(x) + t > 1$). If $x_t \in \mu$ or $x_t q \mu$, then we write $x_t \in \vee q \mu$. The symbol $\overline{\in \vee q}$ means $\in \vee q$ does not hold.

Using the notion of "belongingness $(\in)$" and "quasi-coincidence (q)" of fuzzy points with fuzzy sets, the concept of (α, β)-fuzzy subgroup, where α, β are any two of $\{\in, q, \in \vee q, \in \wedge q\}$ with $\alpha \neq \in \wedge q$, is introduced in [12]. It is noteworthy that the most viable generalization of Rosenfeld's fuzzy subgroup is the notion of $(\in, \in \vee q)$-fuzzy subgroup. The detailed study of the $(\in, \in \vee q)$-fuzzy subgroup has been considered in [13]. In [87], Davvaz and Corsini extended the concept of $(\in, \in \vee q)$-fuzzy subgroups to the concept of $(\in, \in \vee q)$-fuzzy subpolygroups in the following way:

Definition 2.4.1. A fuzzy subset μ of a polygroup P is said to be an $(\in, \in \vee q)$*-fuzzy subpolygroup* of P if for all $t, r \in (0, 1]$ and $x, y \in P$,

(1) $x_t, y_r \in \mu$ implies $z_{t \wedge r} \in \vee q \mu$, for all $z \in x \cdot y$,
(2) $x_t \in \mu$ implies $(x^{-1})_t \in \vee q \mu$.

We note that if μ is a fuzzy subpolygroup of P according to Definition 2.3.1, then μ is an $(\in, \in \vee q)$-fuzzy subpolygroup of P according to Definition 2.4.1. But, the converse is not true shown by the following example.

EXAMPLE 28 Let $P = \{e, a, b\}$ be the polygroup defined by the multiplication table:

$\cdot$	e	a	b
e	e	a	b
a	a	$\{e,b\}$	$\{a,b\}$
b	b	$\{a,b\}$	$\{e,a\}$

If $\mu : P \to [0,1]$ is defined by

$$\mu(e) = 0.8, \quad \mu(a) = 0.7, \quad \mu(b) = 0.6,$$

then it is easy to see that μ is an $(\in, \in \vee q)$-fuzzy subpolygroup of P, but is not a fuzzy subpolygroup of P.

Proposition 2.4.2. *Conditions (1) and (2) in Definition 2.4.1 are equivalent to the following conditions, respectively.*

(1′) $\mu(x) \wedge \mu(y) \wedge 0.5 \leq \bigwedge\limits_{z \in x \cdot y} \mu(z)$, *for all* $x, y \in P$.

(2′) $\mu(x) \wedge 0.5 \leq \mu(x^{-1})$, *for all* $x \in P$.

Proof. $(1 \Rightarrow 1')$: Suppose that $x, y \in P$. We consider the following cases:

(a) $\mu(x) \wedge \mu(y) < 0.5$,
(b) $\mu(x) \wedge \mu(y) \geq 0.5$.

Case a: Assume that there exists $z \in x \cdot y$ such that $\mu(z) < \mu(x) \wedge \mu(y) \wedge 0.5$, which implies that $\mu(z) < \mu(x) \wedge \mu(y)$. Choose t such that $\mu(z) < t < \mu(x) \wedge \mu(y)$. Then, $x_t, y_t \in \mu$, but $z_t \overline{\in \vee q} \mu$ which contradicts (1).

Case b: Assume that $\mu(z) < 0.5$ for some $z \in x \cdot y$. Then, $x_{0.5}, y_{0.5} \in \mu$, but $z_{0.5} \overline{\in \vee q} \mu$, a contradiction. Hence, $(1')$ holds.

$(2 \Rightarrow 2')$: Suppose that $x \in P$, we consider the following cases:
(a) $\mu(x) < 0.5$,
(b) $\mu(x) \geq 0.5$.

Case a: Assume that $\mu(x) = t < 0.5$ and $\mu(x^{-1}) = r < \mu(x)$. Choose s such that $r < s < t$ and $r + s < 1$. Then, $x_s \in \mu$, but $(x^{-1})_s \overline{\in \vee q} \mu$ which contradicts (2). So, $\mu(x^{-1}) \geq \mu(x) = \mu(x) \wedge 0.5$.

Case b: Let $\mu(x) \geq 0.5$. If $\mu(x^{-1}) < \mu(x) \wedge 0.5$, then $x_{0.5} \in \mu$, but $(x^{-1})_{0.5} \overline{\in \vee q} \mu$, which contradicts (2). So, $\mu(x^{-1}) \geq \mu(x) \wedge 0.5$.

$(1' \Rightarrow 1)$: Let $x_t, y_r \in \mu$. Then, $\mu(x) \geq t$ and $\mu(y) \geq r$. For every $z \in x \cdot y$ we have

$$\mu(z) \geq \mu(x) \wedge \mu(y) \wedge 0.5 \geq t \wedge r \wedge 0.5.$$

If $t \wedge r > 0.5$, then $\mu(z) \geq 0.5$ which implies that $\mu(z) + t \wedge r > 1$.
If $t \wedge r \leq 0.5$, then $\mu(z) \geq t \wedge r$.
Therefore, $z_{t \wedge r} \in \vee q \mu$, for all $z \in x \cdot y$.

$(2' \Rightarrow 2)$: Let $x_t \in \mu$. Then, $\mu(x) \geq t$. Now, we have $\mu(x^{-1}) \geq \mu(x) \wedge 0.5 \geq t \wedge 0.5$, which implies that $\mu(x^{-1}) \geq t$ or $\mu(x^{-1}) \geq 0.5$ according as $t \leq 0.5$ or $t > 0.5$. Therefore, $(x^{-1})_t \in \vee q \mu$. ∎

By Definition 2.4.1 and Proposition 2.4.2, we immediately obtain:

Corollary 2.4.3. *A fuzzy subset μ of a polygroup P is an $(\in, \in \vee q)$-fuzzy subpolygroup of P if and only if the conditions $(1')$ and $(2')$ in Proposition 2.4.2 hold.*

If μ is an $(\in, \in \vee q)$-fuzzy subpolygroup of a polygroup P, then it is easy to see that $\mu(e) \geq \mu(x) \wedge 0.5$, for all $x \in P$.

Let P be a polygroup and χ_K be the characteristic function of a subset K of P. Then, it is not difficult to see that χ_K is an $(\in, \in \vee q)$-fuzzy subpolygroup if and only if K is a subpolygroup.

Now, we characterize $(\in, \in \vee q)$-fuzzy subpolygroups by their level subpolygroups.

Theorem 2.4.4. *Let μ be an $(\in, \in \vee q)$-fuzzy subpolygroup of P. Then, for all $0 < t \leq 0.5$, μ_t is a empty set or a subpolygroup of P. Conversely, if μ is a fuzzy subset of P such that μ_t $(\neq \emptyset)$ is a subpolygroup of P, for all $0 < t \leq 0.5$, then μ is an $(\in, \in \vee q)$-fuzzy subpolygroup of P.*

Proof. Let μ be an $(\in, \in \vee q)$-fuzzy subpolygroup of P and $0 < t \leq 0.5$. Let $x, y \in \mu_t$. Then, $\mu(x) \geq t$ and $\mu(y) \geq t$. Now,

$$\bigwedge_{z \in x \cdot y} \mu(z) \geq \mu(x) \wedge \mu(y) \wedge 0.5 \geq t \wedge 0.5 = t.$$

Therefore, for every $z \in x \cdot y$ we have $\mu(z) \geq t$ or $z \in \mu_t$, so $x \cdot y \subseteq \mu_t$. Also, we have $\mu(x^{-1}) \geq \mu(x) \wedge 0.5 \geq t \wedge 0.5 = t$, and so $x^{-1} \in \mu_t$.

Conversely, let μ be a fuzzy subset of P such that μ_t $(\neq \emptyset)$ is a subpolygroup of P, for all $0 < t \leq 0.5$. For every $x, y \in P$, we can write

$$\begin{aligned} \mu(x) &\geq \mu(x) \wedge \mu(y) \wedge 0.5 = t_0, \\ \mu(y) &\geq \mu(x) \wedge \mu(y) \wedge 0.5 = t_0. \end{aligned}$$

Then, $x \in \mu_{t_0}$ and $y \in \mu_{t_0}$, so $x \cdot y \subseteq \mu_{t_0}$. Therefore, for every $z \in x \cdot y$ we have $\mu(z) \geq t_0$ which implies that

$$\bigwedge_{z \in x \cdot y} \mu(z) \geq t_0,$$

and in this way the condition $(1')$ of Proposition 2.4.2 is verified. In order to verify the second condition, let $x \in P$, we can write $\mu(x) \geq \mu(x) \wedge 0.5 = t_0$. Then, $x \in \mu_{t_0}$, and so $x^{-1} \in \mu_{t_0}$. Therefore, $\mu(x^{-1}) \geq \mu(x) \wedge 0.5$. ∎

Naturally, a corresponding result should be considered when μ_t is a subpolygroup of P for all $t \in (0.5, 1]$.

Theorem 2.4.5. *Let μ be a fuzzy subset of a polygroup P. Then, μ_t $(\neq \emptyset)$ is a subpolygroup of P for all $t \in (0.5, 1]$ if and only if*

(1) $\mu(x) \wedge \mu(y) \leq \bigwedge_{z \in x \cdot y} (\mu(z) \vee 0.5)$, *for all $x, y \in P$.*

(2) $\mu(x) \leq \mu(x^{-1}) \vee 0.5$, *for all* $x \in P$.

Proof. If there exist $x, y, z \in P$ with $z \in x \cdot y$ such that

$$\mu(z) \vee 0.5 < \mu(x) \wedge \mu(y) = t,$$

then $t \in (0.5, 1]$, $\mu(z) < t$, $x \in \mu_t$, and $y \in \mu_t$. Since $x, y \in \mu_t$ and μ_t is a subpolygroup, $x \cdot y \subseteq \mu_t$ and $\mu(z) \geq t$, for all $z \in x \cdot y$, which is a contradiction with $\mu(z) < t$. Therefore,

$$\mu(x) \wedge \mu(y) \leq \mu(z) \vee 0.5, \text{ for all } x, y, z \in P \text{ with } z \in x \cdot y,$$

which implies that

$$\mu(x) \wedge \mu(y) \leq \bigwedge_{z \in x \cdot y} (\mu(z) \vee 0.5), \text{ for all } x, y \in P.$$

Hence, (1) holds.

Now, assume that for some $x \in P$, $\mu(x^{-1}) \vee 0.5 < \mu(x) = t$. Then, $t \in (0.5, 1]$, $\mu(x^{-1}) < t$ and $x \in \mu_t$. Since $x \in \mu_t$, we obtain $x^{-1} \in \mu_t$ or $\mu(x^{-1}) \geq t$, which is a contradiction. Hence, (2) holds.

Conversely, assume that $t \in (0.5, 1]$ and $x, y \in \mu_t$. Then,

$$0.5 < t \leq \mu(x) \wedge \mu(y) \leq \bigwedge_{z \in x \cdot y} (\mu(z) \vee 0.5).$$

It follows that for every $z \in x \cdot y$, $0.5 < t \leq \mu(z) \vee 0.5$ and so $t \leq \mu(z)$, which implies that $z \in \mu_t$. Hence, $x \cdot y \subseteq \mu_t$.

Now, let $t \in (0.5, 1]$ and $x \in \mu_t$. Then, using condition (2), we have

$$0.5 < t \leq \mu(x) \leq \mu(x^{-1}) \vee 0.5 = \mu(x^{-1}),$$

and so $x^{-1} \in \mu_t$. Therefore, μ_t is a subpolygroup of P, for all $t \in (0.5, 1]$. ■

Let μ be a fuzzy subset of a polygroup P and

$$J = \{t \mid t \in (0, 1] \text{ and } \mu_t \text{ is the emptyset or a subpolygroup of } P\}.$$

When $J = (0, 1]$, then μ is an ordinary fuzzy subpolygroup of the polygroup P. When $J = (0, 0.5]$, μ is an $(\in, \in \vee q)$-fuzzy subpolygroup of the polygroup P.

In [201], Yuan, Zhang and Ren gave the definition of a fuzzy subgroup with thresholds which is a generalization of Rosenfeld's fuzzy subgroup, and Bhakat and Das's fuzzy subgroup. Davvaz and Corsini [87] extended the concept of a fuzzy subgroup with thresholds to the concept of fuzzy subpolygroup with thresholds in the following way:

Definition 2.4.6. Let $\alpha, \beta \in [0, 1]$ and $\alpha < \beta$. Let μ be a fuzzy subset of a polygroup P. Then, μ is called a *fuzzy subpolygroup with thresholds of* P if

(1) $\mu(x) \wedge \mu(y) \wedge \beta \leq \bigwedge_{z \in x \cdot y} (\mu(z) \vee \alpha)$, for all $x, y \in P$.

(2) $\mu(x) \wedge \beta \leq \mu(x^{-1}) \vee \alpha$, for all $x \in P$.

If μ is a fuzzy subpolygroup with thresholds of P, then we can conclude that μ is an ordinary fuzzy subpolygroup when $\alpha = 0$, $\beta = 1$, while μ is an $(\in, \in \vee q)$-fuzzy subpolygroup when $\alpha = 0$, $\beta = 0.5$.

Now, we characterize fuzzy subpolygroups with thresholds by their level subpolygroups.

Theorem 2.4.7. *A fuzzy subset μ of a polygroup P is a fuzzy subpolygroup with thresholds of P if and only if μ_t $(\neq \emptyset)$ is a subpolygroup of P, for all $t \in (\alpha, \beta]$.*

Proof. Let μ be a fuzzy subpolygroup with thresholds of P and $t \in (\alpha, \beta]$. Let $x, y \in \mu_t$. Then, $\mu(x) \geq t$ and $\mu(y) \geq t$. Now,

$$\bigwedge_{z \in x \cdot y} (\mu(z) \vee \alpha) \geq \mu(x) \wedge \mu(y) \wedge \beta \geq t \wedge \beta = t > \alpha.$$

So, for every $z \in x \cdot y$ we have $\mu(z) \vee \alpha \geq t > \alpha$ which implies that $\mu(z) \geq t$ and $z \in \mu_t$. Hence, $x \cdot y \subseteq \mu_t$. Now, let $x \in \mu_t$. Then, $\mu(x^{-1}) \vee \alpha \geq \mu(x) \wedge \beta \geq t > \alpha$. So, $\mu(x^{-1}) \geq t$ and $x^{-1} \in \mu_t$. Therefore, μ_t is a subpolygroup of P, for all $t \in (\alpha, \beta]$.

Conversely, let μ be a fuzzy subset of P such that μ_t $(\neq \emptyset)$ is a subpolygroup of P, for all $t \in (\alpha, \beta]$. If there exist $x, y, z \in P$ with $z \in x \cdot y$ such that

$$\mu(z) \vee \alpha < \mu(x) \wedge \mu(y) \wedge \beta = t,$$

then $t \in (\alpha, \beta]$, $\mu(z) < t$, $x \in \mu_t$ and $y \in \mu_t$. Since μ_t is a subpolygroup of P and $x, y \in \mu_t$, it follows that $x \cdot y \subseteq \mu_t$. Hence, $\mu(z) \geq t$, for all $z \in x \cdot y$. This is a contradiction with $\mu(z) < t$. Therefore,

$$\mu(x) \wedge \mu(y) \wedge \beta \leq \mu(z) \vee \alpha, \text{ for all } x, y, z \in P \text{ with } z \in x \cdot y,$$

which implies that

$$\mu(x) \wedge \mu(y) \wedge \beta \leq \bigwedge_{z \in x \cdot y} (\mu(z) \vee \alpha) \text{ for all } x, y \in P.$$

Hence, the condition (1) of Definition 2.4.6 holds.

Now, assume that there exists $x_0 \in P$ such that $t = \mu(x_0) \wedge \beta > \mu(x_0^{-1}) \vee \alpha$. Then, $x_0 \in \mu_t$, $t \in (\alpha, \beta]$ and $\mu(x_0^{-1}) < t$. Since μ_t is a subpolygroup of P, it follows that $\mu(x_0^{-1}) \geq t$. This is a contradiction with $\mu(x_0^{-1}) < t$. Therefore, $\mu(x) \wedge \beta \leq \mu(x^{-1}) \vee \alpha$, for any $x \in P$. Hence, the second condition of Definition 2.4.6 holds. ∎

Now, we give the definition of $(\in, \in \vee q)$-fuzzy normal subpolygroup of a polygroup, which is a generalization of the notion of fuzzy normal subpolygroup.

Definition 2.4.8. An $(\in, \in \vee q)$-fuzzy subpolygroup μ of a polygroup P is said to be $(\in, \in \vee q)$-*fuzzy normal* if for every $x, y \in P$ and $t \in (0, 1]$,

$$x_t \in \mu \text{ implies } z_t \in \vee q \mu, \text{ for all } z \in y \cdot x \cdot y^{-1}.$$

Theorem 2.4.9. *For an $(\in, \in \vee q)$-fuzzy subpolygroup μ of P, the following statements are equivalent:*

(1) *μ is an $(\in, \in \vee q)$-fuzzy normal subpolygroup of P.*
(2) $\bigwedge_{z \in y \cdot x \cdot y^{-1}} \mu(z) \geq \mu(x) \wedge 0.5$, *for all $x, y \in P$.*
(3) $\mu(z) \geq \mu(z') \wedge 0.5$, *for any $z \in x \cdot y$, $z' \in y \cdot x$, and for all $x, y \in P$.*
(4) $\bigwedge_{z \in x^{-1} \cdot y^{-1} \cdot x \cdot y} \mu(z) \geq \mu(x) \wedge 0.5$, *for all $x, y \in P$.*

Proof. ($1 \Rightarrow 2$): Suppose that μ is an $(\in, \in \vee q)$-fuzzy normal subpolygroup of P and $x, y \in P$. We consider the following cases:
(a) $\mu(x) < 0.5$.
(b) $\mu(x) \geq 0.5$.

Case a: Assume that there exists $z \in y \cdot x \cdot y^{-1}$ such that $\mu(z) < \mu(x) \wedge 0.5$, which implies that $\mu(z) < \mu(x)$. Choose t such that $\mu(z) < t < \mu(x)$. Then, $x_t \in \mu$, but $z_t \overline{\in \vee q} \mu$ which contradicts (1).

Case b: Assume that $\mu(z) < 0.5$, for some $z \in y \cdot x \cdot y^{-1}$. Then, we have $\mu(z) < 0.5 \leq \mu(x)$, which implies that $x_{0.5} \in \mu$, but $z_{0.5} \overline{\in \vee q} \mu$, a contradiction. Hence, (2) holds.

($2 \Rightarrow 1$): Suppose that $x_t \in \mu$ and $x, y \in P$. Then, $\mu(x) \geq t$. For every $z \in y \cdot x \cdot y^{-1}$, we have $\mu(z) \geq \mu(x) \wedge 0.5 \geq t \wedge 0.5$ which implies that $\mu(z) \geq t$ or $\mu(z) \geq 0.5$ according as $t \leq 0.5$ or $t > 0.5$. Therefore, $z_t \in \vee q \mu$.

($2 \Rightarrow 3$): Let $x, y \in P$, for every $z \in x \cdot y$ and $z' \in y \cdot x$, we obtain $x \in y^{-1} \cdot z'$, which implies that $z \in y^{-1} \cdot z' \cdot y$. Now, by using (2), we have $\mu(z) \geq \mu(z') \wedge 0.5$.

($3 \Rightarrow 2$): Suppose that there exist $x, y, z \in P$ with $z \in y \cdot x \cdot y^{-1}$ such that $\mu(z) < \mu(x) \wedge 0.5$. Since $z \in y \cdot x \cdot y^{-1}$, there exists $a \in x \cdot y^{-1}$ such that $z \in y \cdot a$. Since $a \in x \cdot y^{-1}$, $x \in a \cdot y$. Therefore, by (3), we have $\mu(x) \wedge 0.5 \leq \mu(z)$, which is a contradiction.

($2 \Rightarrow 4$): Let $x, y \in P$. For every $z \in x^{-1} \cdot y^{-1} \cdot x \cdot y$, there exists $z' \in y^{-1} \cdot x \cdot y$ such that $z \in x^{-1} \cdot z'$. Since μ is an $(\in, \in \vee q)$-fuzzy subpolygroup, then $\mu(z) \geq \mu(x^{-1}) \wedge \mu(z') \wedge 0.5$. By (2), we have $\mu(z) \geq \mu(x^{-1}) \wedge \mu(x) \wedge 0.5$, and so $\mu(z) \geq \mu(x) \wedge 0.5$. Therefore,

$$\bigwedge_{z \in x^{-1} \cdot y^{-1} \cdot x \cdot y} \mu(z) \geq \mu(x) \wedge 0.5.$$

($4 \Rightarrow 2$): Let $x, y \in P$. For every $z \in y \cdot x \cdot y^{-1}$, we have $z \in x \cdot x^{-1} \cdot y \cdot x \cdot y^{-1}$. Hence, there exists $z' \in x^{-1} \cdot y \cdot x \cdot y^{-1}$ such that $z \in x \cdot z'$. So,

$$\mu(z) \geq \mu(z') \wedge \mu(x) \wedge 0.5 \geq \mu(x) \wedge 0.5 \wedge \mu(x) \wedge 0.5 = \mu(x) \wedge 0.5.$$

Therefore, $\bigwedge_{z \in y \cdot x \cdot y^{-1}} \mu(z) \geq \mu(x) \wedge 0.5$. ■

Corollary 2.4.10. *Let μ be an $(\in, \in \vee q)$-fuzzy normal subpolygroup of P. Then, for all $x, y \in P$, we have*

$$\left(\bigwedge_{z \in x \cdot y} \mu(z)\right) \wedge 0.5 = \left(\bigwedge_{z' \in y \cdot x} \mu(z')\right) \wedge 0.5.$$

Proof. Suppose that $x, y \in P$. For every $z \in x \cdot y$, we have $z \in y^{-1} \cdot (y \cdot x) \cdot y$. So, there exists $z' \in y \cdot x$ such that $z \in y^{-1} \cdot z' \cdot y$. Since μ is $(\in, \in \vee q)$-fuzzy normal, we have $\mu(z) \geq \mu(z') \wedge 0.5$, which implies that $\mu(z) \geq \left(\bigwedge_{z' \in y \cdot x} \mu(z')\right) \wedge 0.5$, and so $\bigwedge_{z \in x \cdot y} \mu(z) \geq \left(\bigwedge_{z' \in y \cdot x} \mu(z')\right) \wedge 0.5$. Therefore,

$$\left(\bigwedge_{z \in x \cdot y} \mu(z)\right) \wedge 0.5 \geq \left(\bigwedge_{z' \in y \cdot x} \mu(z')\right) \wedge 0.5.$$

Similarly, we obtain

$$\left(\bigwedge_{z \in x \cdot y} \mu(z)\right) \wedge 0.5 \leq \left(\bigwedge_{z' \in y \cdot x} \mu(z')\right) \wedge 0.5.$$

■

Theorem 2.4.11. *Let μ be an $(\in, \in \vee q)$-fuzzy normal subpolygroup of P. Then, μ_t $(\neq \emptyset)$ is a normal subpolygroup of P, for all $t \in (0, 0.5]$. Conversely, if μ is a fuzzy subset of P such that μ_t $(\neq \emptyset)$ is a normal subpolygroup of P for all $t \in (0, 0.5]$, then μ is an $(\in, \in \vee q)$-fuzzy normal.*

Proof. First, let μ be an $(\in, \in \vee q)$-fuzzy normal subpolygroup of P. Then, μ_t is a subpolygroup of P by Theorem 2.4.4. Now, we show that μ_t is normal. Assume that $x \in \mu_t$ and $y \in P$. Then, for every $z \in y \cdot x \cdot y^{-1}$, we have

$$\begin{aligned} \mu(z) &\geq \mu(x) \wedge 0.5 \text{ (since } \mu \text{ is } (\in, \in \vee q) - \text{fuzzy normal)} \\ &\geq t \wedge 0.5 = t \end{aligned}$$

and so $z \in \mu_t$. Therefore, $y \cdot x \cdot y^{-1} \subseteq \mu_t$, i.e., μ_t is normal, for all $t \in (0, 0.5]$.

Conversely, let μ be a fuzzy subset of P such that μ_t is a normal subpolygroup of P, for all $t \in (0, 0.5]$. Then, μ is an $(\in, \in \vee q)$-fuzzy subpolygroup of P by Theorem 2.4.4. Now, we show that μ is $(\in, \in \vee q)$-fuzzy normal. Assume that $x, y \in P$, we have $\mu(x) \geq \mu(x) \wedge 0.5 = t_0$, hence $x \in \mu_{t_0}$. Since μ_{t_0} is normal, we have $y \cdot x \cdot y^{-1} \subseteq \mu_{t_0}$. So,

$$z \in \mu_{t_0}, \text{ for all } z \in y \cdot x \cdot y^{-1},$$

which implies that

$$\mu(z) \geq t_0, \text{ for all } z \in y \cdot x \cdot y^{-1}.$$

Therefore,

$$\bigwedge_{z \in y \cdot x \cdot y^{-1}} \mu(z) \geq \mu(x) \wedge 0.5.$$

■

Definition 2.4.12. Let μ be an $(\in, \in \vee q)$-fuzzy subpolygroup of P. For any $x \in P$, $\overline{\mu_x}$ and $\underline{\mu_x}$ are define by

$$\overline{\mu_x}(a) = \left(\bigwedge_{z \in a \cdot x^{-1}} \mu(z) \right) \wedge 0.5,$$

$$\underline{\mu_x}(a) = \left(\bigwedge_{z \in x^{-1} \cdot a} \mu(z) \right) \wedge 0.5,$$

for all $a \in P$, and are called $(\in, \in \vee q)$-*fuzzy left* and *fuzzy right cosets of* P *determined by* x *and* μ.

The following example shows that for an $(\in, \in \vee q)$-fuzzy subpolygroup μ of a polygroup P, $(\in, \in \vee q)$-fuzzy left coset need not be equal to corresponding $(\in, \in \vee q)$-fuzzy right coset.

EXAMPLE 29 Let P be the polygroup defined in Example 28, and let

$$S_3 = \{e,\ (1\ 2),\ (1\ 3),\ (2\ 3),\ (1\ 2\ 3),\ (1\ 3\ 2)\}$$

be the symmetric group of degree 3. We consider the polygroup $S_3[P]$ (the extension of S_3 by P). We define $\mu : S_3[P] \to [0, 1]$ by

$$\begin{aligned} &\mu(e) = 0.8, \\ &\mu(a) = 0.3, \\ &\mu(b) = 0.3, \\ &\mu((12)) = 0.8, \\ &\mu(x) = 0.4, \text{ for all } x \in S_3 - \{e, (12)\}. \end{aligned}$$

Then, it is no difficult to see that μ is an $(\in, \in \vee q)$-fuzzy subpolygroup of $S_3[P]$. We have

$$\overline{\mu_{(1\ 3\ 2)}}((1\ 3)) = 0.4 \text{ and } \underline{\mu_{(132)}}((13)) = 0.5.$$

Hence, $\overline{\mu_{(1\ 3\ 2)}} \neq \underline{\mu_{(1\ 3\ 2)}}$. This happens because μ is not an $(\in, \in \vee q)$-fuzzy normal subpolygroup of $S_3[P]$.

Proposition 2.4.13. *Let* μ *be an* $(\in, \in \vee q)$*-fuzzy normal subpolygroup of* P*. Then,*

$$\overline{\mu_x} = \underline{\mu_x}, \text{ for all } x \in P.$$

Proof. It is straightforward. ■

Proposition 2.4.14. *Let $x, y \in P$ and for any $a \in P$, $\overline{\mu_x}(a) = \overline{\mu_y}(a)$. Then, for any non-empty subset S of P, we have*

$$\left(\bigwedge_{a \in S \cdot x^{-1}} \mu(a)\right) \wedge 0.5 = \left(\bigwedge_{a \in S \cdot y^{-1}} \mu(a)\right) \wedge 0.5.$$

Proof. We have

$$\left(\bigwedge_{a \in S \cdot x^{-1}} \mu(a)\right) \wedge 0.5 \leq \left(\bigwedge_{a \in s \cdot x^{-1}} \mu(a)\right) \wedge 0.5 = \left(\bigwedge_{a \in s \cdot y^{-1}} \mu(a)\right) \wedge 0.5,$$

for all $s \in S$. Therefore,

$$\left(\bigwedge_{a \in S \cdot x^{-1}} \mu(a)\right) \wedge 0.5 \leq \bigwedge_{s \in S} \left(\left(\bigwedge_{a \in s \cdot y^{-1}} \mu(a)\right) \wedge 0.5\right) = \left(\bigwedge_{a \in S \cdot y^{-1}} \mu(a)\right) \wedge 0.5.$$

The proof of converse inequality is similar to the above one. ■

Theorem 2.4.15. *Let μ be an $(\in, \in \vee q)$-fuzzy normal subpolygroup of P and $\mu_{0.5} \neq \emptyset$. Then, for all $a, b \in P$,*

$$\overline{\mu_a} = \overline{\mu_b} \Leftrightarrow \mu_{0.5} \cdot a = \mu_{0.5} \cdot b.$$

Proof. Assume that $a, b \in P$ and $\overline{\mu_a} = \overline{\mu_b}$. Since $\mu_{0.5} \neq \emptyset$, $\mu(e) \geq 0.5$. Since $e \in b^{-1} \cdot b$ and μ is normal, for every $z \in b \cdot b^{-1}$ we have $\mu(z) \geq \mu(e) \wedge 0.5 = 0.5$. Thus,

$$\begin{aligned}
\bigwedge_{z \in b \cdot b^{-1}} \mu(z) \geq 0.5 &\Rightarrow \left(\bigwedge_{z \in b \cdot b^{-1}} \mu(z)\right) \wedge 0.5 \geq 0.5 \\
&\Rightarrow \overline{\mu_b}(b) \geq 0.5 \\
&\Rightarrow \overline{\mu_a}(b) \geq 0.5 \\
&\Rightarrow \left(\bigwedge_{z \in b \cdot a^{-1}} \mu(z)\right) \wedge 0.5 \geq 0.5 \\
&\Rightarrow \bigwedge_{z \in b \cdot a^{-1}} \mu(z) \geq 0.5.
\end{aligned}$$

Hence, there exists $z_0 \in b \cdot a^{-1}$ such that $\mu(z_0) \geq 0.5$. Since μ is normal, for all $z \in a^{-1} \cdot b$, we have $\mu(z) \geq \mu(z_0) \wedge 0.5 = 0.5$ and so $z \in \mu_{0.5}$. Hence, $a^{-1} \cdot b \subseteq \mu_{0.5}$. Since $\mu_{0.5}$ is a normal subpolygroup of P, $\mu_{0.5} \cdot a = \mu_{0.5} \cdot b$.

Conversely, suppose that $\mu_{0.5} \cdot a = \mu_{0.5} \cdot b$. Then, for every $x \in P$, we have $x \cdot a^{-1} \cdot \mu_{0.5} = x \cdot b^{-1} \cdot \mu_{0.5}$. Also, we have

$$\begin{aligned}
\overline{\mu_a}(x) &= \left(\bigwedge_{z \in x \cdot a^{-1}} \mu(z)\right) \wedge 0.5, \\
\overline{\mu_b}(x) &= \left(\bigwedge_{z \in x \cdot b^{-1}} \mu(z)\right) \wedge 0.5 = \left(\bigwedge_{z \in b^{-1} \cdot x} \mu(z)\right) \wedge 0.5. \qquad (\star)
\end{aligned}$$

Now, we put

$$r = \bigwedge_{z \in x \cdot a^{-1}} \mu(z) \text{ and } s = \bigwedge_{z \in x \cdot b^{-1}} \mu(z).$$

We show that ($r < 0.5$ and $s < 0.5$) or ($r \geq 0.5$ and $s \geq 0.5$). In order to show this, assume that $r < 0.5$ and $s \geq 0.5$. Since $r < 0.5$, there exists $z_0 \in x \cdot a^{-1}$ such that $\mu(z_0) < 0.5$. This implies that $z_0 \notin \mu_{0.5}$ and so $x \cdot a^{-1} \not\subseteq \mu_{0.5}$. Hence, $x \cdot a^{-1} \cdot \mu_{0.5} \neq \mu_{0.5}$. Since $s \geq 0.5$, for every $z \in x \cdot b^{-1}$, $\mu(z) \geq 0.5$. This implies that $x \cdot b^{-1} \subseteq \mu_{0.5}$ and so $x \cdot b^{-1} \cdot \mu_{0.5} = \mu_{0.5}$. Therefore, $x \cdot a^{-1} \cdot \mu_{0.5} \neq x \cdot b^{-1} \cdot \mu_{0.5}$, which is a contradiction. Similarly, we can show that it is not possible to have $r \geq 0.5$ and $s < 0.5$.

Now, if $r \geq 0.5$ and $s \geq 0.5$, then $\overline{\mu_a}(x) = 0.5 = \overline{\mu_b}(x)$, for all $x \in P$.

If $r < 0.5$ and $s < 0.5$, then $\overline{\mu_a}(x) = r$ and $\overline{\mu_b}(x) = s$. In this case, we show that $r = s$. By ($\star$), clearly, we have

$$s = \bigwedge_{z \in x \cdot b^{-1}} \mu(z) = \bigwedge_{z \in b^{-1} \cdot x} \mu(z).$$

Since $r < 0.5$, $\bigwedge\limits_{z \in x \cdot a^{-1}} \mu(z) < 0.5$. So, there exists $z_0 \in x \cdot a^{-1}$ such that $\mu(z_0) < 0.5$. Hence,

$$r = \bigwedge_{z \in x \cdot a^{-1}} \mu(z) = \bigwedge_{\substack{z \in x \cdot a^{-1} \\ \mu(z) < 0.5}} \mu(z).$$

Similarly, we obtain

$$s = \bigwedge_{z \in b^{-1} \cdot x} \mu(z) = \bigwedge_{\substack{z \in b^{-1} \cdot x \\ \mu(z) < 0.5}} \mu(z).$$

Now, we show that if $z_0 \in x \cdot a^{-1}$, $z_0' \in b^{-1} \cdot x$, $\mu(z_0) < 0.5$ and $\mu(z_0') < 0.5$, then $\mu(z) = \mu(z_0)$. This implies that $r = s$.

Since $\mu_{0.5} \cdot a = \mu_{0.5} \cdot b$ and $b \in \mu_{0.5} \cdot b$, we have $b \in \mu_{0.5} \cdot a$. Hence, there exists $y \in \mu_{0.5}$ such that $b \in y \cdot a$ and so $b^{-1} \in a^{-1} \cdot y^{-1}$. Since $z_0 \in x \cdot a^{-1}$, we have $x \in z_0 \cdot a$. Thus, $z_0' \in a^{-1} \cdot y^{-1} \cdot z_0 \cdot a$. Hence, there exists $c \in y^{-1} \cdot z_0$ such that $z_0' \in a^{-1} \cdot c \cdot a$. Thus,

$$\mu(z_0') \geq \mu(c) \wedge 0.5 \geq \mu(y^{-1}) \wedge \mu(z_0) \wedge 0.5.$$

Since $\mu(y^{-1}) \geq 0.5$ and $\mu(z_0) < 0.5$, we obtain $\mu(z_0') \geq \mu(z_0)$. Similarly, we obtain $\mu(z_0) \geq \mu(z_0')$. ∎

Theorem 2.4.16. *Let μ be an $(\in, \in \vee q)$-fuzzy normal subpolygroup of P and $\mu_{0.5} \neq \emptyset$. Let P/μ be the set of all $(\in, \in \vee q)$-fuzzy left coset of μ in P. Then, P/μ is a polygroup if the hyperoperation is defined by*

$$\overline{\mu_x} \odot \overline{\mu_y} = \{\overline{\mu_z} \mid z \in \mu_{0.5} \cdot x \cdot y\}$$

and $(\overline{\mu_x})^{-I} = \overline{\mu_{x^{-1}}}$.

Proof. It is straightforward. ■

Naturally, a corresponding result should be considered when μ_t is a normal subpolygroup of P, for all $t \in (0.5, 1]$.

Theorem 2.4.17. *Let μ be a fuzzy subset of a polygroup P. Then, μ_t $(\neq \emptyset)$ is a normal subpolygroup of P for all $t \in (0.5, 1]$, if and only if the conditions (1) and (2) in Theorem 2.4.5 and the following condition hold:*

$$\mu(x) \leq \bigwedge_{z \in y \cdot x \cdot y^{-1}} (\mu(z) \vee 0.5), \text{ for all } x, y \in P.$$

Proof. The proof is similar to the proof of Theorem 2.4.11. ■

Theorem 2.4.18. *Let $\alpha, \beta \in [0, 1]$, $\alpha < \beta$ and μ be a fuzzy subset of P. Then, μ_t $(\neq \emptyset)$ is a normal subpolygroup of P for all $t \in (\alpha, \beta]$ if and only if μ is a fuzzy subpolygroup with thresholds of P and the following condition holds:*

$$\bigwedge_{z \in y \cdot x \cdot y^{-1}} (\mu(z) \vee \alpha) \geq \mu(x) \wedge \beta, \tag{2.1}$$

for all $x, y \in P$.

Proof. First, let μ be a fuzzy subpolygroup with thresholds of P and condition (2.1) holds. Then, μ_t is a subpolygroup of P by Theorem 3.4.13. Now, we show that μ_t is normal. Assume that $x \in \mu_t$ and $y \in P$. Then, for every $z \in y \cdot x \cdot y^{-1}$, we have

$$\mu(z) \vee \alpha \geq \mu(x) \wedge \beta = t \wedge \beta \geq t > \alpha.$$

So, for every $z \in y \cdot x \cdot y^{-1}$ we have $\mu(z) \vee \alpha \geq t > \alpha$, which implies that $\mu(z) \geq t$ and $z \in \mu_t$. Hence, $y \cdot x \cdot y^{-1} \subseteq \mu_t$, i.e., μ_t is normal for all $t \in (\alpha, \beta]$.

Conversely, let μ be a fuzzy subset of P such that μ_t $(\neq \emptyset)$ is a normal subpolygroup of P, for all $t \in (\alpha, \beta]$. Then, μ is a fuzzy subpolygroup with thresholds of P by Theorem 3.4.13. Now, we verify the condition (2.1). If there exist $x, y, z \in P$ with $z \in y \cdot x \cdot y^{-1}$ such that $\mu(z) \vee \alpha < \mu(x) \wedge \beta = t$, then $t \in (\alpha, \beta]$, $\mu(z) < t$ and $x \in \mu_t$. Since μ_t is normal, so $y \cdot x \cdot y^{-1} \subseteq \mu_t$. Hence, $\mu(z) \geq t$, for all $z \in y \cdot x \cdot y^{-1}$. This is a contradiction with $\mu(z) < t$. Therefore,

$$\mu(z) \vee \alpha \geq \mu(x) \wedge \beta \text{ for all } x, y \in P \text{ with } z \in y \cdot x \cdot y^{-1},$$

and so (2.1) holds. ■

Logic started as the study of language in arguments and persuasion, and it can be used to judge the correctness of chain of reasonings. In a mathematical proof, for instance, the goal of the theory is to reduce the principles of reasoning to a code. The "truth" or "falsity" assigned to a Proposition is its truth-value. In fuzzy logic , a Proposition may be true or false, or an intermediate truth-value may be also true. The sentence "Kenneth Young is a bad guy" is an example of a fuzzy Proposition. In practice, one would subdivide the unit interval into finer divisions or work with a continuous truth-domain. In daily conversation and mathematics, sentences are connected with words such as "and"," or", "if-then (or implies)", and "if and only if" etc. These words are called *connectives.* A sentence which is modified by the word "not" is called the *negation* of the original sentence. The word "and" is used to join two sentences to form the "conjunction" of the two sentences. The word "or" is used to join two sentences to form the "disjunction" of the two sentences. From two sentences , we may construct one of the form "If ... then ...". This is called an " implication ".

Some operators such as $\wedge, \vee, \neg, \rightarrow$ in fuzzy logic can also be defined by using the truth tables, and the extension principle can also be applied to derive definitions of the operators. In fuzzy logic, the truth value of a fuzzy Proposition P is denoted by $[P]$. In the following, we display some of the fuzzy logical and corresponding set-theoretical notions which are frequently used in this section:

$[x \in \mu] = \mu(x)$;
$[x \notin \mu] = 1 - \mu(x)$;
$[P \wedge Q] = \min\{[P], [Q]\}$;
$[P \vee Q] = \max\{[P], [Q]\}$;
$[P \rightarrow Q] = \min\{1, 1 - [P] + [Q]\}$;
$[\forall x P(x)] = \inf[P(x)]$;
$\models P$ if and only if $[P] = 1$ for all valuations.

The truth valuation rules given above are those in the Łukasiewicz system of continuous-valued logic.

Definition 2.4.19. A function $I : [0,1] \times [0,1] \rightarrow [0,1]$ is called *fuzzy implication* if it is monotonic with respect to both variables (separately) and fulfills the binary implication truth table:

$$I(0,0) = I(0,1) = I(1,1) = 1, \quad I(1,0) = 0.$$

By monotonicity

$$I(0,x) = I(x,1) = 1 \quad \text{for all } x \in [0,1],$$

where I is decreasing with respect to the first variable ($I(1,0) < I(0,0)$) and I is increasing with respect to the second variable ($I(1,0) < I(1,1)$).

In practice, various implication operators can be defined. A selection of the most important multi-valued implications in the following table is shown, where α denotes the degree of truth (or degree of membership) of the premise, and we use β to denote the respective values for the consequence, and I is the resulting degree of truth for the implication:

Name	Definition of Implication Operators
Zadeh	$I_m(\alpha,\beta) = \max\{1-\alpha, \min\{\alpha,\beta\}\}$
Łukasiewicz	$I_a(\alpha,\beta) = \min\{1, 1-\alpha+\beta\}$
Standard Star(Gödel)	$I_g(\alpha,\beta) = \begin{cases} 1 & \text{if} \quad \alpha \leq \beta \\ \beta & \text{if} \quad \alpha > \beta \end{cases}$
Contraposition of Gödel	$I_{cg}(\alpha,\beta) = \begin{cases} 1 & \text{if} \quad \alpha \leq \beta \\ 1-\alpha & \text{if} \quad \alpha > \beta \end{cases}$
Gaines-Rescher	$I_{gr}(\alpha,\beta) = \begin{cases} 1 & \text{if} \quad \alpha \leq \beta \\ 0 & \text{if} \quad \alpha > \beta \end{cases}$
Kleene-Dienes	$I_b(\alpha,\beta) = \max\{1-\alpha, \beta\}$
Goguen	$I_{gg}(\alpha,\beta) = \begin{cases} 1 & \text{if} \quad \alpha \leq \beta \\ \frac{\beta}{\alpha} & \text{if} \quad \alpha > \beta \end{cases}$
Reichenbach	$I_{RC}(\alpha,\beta) = 1-\alpha+\alpha\beta$

The quality of the above implication operators could be evaluated either empirically or axiomatically. We now denote the set of all fuzzy implications by FI. Then, the sup-min composition of fuzzy relations $I, J \in FI$ can be expressed by the relation $I \circ J$,

$$(I \circ J)(x,z) = \bigvee_{y\in[0,1]} (I(x,y) \wedge J(y,z)), \quad \text{for } x,z \in [0,1].$$

By using Łukasiewicz and Kleene-Dienes implications, we can easily see that

$$(I_a \circ I_b)(1,0) = (I_b \circ I_a)(1,0) = \bigvee\nolimits_{y\in[0,1]}(y \wedge (1-y)) = 0.5 > 0.$$

Therefore, the composition of fuzzy implications is not a fuzzy implication. It is not also difficult to see the following: $I \circ J \in FI \ \Leftrightarrow\ (I \circ J)(1,0) = 0$.

In the following definition, we consider the implication operators in Łukasiewicz system of continuous-valued logic.

Definition 2.4.20. Let a fuzzy subset μ of a polygroup P satisfies:

(1) For any $x, y \in P$, $\models \ [\ [x \in \mu] \wedge [y \in \mu] \to [\forall z \in x \cdot y,\ z \in \mu]\ ]$.
(2) For any $x \in P$, $\models \ [[\ x \in \mu] \to [x^{-1} \in \mu]]$.

Then, μ is called a *fuzzifying subpolygroup of* P.

Clearly, Definition 2.4.20 is equivalent to Definition 2.3.1. Therefore, a fuzzifying subpolygroup is an ordinary fuzzy subpolygroup.

Now, we consider the concept of t-tautology, i.e.,

$$\models_t P \text{ if and only if } [P] \geq t \text{ for all valuations.}$$

We define the concept of implication-based fuzzy subpolygroup in the following way.

Definition 2.4.21. Let μ be a fuzzy subset of a polygroup P and $t \in (0,1]$ be a fixed number. If

(1) for any $x, y \in P$, $\models_t \; [\,[x \in \mu] \wedge [y \in \mu] \to [\forall z \in x \cdot y,\; z \in \mu]\,]$,
(2) for any $x \in P$, $\models_t \; [\,[x \in \mu] \to [x^{-1} \in \mu]]$,

then μ is called a *t-implication-based fuzzy subpolygroup of P.*

Now, let I be an implication operator. Then the following result follows.

Corollary 2.4.22. *μ is a t-implication-based fuzzy subpolygroup of a polygroup P if and only if*

(1) $I_a(\mu(x) \wedge \mu(y), \bigwedge\limits_{z \in x \cdot y} \mu(z)) \geq t$, *for all* $x, y \in P$.
(2) *For any* $x \in P$, $I_a(\mu(x), \mu(x^{-1})) \geq t$.

2.5 Roughness in Polygroups

Let U be any non-empty set and let $\mathcal{B}$ be a complete subalgebra of the Boolean algebra $\mathcal{P}(U)$ of subsets of U. The pair $(U, \mathcal{B})$ is called a *rough universe.* Let $\mathcal{R}$ be the relation defined as follows:

$$A = (A_L, A_U) \in \mathcal{R} \text{ if and only if } A_L, A_U \in \mathcal{B},\; A_L \subseteq A_U.$$

The elements of $\mathcal{R}$ are called *rough sets.*

Let $A = (A_L, A_U)$ and $B = (B_L, B_U)$ be any two rough sets. Then,

$$A \cup B = (A_L \cup B_L, A_U \cup B_U), \quad A \cap B = (A_L \cap B_L, A_U \cap B_U),$$

and $A \subseteq B$ if and only if $A_L \subseteq B_L$ and $A_U \subseteq B_U$.

The theory of rough sets can be developed in at least two ways: constructive and axiomatic approaches. They are complementary to each other. The constructive method is more suitable for practical applications of rough sets, while the axiomatic method is appropriate for studding the structure of rough set algebra.

Let N be a normal subpolygroup of a polygroup P, and X be a non-empty subset of P. In [70], Davvaz considered the *lower* and *upper approximations* of X with respect to N as follows:

$$N_*(X) = \{x \in P \mid xN \subseteq X\}$$

and

$$N^*(X) = \{x \in P \mid xN \cap X \neq \emptyset\}.$$

The indiscernibility relation enables us to divide objects of the polygroup P into three disjoint sets with respect to any subset $X \subseteq P$:

(1) The objects which surely are in X.
(2) The objects which are surely not in X.
(3) The objects which possibly are in X.

The objects in class 1 form the lower approximation of X, and the objects of type 1 and 3 form together its upper approximation.

The *lower* and *upper approximations* can be presented in an equivalent form as shown below:

$$\underline{N}(X) = \{xN \in P/N \mid xN \subseteq X\}$$

and

$$\overline{N}(X) = \{xN \in P/N \mid xN \cap X \neq \emptyset\}.$$

In this case, $\underline{N}(X)$ and $\overline{N}(X)$ are subsets of P/N.

It is straightforward to show that every subpolygroup of P/N is of the form X/N, where X is a subpolygroup of P that contains N. Now, we collect in the following proposition the basic property of approximations.

Proposition 2.5.1. *Let N be a normal subpolygroup of a polygroup P. If X is a subpolygroup of P such that $N \subseteq X$, then*

$$\underline{N}(X) \subseteq X/N \subseteq \overline{N}(X).$$

Proof. The proof follows from their definitions. ■

Therefore, $N(X) = (\underline{N}(X), \overline{N}(X))$ is a rough set.

Proposition 2.5.2. *Let N be a normal subpolygroup of a polygroup P. If X is a subpolygroup of P, then $\overline{N}(X)$ is a subpolygroup of P/N.*

Proof. Suppose that aN and bN are any elements of $\overline{N}(X)$. Then, there exist elements x and y in P such that $x \in aN \cap X$ and $y \in bN \cap X$, and so $x \in aN$, $y \in bN$, $\{x, y\} \subseteq X$. Since X is a subpolygroup of P, $xy \subseteq X$. And since N is a normal subpolygroup of P, $xy \subseteq (aN)(bN) = abN$. Now, for every $cN \in aN \odot bN$ we have $xy \subseteq cN$. Therefore, $cN \cap X \neq \emptyset$, and so $cN \in \overline{N}(X)$ which yields $aN \odot bN \subseteq \overline{N}(X)$. Therefore, the first condition of the definition is satisfies. Now, we prove the second condition. Assume that aN is any element of $\overline{N}(X)$, then there exists $x \in aN \cap X$, so $x \in aN$ and $x \in X$. Since X is a subpolygroup of P, then $x^{-1} \in X$. Since $x \in aN$, then there exists $n \in N$ such that $x \in an$, and since N is a normal subpolygroup of P and $n^{-1} \in N$, we have $x^{-1} \in (an)^{-1} = n^{-1}a^{-1} \in Na^{-1} = a^{-1}N$. Therefore, $x^{-1} \in a^{-1}N \cap X$ which implies that $a^{-1}N \cap X \neq \emptyset$, and so $a^{-1}N = (aN)^{-I} \in \overline{N}(X)$. This completes the proof. ■

Proposition 2.5.3. *Let N be a normal subpolygroup of a polygroup P. If X is a subpolygroup of P such that $N \subseteq X$, then $\underline{N}(X)$ is a subpolygroup of P/N.*

Proof. Assume that aN and bN are any elements of $\underline{N}(X)$. Then, $aN \subseteq X$ and $bN \subseteq X$. Therefore, we have $abN = (aN)(bN) \subseteq XX \subseteq X$, and so for every $c \in abN$, we have $cN \subseteq X$. This implies that $aN \odot bN \subseteq \underline{N}(X)$. Therefore, the first condition of the definition is satisfies. Now, we prove the second condition. Let aN be any element of $\underline{N}(X)$. Then, $a = ae \in aN \subseteq X$. Assuming that X is a subpolygroup of P, it follows $a^{-1} \in X$. Therefore, we have $a^{-1}N \subseteq XX \subseteq X$ which implies that $a^{-1}N \in \underline{N}(X)$, and so $(aN)^{-I} \in \underline{N}(X)$. Therefore, $\underline{N}(X)$ is a subpolygroup of P/N. ■

Definition 2.5.4. Let N be a normal subpolygroup of a polygroup P and $A = (A_L, A_U)$ be a rough set in P/N. If A_L, A_U are subpolygroups of P/N, then we call $A = (A_L, A_U)$ a *quotient rough polygroup.*

Theorem 2.5.5. *Let N be a normal subpolygroup of a polygroup P. If X is a subpolygroup of P that contains N, then $N(X)$ is a quotient rough polygroup.*

Proof. The proof follows from Proposition 2.5.2 and 2.5.3. ■

The study of properties of rough sets on a polygroup is a meaningful research topic of rough sets theory. The existing research of rough sets on a polygroup are mainly concerned with crisp sets. In this section we concentrate our study on algebraic properties of fuzzy rough sets with respect to a polygroup. We introduce a new approach to the combination of fuzzy sets and rough sets. The combination is based on membership functions. It provides a new perspective to the theories of fuzzy sets, rough sets and algebraic hyperstructures.

Let $S = (S_L, S_U)$ be a rough set. A *fuzzy rough set* $A = (A_L, A_U)$ in S is characterized by a pair of maps

$$\mu_{A_L} : S_L \to [0,1], \quad \mu_{A_U} : S_U \to [0,1]$$

with the property that $\mu_{A_L}(x) \leq \mu_{A_U}(x)$, for all $x \in S_L$. For any two fuzzy rough sets $A = (A_L, A_U)$ and $B = (B_L, B_U)$ in S we define

(1) $A \subseteq B$ if and only if $\mu_{A_L}(x) \leq \mu_{B_L}(x)$, for all $x \in S_L$, and $\mu_{A_U}(y) \leq \mu_{B_U}(y)$, for all $y \in S_U$.
(2) $C = A \cup B$ if and only if $\mu_{C_L}(x) = \mu_{A_L}(x) \vee \mu_{B_L}(x)$, for all $x \in S_L$, and $\mu_{C_U}(y) = \mu_{A_U}(y) \vee \mu_{B_U}(y)$, for all $y \in S_U$.
(3) $D = A \cap B$ if and only if $\mu_{D_L}(x) = \mu_{A_L}(x) \wedge \mu_{B_L}(x)$, for all $x \in S_L$, and $\mu_{D_U}(y) = \mu_{A_U}(y) \wedge \mu_{B_U}(y)$, for all $y \in S_U$.

Now, assume that $N(X) = (\underline{N}(X), \overline{N}(X))$ is a rough polygroup. The difference $\widehat{N(X)} = \overline{N}(X) - \underline{N}(X)$ is called the *boundary region* of X. In the case when $\widehat{N(X)} = \emptyset$, the set X is said to be *exact*. Let $A = (A_L, A_U)$ be a fuzzy rough set of $N(X)$. We define $\mu_{A_L^*} : \overline{N}(X) \to L$ as follows:

$$\mu_{A_L^*}(aN) = \begin{cases} \mu_{A_L}(aN) & \text{if } aN \in \underline{N}(X) \\ 0 & \text{if } aN \in \widehat{N(X)}. \end{cases}$$

Definition 2.5.6. Let $N(X)$ be a rough polygroup. An *interval-valued L-fuzzy subset* F is given by

$$F = \{(aN, [\mu_{A_L^*}(aN), \mu_{A_U}(aN)]) \mid aN \in \overline{N}(X)\},$$

where (A_L, A_U) is a fuzzy rough set in $N(X)$.

Suppose that $\widetilde{\mu_F}(aN) = [\mu_{A_L^*}(aN), \mu_{A_U}(aN)]$ for all $aN \in \overline{N}(X)$, and $D(L)$ denotes the family of all closed subintervals of L. If $\mu_{A_L^*}(aN) = \mu_{A_U}(aN) = t$ where $0 \leq t \leq 1$, then we have $\widetilde{\mu_F}(aN) = [t, t]$ which we also assume, for the sake of convenience, to belong to $D(L)$. Thus, $\widetilde{\mu_F}(aN) \in D(L)$ for all $aN \in \overline{N}(X)$.

Definition 2.5.7. Let $D_1 = [a_1, b_1]$, $D_2 = [a_2, b_2]$ be elements of $D(L)$. Then, we define

$$\text{rmax}(D_1, D_2) = [a_1 \vee a_2,\ b_1 \vee b_2],$$

$$\text{rmin}(D_1, D_2) = [a_1 \wedge a_2,\ b_1 \wedge b_2].$$

We call $D_2 \leq D_1$ if and only if $a_2 \leq a_1$ and $b_2 \leq b_1$.

Definition 2.5.8. Let $N(X)$ be a rough polygroup. A fuzzy rough set $A = (A_L, A_U)$ in $N(X)$ is called a *fuzzy rough subpolygroup* if for all $xN, yN \in \overline{N}(X)$, the following hold.

(1) $\text{rmin}(\widetilde{\mu_F}(xN), \widetilde{\mu_F}(yN)) \leq \widetilde{\mu_F}(zN)$, for all $zN \in xN \odot yN$.
(2) $\mu_{A_L^*}(xN) \leq \mu_{A_L^*}(x^{-1}N)$, $\mu_{A_U}(xN) \leq \mu_{A_U}(x^{-1}N)$.

Lemma 2.5.9. *Let $N(X)$ be a rough polygroup. If $A = (A_L, A_U)$ and $B = (B_L, B_U)$ are two fuzzy rough subpolygroups of $N(X)$, then $A \cap B$ is a fuzzy rough subpolygroup of $N(X)$.*

Proof. It is straightforward. ■

Definition 2.5.10. Let $N(X)$ be a rough polygroup and $A = (A_L, A_U)$ be a fuzzy rough set of $N(X)$. We define

$$\underline{A}_t = \{xN \mid xN \in \underline{N}(X),\ \mu_{A_L}(xN) \geq t\},$$

$$\overline{A}_t = \{xN \mid xN \in \overline{N}(X),\ \mu_{A_U}(xN) \geq t\}.$$

Then, $(\underline{A}_t, \overline{A}_t)$ is called a *level rough set*.

Theorem 2.5.11. *Let $N(X)$ be a quotient rough polygroup and $A = (A_L, A_U)$ be a fuzzy rough set of $N(X)$. Then, A is a fuzzy rough subpolygroup of $N(X)$ if and only if for every $0 \le t \le 1$, $(\underline{A}_t, \overline{A}_t)$ is a quotient rough subpolygroup of $N(X)$.*

Proof. Suppose that for every $0 \le t \le 1$, $(\underline{A}_t, \overline{A}_t)$ is a rough subpolygroup of $N(X)$. For every $xN, yN \in \overline{N}(X)$ we must prove all the conditions in Definition 2.5.8. We put

$$\mathrm{rmin}(\widetilde{\mu_F}(xN), \widetilde{\mu_F}(yN)) = [t_0, t_1].$$

Then,

$$\mu_{A_L^*}(xN) \wedge \mu_{A_L^*}(yN) = t_0, \qquad \mu_{A_U}(xN) \wedge \mu_{A_U}(yN) = t_1.$$

We obtain $xN \in \overline{A}_{t_1}, yN \in \overline{A}_{t_1}$. Then, $xN \odot yN \subseteq \overline{A}_{t_1}$, and so for every $zN \in xN \odot yN$ we have $zN \in \overline{A}_{t_1}$. Thus, $\mu_{A_U}(zN) \ge t_1$. On the other hand, if $xN \in \widehat{N(X)}$ or $yN \in \widehat{N(X)}$ then $t_0 = 0$ and for $zN \in xN \odot yN$ we have $\mu_{A_L^*}(zN) \ge 0 = t_0$. If $xN \notin \widehat{N(X)}$ and $yN \notin \widehat{N(X)}$, then $\mu_{A_L^*}(xN) = \mu_{A_L}(xN)$, $\mu_{A_L^*}(yN) = \mu_{A_L}(yN)$, so $\mu_{A_L}(xN) \wedge \mu_{A_L}(yN) = t_0$. Thus, $xN \in \underline{A}_{t_0}$, $yN \in \underline{A}_{t_0}$. Therefore, for $zN \in xN \odot yN$ we obtain $zN \in \underline{A}_{t_0}$ which implies that $\mu_{A_L^*}(zN) \ge t_0$. Therefore,

$$\widetilde{\mu_F}(zN) = [\mu_{A_L^*}(zN), \mu_{A_U}(zN)] \ge [t_0, t_1]$$

and so $\widetilde{\mu_F}(zN) \ge rmin\{\widetilde{\mu_F}(xN), \widetilde{\mu_F}(yN)\}$. In order to verify the second condition, we know $xN \in \overline{A}_{\mu_{A_U}(xN)}$, so $(xN)^{-I} \in \overline{A}_{\mu_{A_U}(xN)}$ which implies that $\mu_{A_U}(x^{-1}N) \ge \mu_{A_U}(xN)$.

Now, if $xN \in \widehat{N(X)}$, then $\mu_{A_L^*}(xN) = 0$. If $xN \in \underline{N}(X)$, then $xN \in \underline{A}_{\mu_{A_L}(xN)}$, so $(xN)^{-I} \in \underline{A}_{\mu_{A_L}(xN)}$ which implies that $\mu_{A_L}(x^{-1}N) \ge \mu_{A_L}(xN)$. Thus, we obtain $\mu_{A_L^*}(x^{-1}N) \ge \mu_{A_L^*}(xN)$. Therefore, A is a fuzzy rough subpolygroup of $N(X)$.

Conversely, assume that $A = (A_L, A_U)$ is a fuzzy rough subpolygroup of $N(X)$. We show that for every $0 \le t \le 1$, $\underline{A}_t$ and $\overline{A}_t$ are subpolygroups. For every $xN, yN \in \underline{A}_t$, we have $\mu_{A_L}(xN) \ge t$, $\mu_{A_L}(yN) \ge t$, so

$$\mathrm{rmin}(\widetilde{\mu_F}(xN), \widetilde{\mu_F}(yN)) \ge [t, \mu_{A_U}(xN) \wedge \mu_{A_U}(yN)].$$

Therefore, for every $zN \in xN \odot yN$ we have

$$\widetilde{\mu_F}(zN) \ge [t, \mu_{A_L}(xN) \wedge \mu_{A_L}(yN)],$$

then $\mu_{A_L^*}(zN) \ge t$. Since $xN, yN \in \underline{N}(X)$, we have $zN \in \underline{N}(X)$ and so $\mu_{A_L}(zN) \ge t$. Therefore, $xN \odot yN \subseteq \underline{A}_t$. It is easy to see that $x^{-1}N \in \underline{A}_t$.

Now, let $xN, yN \in \overline{A}_t$. Then, we have $\mu_{A_U}(xN) \ge t$, $\mu_{A_U}(yN) \ge t$, so

$$\mathrm{rmin}(\widetilde{\mu_F}(xN), \widetilde{\mu_F}(yN)) \ge [0, t].$$

Therefore, for every $zN \in xN \odot yN$ we have $\widetilde{\mu_F}(zN) \ge [0, t]$. Hence, $\mu_{A_U}(zN) \ge t$. Thus, $xN \odot yN \subseteq \overline{A}_t$. It is easy to see that $x^{-1}N \in \overline{A}_t$. This completes the proof. ■

2.6 F-Polygroups

Throughout this section, I is the unit interval $[0,1]$ and we denote the set of all fuzzy subsets of A by I^A. If $\mu \in I^A$, then by $supp(\mu)$ we mean the set $\{x \in A \mid \mu(x) \neq 0\}$. If $a \in A$ and $t \in I$, then by a *fuzzy point* a_t of A we mean the fuzzy subset of A as follows:

$$a_t(x) = \begin{cases} t \text{ if } x = a \\ 0 \text{ otherwise.} \end{cases}$$

For any subset A of X, we denote χ_A the characteristic function of A. Let A be a non-empty set and $I_*^A = I^A \backslash \{0\}$. Then, by an *$F$-hyperoperation* "$*$" on A we mean a function from $A \times A$ to I_*^A, in other words for any $a, b \in A$, $a * b$ is a non-empty fuzzy subset of A. If $\mu, \eta \in I_*^A$, then $\mu * \eta$ is defined by

$$\mu * \eta = \bigcup_{x \in supp(\mu), y \in supp(\eta)} x * y.$$

Let $\mu \in I_*^A$, $B \in \mathcal{P}^*(A)$ and $a \in A$. Then,

(1) $a * \mu$ and $\mu * a$ denote $\chi_{\{a\}} * \mu$ and $\mu * \chi_{\{a\}}$, respectively.
(2) $a * B$, $B * a$, $\mu * B$ and $B * \mu$ denote $\chi_{\{a\}} * \chi_B$, $\chi_B * \chi_{\{a\}}$, $\mu * \chi_B$ and $\chi_B * \mu$, respectively.

The main references for this section are [207, 208].

Definition 2.6.1. [207] Let $\mathcal{F}$ be a non-empty set and $*$ be an F-hyperoperation on $\mathcal{F}$. Then, $(\mathcal{F}, *)$ is called an *F-polygroup* if

(1) $(x * y) * z = x * (y * z)$, for all $x, y, z \in \mathcal{F}$.
(2) There exists an element $e \in \mathcal{F}$ such that $x \in supp(x * e \cap e * x)$, for all $x \in \mathcal{F}$ (in this case we say that e is an *F-identity* element of $\mathcal{F}$).
(3) For each $x \in \mathcal{F}$, there exists a unique element $x' \in \mathcal{F}$ such that $e \in supp(x * x' \cap x' * x)$ (x' is called the *F-inverse* of x and is denoted by x^{-1}).
(4) $z \in supp(x * y) \Rightarrow x \in supp(z * y^{-1}) \Rightarrow y \in supp(x^{-1} * z)$, for all $x, y, z \in \mathcal{F}$.

EXAMPLE 30 (1) Let $\mathcal{F} = \{a, b\}$. Then, the following table denotes an F-polygroup structure on $\mathcal{F}$,

$*$	a	b
a	$\frac{a}{0.7}, \frac{b}{0}$	$\frac{a}{0}, \frac{b}{0.3}$
b	$\frac{a}{0}, \frac{b}{0.3}$	$\frac{a}{0.7}, \frac{b}{0}$

(2) Let t be an arbitrary element of $I \backslash \{0\}$ and G be a group such that $x^2 = e$, for all $x \in G$. Then, it is easy to see that the F-hyperoperation $*$ which is defined as

$$(x * y)(z) = e_t(xyz), \text{ for all } x, y, z \in G$$

induces an F-polygroup structure on G, where e_t is a fuzzy point of G.

Notice that for $\mu_1, \mu_2 \in I_*^{\mathcal{F}}$, if $\mu_1(t) > 0$, then $t * \mu_2 \subseteq \mu_1 * \mu_2$. Similarly, if $\mu_2(t) > 0$, then $\mu_1 * t \subseteq \mu_1 * \mu_2$. In particular, $e * \mu_2 \subseteq (x * x^{-1}) * \mu_2$, for all $x \in \mathcal{F}$. If $\mu_1 \subseteq \mu_2$, then $\mu_1 * x \subseteq \mu_2 * x$ and $x * \mu_1 \subseteq x * \mu_2$, for all $x \in \mathcal{F}$.

Corollary 2.6.2. *Let $\mu \in I_*^{\mathcal{F}}$ and $z \in \mathcal{F}$. Then,*

$$supp(\mu * z) = \bigcup_{t \in supp(\mu)} supp(t * z) \quad and \quad supp(z * \mu) = \bigcup_{t \in supp(\mu)} supp(z * t).$$

Corollary 2.6.3. *Let $\mu_1, \mu_2 \in I_*^{\mathcal{F}}$. Then,*

$$supp(\mu_1 * \mu_2) = \bigcup_{\substack{x \in supp(\mu_1) \\ y \in supp(\mu_2)}} supp(x * y).$$

Corollary 2.6.4. *Let $(\mathcal{F}, *)$ be an F-polygroup. Then,*

(1) *$e^{-1} = e$ and e is unique. Moreover, $supp(e * e) = \{e\}$.*
(2) *$(x^{-1})^{-1} = x$, for all $x \in \mathcal{F}$.*
(3) $\bigcup_{x \in supp(\mu_1)} x * \mu_2 = \mu_1 * \mu_2 = \bigcup_{y \in supp(\mu_2)} \mu_1 * y$, *for all $\mu_1, \mu_2 \in I_*^{\mathcal{F}}$.*
(4) *$(\mu_1 * \mu_2) * \mu_3 = \mu_1 * (\mu_2 * \mu_3)$, for all $\mu_1, \mu_2, \mu_3 \in I_*^{\mathcal{F}}$.*

Theorem 2.6.5. (1) *Let $(A, \circ)$ be a polygroup. Then, $(A, *)$ is an F-polygroup, where*

$$x * y = \chi_{x \circ y}, \quad for\ all\ x, y \in A$$

($*$ is called the F-hyperoperation induced by $\circ$).

(2) *Let $(A, *)$ be an F-polygroup and $x * e = e * x = \chi_{\{x\}}$, for all $x \in A$. Then, $(A, \odot)$ is a polygroup, where*

$$x \odot y = supp(x * y), \quad for\ all\ x, y \in A$$

($\odot$ is called the hyperoperation extracted from $*$).

Proof. It is straightforward. ■

If $(x * e)(x) = (e * x)(x) = 1$, for all $x \in \mathcal{F}$, then we say that e is of *degree* 1. Let $(A, \circ)$ be a polygroup. Then, the F-identity of F-polygroup $(A, *)$, which is defined in Theorem 2.6.5, is of degree 1.

Definition 2.6.6. Let $\mathcal{F}_1, \mathcal{F}_2$ be two F-polygroups and $f : \mathcal{F}_1 \to \mathcal{F}_2$ be a function such that $f(e) = e'$.

(1) f is called a *homomorphism* if $f(x * y) \leq f(x) *' f(y)$, for all $x, y \in \mathcal{F}_1$.
(2) f is called a *strong homomorphism* if $f(x * y) = f(x) *' f(y)$, for all $x, y \in \mathcal{F}_1$.

If f is a homomorphism, then we have $(f(x))^{-1} = f(x^{-1})$, for all $x \in \mathcal{F}_1$.

Theorem 2.6.7. *Let $\mathcal{F}_1, \mathcal{F}_2$ be two F-polygroups and $\mathcal{F} = \mathcal{F}_1 \times \mathcal{F}_2 = \{(x, y) \mid x \in \mathcal{F}_1,\ y \in \mathcal{F}_2\}$. We define $\otimes$ as follows:*

$$\otimes : \mathcal{F} \times \mathcal{F} \to I_*^{\mathcal{F}}$$
$$((x_1, y_1), (x_2, y_2)) \mapsto (x_1, y_1) \otimes (x_2, y_2),$$

where $((x_1, y_1) \otimes (x_2, y_2))(a, b) = \min\{(x_1 * x_2)(a),\ (y_1 * y_2)(b)\}$, for all $(a, b) \in \mathcal{F}_1 \times \mathcal{F}_2$. Then, $(\mathcal{F}, \otimes)$ is an F-polygroup.

Proof. It is straightforward. ∎

Definition 2.6.8. Let K be a non-empty subset of $\mathcal{F}$. Then, K is called an *F-subpolygroup* if

(1) $supp(x * y) \subseteq K$, for all $x, y \in K$.
(2) $x \in K$ implies $x^{-1} \in K$.

Notice that condition (1) of the above definition is equivalent to $x * y \leq \chi_K$, for all $x, y \in K$.

Proposition 2.6.9. *If K is an F-subpolygroup of $\mathcal{F}$, then $(K, *')$ is an F-polygroup, where by $*'$ we mean the restriction of $*$ to $K \times K$.*

Proof. The proof follows easily from Definitions 2.6.1 and 2.6.8. ∎

The following example shows that the converse of Theorem 2.6.9 is not true.

EXAMPLE 31 Let $\mathcal{F} = \{e, a, b\}$. Then, the following table shows an F-polygroup structure on $\mathcal{F}$.

$*$	e	a	b
e	$\frac{e}{0.7}, \frac{a}{0}, \frac{b}{0}$	$\frac{e}{0}, \frac{a}{0.7}, \frac{b}{0}$	$\frac{e}{0}, \frac{a}{0}, \frac{b}{0.7}$
a	$\frac{e}{0}, \frac{a}{0.7}, \frac{b}{0}$	$\frac{e}{0}, \frac{a}{0.7}, \frac{b}{0}$	$\frac{e}{0.7}, \frac{a}{0.7}, \frac{b}{0.7}$
b	$\frac{e}{0}, \frac{a}{0}, \frac{b}{0.7}$	$\frac{e}{0.7}, \frac{a}{0.7}, \frac{b}{0.7}$	$\frac{e}{0}, \frac{a}{0}, \frac{b}{0.7}$

Now, suppose that $K = \{a\}$. Then, $(K, *')$ is an F-polygroup, where $*'$ is the restriction of $*$ to $K \times K$. But, K is not an F-subpolygroup of $\mathcal{F}$, because $a^{-1} = b$ in $\mathcal{F}$ and $b \notin K$.

Let K be a non-empty subset of $\mathcal{F}$. It is easy to see that K is an F-subpolygroup of $\mathcal{F}$ if and only if $supp(x * y^{-1}) \subseteq K$, for all $x, y \in K$.

Definition 2.6.10. Let N be an F-subpolygroup of $\mathcal{F}$.

(1) N is said to be *weak normal* in $\mathcal{F}$ if $x * N * x^{-1} \leq \chi_N$, for all $x \in \mathcal{F}$. Notice that this condition is equivalent to $supp(x * N * x^{-1}) \subseteq N$, for all $x \in \mathcal{F}$.
(2) N is said to be *normal* in $\mathcal{F}$ if $x * N * x^{-1} = \chi_N$, for all $x \in \mathcal{F}$.

The following example shows a weak normal F-subpolygroup which is not normal.

EXAMPLE 32 We consider Example 31. Let $N = \mathcal{F}$. Then, it is obvious that N is weak normal. Since $(x * N * x^{-1})(a) \leq 0.7 < 1$, for all $x, a \in \mathcal{F}$, we conclude that N is not normal.

Now, we give an example of a normal F-subpolygroup.

EXAMPLE 33 We consider a group G with order greater than 1 and $x^2 = e$, for all $x \in G$. Let $t \in (0, 1)$. We define the F-hyperoperation $*$ on G by

$$(x * y)(z) = \begin{cases} e_1(xyz) & \text{if } z = e \\ e_t(xyz) & \text{if } z \neq e, \end{cases}$$

where e_1, e_t are fuzzy points of G. After some manipulations it can be seen that $(G, *)$is an F-polygroup. Now, suppose that $N = \{e\}$. It is easy to see that N is an F-subpolygroup of G. We prove that $x * N * x^{-1} = \chi_N$, for all $x \in G$. Let $x \in G$ be arbitrary. Then, $z \in supp(x * e) \Leftrightarrow zx = e \Leftrightarrow z = x$. Thus, $supp(x * e) = \{x\}$. So, we obtain $x * e * x = x * x$, for all $x \in G$. Now, let $y \in G$ and $(x * e * x)(y) > 0$. Then, $(x * x)(y) > 0$ and so $y = e$. Therefore, $x * N * x^{-1} \leq \chi_N$, for all $x \in G$. On the other hand, we have $(x * N * x)(e) = (x * x)(e) = 1$, for all $x \in G$. Consequently, $\chi_N \leq x * N * x^{-1}$, for all $x \in G$.

Notice that in the above example, the identity is not of degree 1, because if $x \neq e$, then $(x * e)(x) = t \neq 1$.

Theorem 2.6.11. *Let $e \in \mathcal{F}$ be of degree 1. Then, N is a weak normal F-subpolygroup of $\mathcal{F}$ if and only if N is a normal F-subpolygroup of $\mathcal{F}$.*

Proof. Suppose that N is a weak normal F-subpolygroup of $\mathcal{F}$. First, we show that $N * e = e * N = \chi_N$. In order to show this, assume that $y \in \mathcal{F}$ is arbitrary. If $y \in N$, then $(y * e)(y) = 1$, we conclude that

$$(N * e)(y) = \sup_{x \in N} (x * e)(y) = 1 = \chi_N(y).$$

If $y \notin N$, then since N is an F-subpolygroup of $\mathcal{F}$, we have $(x * e)(y) = 0$, for all $x \in N$. Therefore,

$$(N * e)(y) = \sup_{x \in N} (x * e)(y) = 0 = \chi_N(y).$$

Thus, $N * e = \chi_N$. Similarly, $e * N = \chi_N$. Now, we prove that $x * N = N * x$, for all $x \in \mathcal{F}$. Suppose that $x \in \mathcal{F}$ is arbitrary. By hypothesis we have $x * N * x^{-1} * x \leq N * x$. Therefore,

$$x * N = x * N * e \leq x * N * x^{-1} * x \leq N * x.$$

Similarly, $N * x \leq x * N$. Finally, in order to complete the proof, it is enough to show that $\chi_N \leq x * N * x^{-1}$, for all $x \in \mathcal{F}$. Let $y \in \mathcal{F}$. If $y \notin N$, then it is obvious that $\chi_N(y) = 0 \leq (x * N * x^{-1})(y)$. Otherwise,

$$\begin{aligned}\chi_N(y) = 1 = (e * y)(y) &\leq (x * x^{-1} * y)(y) \\ &\leq (x * x^{-1} * N)(y) \\ &= (x * N * x^{-1})(y). \blacksquare\end{aligned}$$

Lemma 2.6.12. *Let* $\mu \in I_*^{\mathcal{F}}$. *Then,* $x * \mu * y = \bigcup_{a \in supp(\mu)} x * a * y$, *for all* $x, y \in \mathcal{F}$.

Proof. It is straightforward. ■

Lemma 2.6.13. *Let* $n \in \mathbb{N}$ *and* $w, a_1, \ldots, a_n \in \mathcal{F}$. *If* $w \in supp(a_1 * \ldots * a_n)$, *then* $w^{-1} \in supp(a_n^{-1} * \ldots * a_1^{-1})$.

Proof. We can prove the lemma by induction on n. ■

Theorem 2.6.14. *Let* H, K *be* F*-subpolygroups of* $\mathcal{F}$ *and*

$$H \odot K = \bigcup_{x \in H, y \in K} supp(x * y).$$

Then,

(1) $H \odot K$ *is an* F*-subpolygroup of* $\mathcal{F}$ *if and only if* $H \odot K = K \odot H$.
(2) *If* K *is a weak normal* F*-subpolygroup of* $\mathcal{F}$, *then* $H \odot K$ *is an* F*-subpolygroup of* $\mathcal{F}$.
(3) *If* H, K *are weak normal* F*-subpolygroups of* $\mathcal{F}$, *then* $H \odot K$ *is a weak normal* F*-subpolygroup of* $\mathcal{F}$.

Proof. (1) Suppose that $H \odot K$ is an F-subpolygroup of $\mathcal{F}$ and $w \in H \odot K$ be an arbitrary element. Then, $w^{-1} \in H \odot K$. Therefore, there exist $h \in H$ and $k \in K$ such that $w^{-1} \in supp(h * k)$. Thus, by Lemma 2.6.13, we have $w \in supp(k^{-1} * h^{-1})$. Consequently, $w \in K \odot H$, and hence $H \odot K \subseteq K \odot H$. Similarly, $K \odot H \subseteq H \odot K$. That is $H \odot K = K \odot H$.

Conversely, suppose that $H \odot K = K \odot H$. Since $e \in supp(e * e) \subseteq H \odot K$, $H \odot K$ is non-empty. Let $a, b \in H \odot K$. Then, there exist $k, k' \in K$ and $h, h' \in H$ such that $a \in supp(h * k)$ and $b \in supp(h' * k')$. So,

$$a * b^{-1} \leq h * k * k'^{-1} * h'^{-1} \leq h * K * h'^{-1} = \bigcup_{z \in supp(K * h'^{-1})} (h * z).$$

Now, let $s \in supp(a * b^{-1})$. Then, there exists $z' \in supp(K * h'^{-1})$ such that $s \in supp(h * z')$. Therefore, $z' \in supp(k * h'^{-1})$, for some $k \in K$. So, $(h * z')(s) \leq (h * (k * h'^{-1}))(s)$, and there exists $r \in supp(k * h'^{-1})$ such that

$s \in supp(h * r)$. Consequently, $r \in H \odot K = K \odot H$. Thus, $r \in supp(h_2 * k_2)$, for some $h_2 \in H$ and $k_2 \in K$. Hence,

$$0 < (h * r)(s) \leq (h * h_2 * k_2)(s) \leq (H * k_2)(s).$$

Thus, $s \in H \odot K$, which shows that $H \odot K$ is an F-subpolygroup of $\mathcal{F}$.

(2) Suppose that $s \in H \odot K$. Then, there exist $h \in H$ and $k \in K$ such that $s \in supp(h * k)$. Thus, $s \in supp(h * k * h^{-1} * h)$. Since K is a weak normal F-subpolygroup of $\mathcal{F}$, we obtain $s \in supp(K * h) \subseteq K \odot H$. So, $H \odot K \subseteq K \odot H$, and similarly we have $K \odot H \subseteq H \odot K$. Thus, $H \odot K = K \odot H$, which implies that $H \odot K$ is an F-subpolygroup of $\mathcal{F}$, by (1).

(3) Suppose that $x \in \mathcal{F}$ and $b \in supp(x * (H \odot K) * x^{-1})$. Then, by Lemma 2.6.12, we have $b \in supp(x * a * x^{-1})$, for some $a \in H \odot K$. Hence, $a \in supp(h * k)$, for some $h \in H$ and $k \in K$. So, $b \in supp(x * h * k * x^{-1})$. Since H, K are weak normal F-subpolygroups of $\mathcal{F}$, we obtain $b \in supp(x * h * x^{-1} * x * k * x^{-1}) \subseteq H \odot K$. Therefore, $H \odot K$ is a weak normal F-subpolygroups of $\mathcal{F}$. ■

Corollary 2.6.15. *Let $e \in \mathcal{F}$ be of degree 1 and H, K be weak normal F-subpolygroups of $\mathcal{F}$. Then, $H \odot K$ is a normal F-subpolygroup of $\mathcal{F}$.*

Proof. The proof easily follows from Theorem 2.6.11 and Theorem 2.6.14 (3). ■

Lemma 2.6.16. *Let $f : \mathcal{F}_1 \to \mathcal{F}_2$ be a homomorphism of F-polygroups. Then, $f(a * b * c) \leq f(a) * f(b) * f(c)$, for all $a, b, c \in \mathcal{F}_1$.*

Proof. Suppose that $y \in \mathcal{F}_2$ is an arbitrary element. If $f(a * b * c)(y) = 0$, then the assertion holds. Now, assume that $f(a * b * c)(y) \neq 0$. We have

$$\begin{aligned}
f(a * b * c)(y) &= \sup_{x \in f^{-1}(y)} (a * b * c)(x) \\
&= \sup_{x \in f^{-1}(y)} \left(\sup_{z \in supp(a*b)} (z * c)(x) \right) \\
&= \sup_{z \in supp(a*b)} \left(\sup_{x \in f^{-1}(y)} (z * c)(x) \right) \\
&= \sup_{z \in supp(a*b)} f(z * c)(y) \\
&\leq \sup_{z \in supp(a*b)} (f(z) * f(c))(y).
\end{aligned}$$

For every $z \in supp(a * b)$, we have $0 < f(a * b)(f(z)) \leq (f(a) * f(b))(f(z))$. So, $(f(z) * f(c))(y) \leq (f(a) * f(b) * f(c))(y)$, for all $z \in supp(a * b)$. Therefore,

$$f(a * b * c)(y) \leq (f(a) * f(b) * f(c))(y).$$ ■

Definition 2.6.17. Let $f : \mathcal{F}_1 \to \mathcal{F}_2$ be a strong homomorphism of F-polygroups. Then, f is called *zero-invariant* if

$$(x * y)(z) = 0 \Rightarrow f(x * y)(f(z)) = 0, \text{ for all } x, y, z \in \mathcal{F}_1.$$

Theorem 2.6.18. *Let $f : \mathcal{F}_1 \to \mathcal{F}_2$ be a homomorphism of F-polygroups.*

(1) *If K is an F-subpolygroup of $\mathcal{F}_2$, then $f^{-1}(K)$ is an F-subpolygroup of $\mathcal{F}_1$.*
(2) *If K is a weak normal F-subpolygroup of $\mathcal{F}_2$, then $f^{-1}(K)$ is a weak normal F-subpolygroup of $\mathcal{F}_1$.*
(3) *If f is a strong homomorphism and H is an F-subpolygroup of $\mathcal{F}_1$, then $f(H)$ is an F-subpolygroup of $\mathcal{F}_2$.*
(4) *If f is an onto, zero-invariant and strong homomorphism and H is a weak normal F-subpolygroup of $\mathcal{F}_1$, then $f(H)$ is a weak normal F-subpolygroup of $\mathcal{F}_2$.*

Proof. It is straightforward. ■

Corollary 2.6.19. *Let $f : \mathcal{F}_1 \to \mathcal{F}_2$ be a strong homomorphism of F-polygroups, where the identities of $\mathcal{F}_1$ and $\mathcal{F}_2$ are of degree 1.*

(1) *If K is a weak normal F-subpolygroup of $\mathcal{F}_2$, then $f^{-1}(K)$ is a normal F-subpolygroup of $\mathcal{F}_1$.*
(2) *If f is zero-invariant and onto, and H is a weak normal F-subpolygroup of $\mathcal{F}_1$, then $f(H)$ is a normal F-subpolygroup of $\mathcal{F}_2$.*

Proof. The proof immediately follows from Theorems 2.6.18 and 2.6.11. ■

Let $f : \mathcal{F}_1 \to \mathcal{F}_2$ be a homomorphism of F-polygroups. Then, the set $f^{-1}(\{e_2\}) = \{x \in \mathcal{F}_1 \mid f(x) = e_2\}$ is called the *kernel* of f and is denoted by $kerf$. Clearly, $kerf$ is an F-subpolygroup of $\mathcal{F}_1$.

Lemma 2.6.20. *Let N be a normal F-subpolygroup of $\mathcal{F}$. Then,*

(1) *$N * x = x * N$, for all $x \in \mathcal{F}$.*
(2) *$x * N = \chi_N$ if and only if $x \in N$.*
(3) *$N * N = \chi_N$.*

Proof. It is straightforward. ■

Theorem 2.6.21. *Let N be a normal F-subpolygroup of $\mathcal{F}$ and let $\mathcal{F}/N = \{x * N \mid x \in \mathcal{F}\}$. We define the F-hyperoperation $\odot$ on $\mathcal{F}/N$ as follows:*

$$\odot : \mathcal{F}/N \times \mathcal{F}/N \to I_*^{\mathcal{F}/N},$$
$$(x * N, y * N) \mapsto x * N \odot y * N,$$

where $(x * N \odot y * N)(z * N) = (x * y * N)(z)$, for all $z * N \in \mathcal{F}/N$. Then, $(\mathcal{F}/N, \odot)$ is an F-polygroup and is called the quotient F-polygroup.

Proof. It is straightforward. ■

Corollary 2.6.22. *Let N be a weak normal F-subpolygroup of $\mathcal{F}$ and the identity of $\mathcal{F}$ be of degree 1. Then, $(\mathcal{F}/N, \otimes)$ is a polygroup, where $\otimes$ is the hyperoperation extracted from $\odot$.*

Lemma 2.6.23. *Let $\{H_i\}_{i\in I}$ be a family of F-subpolygroups of $\mathcal{F}$. Then, $H = \bigcap\limits_{i\in I} H_i$ is an F-subpolygroup of $\mathcal{F}$.*

Proof. It is straightforward. ∎

Let X be a non-empty subset of $\mathcal{F}$. Then, by the F-subpolygroup $< X >$ *generated* by X, we mean the intersection of all F-subpolygroups of $\mathcal{F}$, containing X.

Theorem 2.6.24. *Let X be a non-empty subset of $\mathcal{F}$ and $e \in X$. Then,*

$$< X >= \bigcup_{n\in\mathbb{N}, a_1,\ldots,a_n \in X\cup X^{-1}} supp(a_1 * \ldots * a_n),$$

where $X^{-1} = \{x^{-1} \mid x \in X\}$.

Proof. The proof follows from Lemma 2.6.13. ∎

Theorem 2.6.25. *Let $C = \{z \in \mathcal{F} \mid z \in supp(x * y * x^{-1} * y^{-1})$, for some $x, y \in \mathcal{F}\}$. Then, $\mathcal{F}' =< C >$ is a weak normal F-subpolygroup of $\mathcal{F}$ called the commutator F-subpolygroup of $\mathcal{F}$. Moreover, if W is a normal F-subpolygroup of $\mathcal{F}$ and $\mathcal{F}' \subseteq W$, then $(\mathcal{F}/W, \odot)$ is a commutative F-polygroup. In particular, if the identity of $\mathcal{F}$ is of degree 1, then $\mathcal{F}/\mathcal{F}'$ is commutative.*

Proof. By Lemma 2.6.23, $\mathcal{F}'$ is an F-subpolygroup of $\mathcal{F}$. Now, we show that $\mathcal{F}'$ is a weak normal F-subpolygroup. In order to do this, let $y \in \mathcal{F}'$ and $x \in \mathcal{F}$ be arbitrary. We have

$$\begin{aligned} x * y * x^{-1} &\leq x * y * x^{-1} * e \leq x * y * x^{-1} * y^{-1} * y \\ &= \bigcup_{a\in supp(x*y*x^{-1}*y^{-1})} a * y \\ &\leq \chi_{\mathcal{F}'}. \end{aligned}$$

Therefore, by Lemma 2.6.12, $\mathcal{F}'$ is a weak normal F-subpolygroup of $\mathcal{F}$.

Now, suppose that W is a normal F-subpolygroup of $\mathcal{F}$ and $\mathcal{F}' \subseteq W$. We prove that $\mathcal{F}/W$ is commutative. By hypothesis, $\chi_{\mathcal{F}'} \leq \chi_W$, which implies that $y^{-1} * x^{-1} * y * x * W \leq \mathcal{F}' * W \leq W * W = \chi_W$, for all $x, y \in \mathcal{F}$. Thus,

$$y * x * W \leq x * x^{-1} * y * x * W \leq x * y * W, \text{ for all } x, y \in \mathcal{F}.$$

Similarly, $x * y * W \leq y * x * W$, for all $x, y \in \mathcal{F}$. Hence, $x * y * W = y * x * W$, for all $x, y \in \mathcal{F}$, which implies that $\mathcal{F}/W$ is commutative. The rest of the proof is clear. ∎

The following corollary follows from Corollary 2.6.22, Theorem 2.6.25 and the definition of $\otimes$.

Corollary 2.6.26. *Let the identity of $\mathcal{F}$ be of degree 1, W be weak normal F-subpolygroup of $\mathcal{F}$ and $\mathcal{F}' \subseteq W$. Then, $(\mathcal{F}/W, \otimes)$ is a commutative polygroup, where $\otimes$ is the hyperoperation extracted from $\odot$.*

Lemma 2.6.27. *Let $a_1, a_2, a_3, a_4 \in \mathcal{F}$ and W be a normal F-subpolygroup of $\mathcal{F}$. Then,*

$$(a_1 * a_2 * a_3 * a_4 * W)(y) = (a_1 * W \odot a_2 * W \odot a_3 * W \odot a_4 * W)(y * W).$$

Proof. It is straightforward. ■

Theorem 2.6.28. *Let $(\mathcal{F}/W, \odot)$ be a commutative F-polygroup. Then, $\mathcal{F}' \subseteq W$.*

Proof. It is enough to show that $C \subseteq W$, where C is defined in Theorem 2.6.25. In order to show this, suppose that $y \in C$ be arbitrary. Then, there exist $a, b \in \mathcal{F}$ such that $y \in supp(a * b * a^{-1} * b^{-1})$. Since $e \in W$, we have $y \in supp(a * b * a^{-1} * b^{-1} * W)$. Hence, by using Lemma 2.6.27 twice, we can show that $y \in W$. ■

Definition 2.6.29. A fuzzy binary relation ρ on a non-empty set X (i.e., $\rho \in I^{X \times X}$) is said to be a *fuzzy similarity relation* if it satisfies for all $x, y, z \in X$:

(1) Reflexivity: $\rho(x, x) = 1$.
(2) Symmetry: $\rho(x, y) = \rho(y, x)$.
(3) Transitivity: $\min\{\rho(x, y), \rho(y, z)\} \leq \rho(x, z)$.

Definition 2.6.30. Let ρ be a fuzzy binary relation between two F-polygroups $\mathcal{F}, \mathcal{F}'$ (i.e., $\rho \in I^{\mathcal{F} \times \mathcal{F}'}$). Then, ρ is called an *FP-relation* if

(1) $(e, e') \in supp(\rho)$, where e and e' are the identity elements of $\mathcal{F}$ and $\mathcal{F}'$, respectively.
(2) $\rho(a, b) \leq \rho(a^{-1}, b^{-1})$, for all $(a, b) \in \mathcal{F} \times \mathcal{F}'$.
(3) $\min\{\rho(a, b), \rho(c, d)\} \leq \rho(x, y)$, for all $x \in supp(a * c)$, $y \in supp(b * d)$ and for all $(a, b), (c, d) \in \mathcal{F} \times \mathcal{F}'$.

Clearly, (2) implies that $\rho(a, b) = \rho(a^{-1}, b^{-1})$, for all $a, b \in \mathcal{F}$.

Theorem 2.6.31. *If ρ is an FP-relation between $\mathcal{F}, \mathcal{F}'$ and if K is an F-subpolygroup of $\mathcal{F}'$, then the subset*

$$H = \{x \in \mathcal{F} \mid (x, y) \in supp(\rho), \ for \ some \ y \in K\}$$

is an F-subpolygroup of $\mathcal{F}$.

Proof. Since $(e, e') \in supp(\rho)$, so $e \in H$. Now, assume that $x \in H$. Then, there exists $y \in K$ such that $\rho(x, y) > 0$. Since $\rho(x, y) \leq \rho(x^{-1}, y^{-1})$ and

$y^{-1} \in K$, we obtain $x^{-1} \in H$. Let $x_1, x_2 \in H$ and $z \in supp(x_1 * x_2)$. Then, (x_1, y_1), $(x_2, y_2) \in supp(\rho)$, for some $y_1, y_2 \in K$. Thus,

$$0 < \min\{\rho(x_1, y_1), \rho(x_2, y_2)\} \leq \rho(z, w),$$

where $w \in supp(y_1 * y_2) \subseteq K$. Hence, $z \in H$. ■

Theorem 2.6.32. *If ρ is a reflexive FP-relation on an F-polygroup $\mathcal{F}$, then ρ is a fuzzy similarity relation on $\mathcal{F}$.*

Proof. It is straightforward. ■

Definition 2.6.33. An FP-relation on an F-polygroup $\mathcal{F}$ which is a fuzzy similarity relation too, is called a *fuzzy congruence relation* on $\mathcal{F}$.

Definition 2.6.34. Let ρ be a fuzzy congruence relation on $\mathcal{F}$. Then, the fuzzy subset $\rho \prec e \succ$ of $\mathcal{F}$ is defined by $\rho \prec e \succ (x) = \rho(e, x)$, for all $x \in \mathcal{F}$.

Definition 2.6.35. Let $\mathcal{F}$ be an F-polygroup and $\mu \in I^{\mathcal{F}}$. Then, μ is called a *fuzzy sub F-polygroup* of $\mathcal{F}$ if

(1) $\min\{\mu(x), \mu(y)\} \leq \mu(z)$, for all $x, y \in \mathcal{F}$ and for all $z \in supp(x * y)$.
(2) $\mu(x) \leq \mu(x^{-1})$, for all $x \in \mathcal{F}$.

Notice that this definition is a generalization of Definition 2.3.1. Clearly, $\mu(x) = \mu(x^{-1})$ and $\mu(x) \leq \mu(e)$, for all $x \in \mathcal{F}$.

Theorem 2.6.36. *If ρ is a fuzzy congruence relation on an F-polygroup $\mathcal{F}$, then the fuzzy subset $\rho \prec e \succ$ is a fuzzy sub F-polygroup of $\mathcal{F}$.*

Proof. Suppose that $x, y \in \mathcal{F}$. Then,

$$\min\{\rho \prec e \succ (x),\ \rho \prec e \succ (y)\} = \min\{\rho(e, x), \rho(e, y)\} \leq \rho(e, z),$$

for all $z \in supp(x * y)$. Also, we have

$$\rho \prec e \succ (x) = \rho(e, x) \leq \rho(e, x^{-1}) = \rho \prec e \succ (x^{-1}).$$

Therefore, $\rho \prec e \succ$ is a fuzzy sub F-polygroup of $\mathcal{F}$. ■

Definition 2.6.37. Let μ be a fuzzy sub F-polygroup of $\mathcal{F}$. Then, μ is said to be *normal* if for all $x, y \in \mathcal{F}$,

$$\mu(z) = \mu(z'), \text{ for all } z \in supp(x * y) \text{ and for all } z' \in supp(y * x).$$

If μ is a fuzzy normal sub F-polygroup of $\mathcal{F}$, then $\mu(z) = \mu(z')$, for all $z, z' \in supp(x * y)$.

Notice that the above definition is a generalization of Definition 2.3.3.

Theorem 2.6.38. *Let μ be a fuzzy sub F-polygroup of $\mathcal{F}$. Then, the following conditions are equivalent:*

(1) *μ is a fuzzy normal sub F-polygroup of $\mathcal{F}$.*
(2) *$\mu(z) = \mu(y)$, for all $x, y \in \mathcal{F}$ and for all $z \in supp(x * y * x^{-1})$.*
(3) *$\mu(z) \geq \mu(y)$, for all $x, y \in \mathcal{F}$ and for all $z \in supp(x * y * x^{-1})$.*
(4) *$\mu(z) \geq \mu(y)$, for all $x, y \in \mathcal{F}$ and for all $z \in supp(x^{-1} * y^{-1} * x * y)$.*

Proof. The proof is similar to the proof of Theorem 2.3.4. ■

Corollary 2.6.39. *If ρ is a fuzzy congruence relation on $\mathcal{F}$, then $\rho \prec e \succ \in I^{\mathcal{F}}$ is a fuzzy normal sub F-polygroup of $\mathcal{F}$.*

Proof. By Theorem 2.6.36, $\rho \prec e \succ$ is a fuzzy sub F-polygroup of $\mathcal{F}$. Now, we suppose that $z \in supp(x * y * x^{-1})$. Then, $z \in supp(a * x^{-1})$, for some $a \in supp(x * y)$. Hence, we have

$$\begin{aligned}\rho \prec e \succ (z) &= \rho(e, z)\\ &\geq \min\left\{\rho(x, a), \rho(x^{-1}, x^{-1})\right\}\\ &= \rho(x, a)\\ &\geq \min\{\rho(x, x), \rho(e, y)\}\\ &= \rho(e, y)\\ &= \rho \prec e \succ (y).\end{aligned}$$

Therefore, by Theorem 2.6.38 (3), $\rho \prec e \succ$ is normal. ■

Corollary 2.6.40. *N is a weak normal F-subpolygroup of $\mathcal{F}$ if and only if χ_N is a fuzzy normal sub F-polygroup of $\mathcal{F}$.*

Corollary 2.6.41. *Let N be a normal F-subpolygroup of $\mathcal{F}$. Then, $x*N*x^{-1}$ is a fuzzy normal sub F-polygroup of $\mathcal{F}$, for all $x \in \mathcal{F}$.*

We know that every fuzzy subgroup of an abelian group is normal. But, the following example shows that this is not true in the case of F-polygroups.

EXAMPLE 34 Let G be Klein's 4-group. Then, $(G, \circ)$ is a polygroup, where the hyperoperation $\circ$ is defined as follows:

$$\begin{aligned}&x \circ y = \{x, y, xy\}, \text{ if } x \neq y^{-1},\ x, y \neq e,\\ &x \circ x^{-1} = x^{-1} \circ x = G, \text{ if } x \neq e,\\ &x \circ e = e \circ x = \{x\}, \text{ for all } x \in G.\end{aligned}$$

Now, let $*$ be the F-hyperoperation induced by $\circ$, i.e., $x * y = \chi_{x \circ y}$, for all $x, y \in G$. Then, $(G, *)$ is an F-polygroup. Let $s, t \in [0, 1]$ such that $s < t$. Define a fuzzy subset μ of G as follows:

$$\mu(a) = \mu(b) = \mu(c) = s,\ \mu(e) = t.$$

Then, it is easy to see that μ is a fuzzy sub F-polygroup of $(G, *)$. But, since $e \in supp(a * a)$, $a \in supp(a * a)$ and $\mu(e) \neq \mu(a)$, we conclude that μ is not normal.

Theorem 2.6.42. *Let $\mathcal{F}$ be a commutative F-polygroup and μ be a normal fuzzy sub F-polygroup of $\mathcal{F}$ such that $\mu(e) = 1$. Define $\rho \in I^{\mathcal{F}\times\mathcal{F}}$ as follows: $\rho(x, y) = \mu(z)$, for some arbitrary element $z \in supp(x * y^{-1})$. Then, ρ is a fuzzy congruence relation and $\mu = \rho \prec e \succ$.*

Proof. We know μ is a normal fuzzy sub F-polygroup of $\mathcal{F}$. Hence, $\mu(z) = \mu(z')$, for all $x, y \in \mathcal{F}$ and for all $z, z' \in supp(x * y^{-1})$. So, ρ is well defined. Clearly, $\rho(e, e) > 0$. Now, for all $(x, y) \in \mathcal{F} \times \mathcal{F}$ we have

$$\begin{aligned}\rho(x, y) &= \mu(w), \text{ where } w \in supp(x * y^{-1}) \\ &= \mu(w^{-1}), \text{ where } w^{-1} \in supp(y * x^{-1}) \\ &= supp(x^{-1} * y), \text{ by commutativity of } * \\ &= \rho(x^{-1}, y^{-1}).\end{aligned}$$

Now, we show that $\min\{\rho(a_1, b_1), \rho(a_2, b_2)\} \le \rho(x, y)$, for all $x \in supp(a_1 * a_2)$ and for all $y \in supp(b_1 * b_2)$. Suppose that $x \in supp(a_1 * a_2)$, $y \in supp(b_1 * b_2)$ and $z \in supp(x * y^{-1})$ be arbitrary. Then, we have $z \in supp(x * y^{-1}) \subseteq supp(a_1 * b_1^{-1} * a_2 * b_2^{-1})$. Thus, $z \in supp(u * v)$, for some $u \in supp(a_1 * b_1^{-1})$ and $v \in supp(a_2 * b_2^{-1})$. Therefore, we obtain

$$\rho(x, y) = \mu(z) \ge \min\{\mu(u), \mu(v)\} = \min\{\rho(a_1, b_1), \rho(a_2, b_2)\}.$$

Consequently, ρ is an FP-relation. Since $\rho(x, x) = \mu(e) = 1$, ρ is reflexive. Hence, by Theorem 2.6.32 and Definition 2.6.33, ρ is a fuzzy congruence relation. Since $\mu(x) = \mu(x^{-1})$, for all $x \in \mathcal{F}$, we obtain $\mu = \rho \prec e \succ$. ∎

Corollary 2.6.43. *Let μ be a fuzzy subset of an abelian F-polygroup $\mathcal{F}$. Then, μ is a fuzzy normal sub F-polygroup of $\mathcal{F}$ and $\mu(e) = 1$ if and only if there exists a fuzzy congruence relation ρ on $\mathcal{F}$ such that $\mu = \rho \prec e \succ$.*

Chapter 3
Fuzzy H_v-Subgroups of H_v-Groups

In [76, 77], Davvaz applied the concept of fuzzy sets to the theory of algebraic hyperstructures and defined fuzzy subhypergroup (respectively, H_v-subgroup) of a hypergroup (resp. H_v-group) which is a generalization of the concept of Rosenfeld's fuzzy subgroup of a group. Since the concept of an H_v-group is a generalization of a hypergroup, in this chapter, we present basic definitions and results concerning fuzzy H_v-subgroups of H_v-groups.

3.1 Fuzzy H_v-Subgroups and Fuzzy Homomorphisms

In this section first we define fuzzy subhypergroup of a hypergroup and then we obtain the relation between a fuzzy subhypergroup and level subhypergroup. This relation is expressed in terms of a necessary and sufficient condition. The main references for this section are [76, 77, 78, 79].

Definition 3.1.1. Let $(H, \cdot)$ be a hypergroup (or H_v-group) and let μ be a fuzzy subset of H. Then, μ is said to be a *fuzzy subhypergroup* (or *H_v-subgroup*) of H if the following axioms hold.

(1) $\min\{\mu(x), \mu(y)\} \leq \inf_{\alpha \in x \cdot y} \{\mu(\alpha)\}$, for all $x, y \in H$.

(2) For all $x, a \in H$ there exists $y \in H$ such that $x \in a \cdot y$ and

$$\min\{\mu(a), \mu(x)\} \leq \mu(y).$$

(3) For all $x, a \in H$ there exists $z \in H$ such that $x \in z \cdot a$ and

$$\min\{\mu(a), \mu(x)\} \leq \mu(z).$$

Condition (2) is called the *left fuzzy reproduction axiom*, while (3) is called the *right fuzzy reproduction axiom*.

EXAMPLE 35 Let $(G, \cdot)$ be a group and μ be a fuzzy subgroup of G. If we define the following hyperoperation on G, $\star : G \times G \to \mathcal{P}^*(G)$ with $x \star y = \{t \mid \mu(t) = \mu(x \cdot y)\}$, then $(G, \star)$ is an H_v-group and μ is a fuzzy H_v-subgroup of G.

B. Davvaz and I. Cristea, *Fuzzy Algebraic Hyperstructures*,
Studies in Fuzziness and Soft Computing 321, DOI: 10.1007/978-3-319-14762-8_3

EXAMPLE 36 Suppose that H is a set and μ is a fuzzy subset of H. We define the hyperoperation $\star : H \times H \to \mathcal{P}^*(H)$ as follows: Assume that $x, y \in H$, if $\mu(x) \leq \mu(y)$, then $y \star x = x \star y = \{t \mid t \in H, \mu(x) \leq \mu(t) \leq \mu(y)\}$. Then, $(H, \star)$ is a hypergroup as well as a join space. If $(H, \cdot)$ is a group and μ is a fuzzy subgroup of H, then μ is a subhypergroup of $(H, \star)$.

Lemma 3.1.2. *Let $(H, \cdot)$ be an H_v-group and μ be a fuzzy H_v-subgroup of H. Then,*

$$\min\{\mu(x_1), \ldots, \mu(x_n)\} \leq \inf_{\alpha \in (\ldots((x_1 \cdot x_2) \cdot x_3) \ldots) \cdot x_n} \{\mu(\alpha)\},$$

for all $x_1, x_2, \ldots, x_n \in H$.

Proof. We shall prove the validity of this lemma by mathematical induction. First, the lemma is clearly true for $n = 2$. To complete the proof we assume the validity of the lemma for $n = k - 1$, that is, we assume that

$$\min\{\mu(x_1), \ldots, \mu(x_{k-1})\} \leq \inf_{r \in (\ldots((x_1 \cdot x_2) \cdot x_3) \ldots) \cdot x_{k-1}} \{\mu(r)\},$$

for all $x_1, x_2, \ldots, x_{k-1} \in H$. Then,

$$\begin{aligned}
\min\{\mu(x_1), \ldots, \mu(x_{k-1}), \mu(x_k)\} &= \min\{\min\{\mu(x_1), \ldots, \mu(x_{k-1})\}, \mu(x_k)\} \\
&\leq \min\left\{ \inf_{r \in (\ldots((x_1 \cdot x_2) \cdot x_3) \ldots) \cdot x_{k-1}} \{\mu(r), \ \mu(x_k)\right\} \\
&= \inf_{r \in (\ldots((x_1 \cdot x_2) \cdot x_3) \ldots) \cdot x_{k-1}} \{\min\{\mu(r), \mu(x_k)\}\} \\
&\leq \inf_{r \in (\ldots((x_1 \cdot x_2) \cdot x_3) \ldots) \cdot x_{k-1}} \left\{ \inf_{\alpha \in r \cdot x_k} \{\mu(\alpha)\} \right\} \\
&\leq \inf_{\alpha \in ((\ldots((x_1 \cdot x_2) \cdot x_3) \ldots) \cdot x_{k-1}) \cdot x_k} \{\mu(\alpha)\}.
\end{aligned}$$

■

Let $(H, \cdot)$ be an H_v-group. An *n-ary hyperproduct* can be defined, induced by $\cdot$, by inserting $n - 2$ parentheses in the sequence of elements $x_1, \ldots, x_n$ in a standard position. Let us denote by $p(x_1, \ldots, x_n)$ such a pattern of $n - 2$ parentheses and by, P_n the set of all such patterns.

Corollary 3.1.3. *Let $(H, \cdot)$ be an H_v-group and μ be a fuzzy H_v-subgroup of H. Then,*

$$\min\{\mu(x_1), \ldots, \mu(x_n)\} \leq \inf\{\mu(\alpha) \mid \alpha \in \bigcup_{p(x_1, \ldots, x_n) \in P_n(x_1, \ldots, x_n)} p(x_1, \ldots, x_n)\}.$$

for all $x_1, x_2, \ldots, x_{k-1} \in H$.

Theorem 3.1.4. *Let $(H, \cdot)$ be an H_v-group and μ be a fuzzy subset of H. Then, μ is a fuzzy H_v-subgroup of H if and only if for every t, $0 \leq t \leq 1$, $\mu_t \neq \emptyset$ is an H_v-subgroup of H.*

Proof. Let μ be a fuzzy H_v-subgroup of H. For every x, y in μ_t we have $\min\{\mu(x), \mu(y)\} \geq t$ and so $\inf_{\alpha \in x \cdot y} \{\mu(\alpha)\} \geq t$. Therefore, for every $\alpha \in x \cdot y$ we have $\alpha \in \mu_t$, so $x \cdot y \subseteq \mu_t$. Hence, for every $a \in \mu_t$ we have $a \cdot \mu_t \subseteq \mu_t$. Now, let $x \in \mu_t$. Then, there exists $y \in H$ such that $x \in a \cdot y$ and $\min\{\mu(a), \mu(x)\} \leq \mu(y)$. From $x \in \mu_t$ and $a \in \mu_t$ we get $\min\{\mu(x), \mu(a)\} \geq t$. So, $y \in \mu_t$, and this proves $\mu_t \subseteq a \cdot \mu_t$.

Conversely, assume that for every t, $0 \leq t \leq 1$, $\mu_t \neq \emptyset$ is an H_v-subgroup of H. For every x, y in H we can write $\mu(x) \geq \min\{\mu(x), \mu(y)\}$ and $\mu(y) \geq \min\{\mu(x), \mu(y)\}$. If we put $t_0 = \min\{\mu(x), \mu(y)\}$, then $x \in \mu_{t_0}$ and $y \in \mu_{t_0}$, so $x \cdot y \subseteq \mu_{t_0}$. Therefore, for every $\alpha \in x \cdot y$ we have $\mu(\alpha) \geq t_0$ implying $\inf_{\alpha \in x \cdot y} \{\mu(\alpha)\} \geq \min\{\mu(x), \mu(y)\}$ and in this way the condition (1) of Definition 3.1.1 is verified. To verify the second condition, if for every $a, x \in H$ we put $t_1 = \min\{\mu(a), \mu(x)\}$, then $x \in \mu_{t_1}$ and $a \in \mu_{t_1}$. So, there exists $y \in \mu_{t_1}$ such that $x \in a \cdot y$. On the other hand, since $y \in \mu_{t_1}$, then $t_1 \leq \mu(y)$ and hence $\min\{\mu(a), \mu(x)\} \leq \mu(y)$. The proof of the third condition of Definition 3.1.1 is similar to the proof of the second condition. ■

The following two corollaries are exactly obtained from Theorem 3.1.4.

Corollary 3.1.5. *Let $(H, \cdot)$ be an H_v-group and μ be a fuzzy H_v-subgroup of H. If $0 \leq t_1 < t_2 \leq 1$, then $\mu_{t_1} = \mu_{t_2}$ if and only if there is no x in H such that $t_1 \leq \mu(x) < t_2$.*

Corollary 3.1.6. *Let $(H, \cdot)$ be an H_v-group and μ be a fuzzy H_v-subgroup of H. If the range of μ is the finite set $\{t_1, t_2, \ldots, t_n\}$, then the set $\{\mu_{t_i} \mid 1 \leq i \leq n\}$ contains all the level H_v-subgroups of μ. Moreover, if $t_1 > t_2 > \ldots > t_n$, then all the level H_v-subgroups μ_{t_i} form the following chain $\mu_{t_1} \subseteq \mu_{t_2} \subseteq \ldots \subseteq \mu_{t_n}$.*

Proposition 3.1.7. *Let $(H, \cdot)$ be an H_v-group and μ be a fuzzy subset of H. Then, μ is a fuzzy H_v-subgroup of H if and only if for every t, $0 \leq t \leq 1$,*

(1) $\mu_t \cdot \mu_t \subseteq \mu_t$,
(2) $a \cdot (H - \mu_t) - (H - \mu_t) \subseteq a \cdot \mu_t$, *for all* $a \in \mu_t$.

Proof. Let μ be a fuzzy H_v-subgroup of H. Then, by Theorem 3.1.4, μ_t is an H_v-subgroup of H, and then it is clear that the three conditions of Definition 3.1.1 are valid.

Conversely, suppose that the three conditions of Definition 3.1.1 hold. Then, by Theorem 3.1.4 it is enough to prove that μ_t is an H_v-subgroup of H. For the proof of the left reproduction axiom, it is enough to show that $\mu_t \subseteq a \cdot \mu_t$, for every $a \in \mu_t$. Assume that there exists $x \in \mu_t$ such that $x \notin a \cdot \mu_t$. Since $x \in \mu_t$, then there exists $b \in H$ such that $x \in a \cdot b$. If $b \in \mu_t$, then $x \in a \cdot b \subseteq a \cdot \mu_t$, which is a contradiction. If $b \in H - \mu_t$, then we have $x \in \{x\} - (H - \mu_t) \subseteq a \cdot b - (H - \mu_t) \subseteq a \cdot (H - \mu_t) - (H - \mu_t) \subseteq a \cdot \mu_t$, again a contradiction. Therefore, $\mu_t - a \cdot \mu_t = \emptyset$ which implies that $\mu_t \subseteq a \cdot \mu_t$. The proof of the right reproduction axiom is similar. ■

Theorem 3.1.8. *Let $(H, \cdot)$ be an H_v-group. Then, every H_v-subgroup of H is a level H_v-subgroup of a fuzzy H_v-subgroup of H.*

Proof. Let A be an H_v-subgroup of H. For a fixed real number c, $0 < c \leq 1$, the fuzzy subset μ is defined as follows

$$\mu(x) = \begin{cases} c \text{ if } x \in A, \\ 0 \text{ otherwise.} \end{cases}$$

We have $A = \mu_c$ and by Theorem 3.1.4, it is enough to prove that μ is a fuzzy H_v-subgroup. This is straightforward and we omit it. ■

Corollary 3.1.9. *Let $(H, \cdot)$ be an H_v-group and A be a non-empty subset of H. Then, a necessary and sufficient condition for A to be an H_v-subgroup is that $A = \mu_{t_0}$, where μ is a fuzzy H_v-subgroup and $0 < t_0 \leq 1$.*

Definition 3.1.10. Let $(H, \cdot)$ be an H_v-group and μ be a fuzzy H_v-subgroup of H. Then, μ is called *right fuzzy closed with respect to H* if for every a, b in H all the x in $b \in a \cdot x$ satisfy $\min\{\mu(b), \mu(a)\} \leq \mu(x)$. We call μ *left fuzzy closed with respect to H* if for all a, b in H all the y in $b \in y \cdot a$ satisfy $\min\{\mu(b), \mu(a)\} \leq \mu(y)$. If μ is left and right fuzzy closed, then μ is called *fuzzy closed.*

Theorem 3.1.11. *If the fuzzy H_v-subgroup μ is right fuzzy closed, then $\mu_t.(H - \mu_t) = H - \mu_t$.*

Proof. If $b \in \mu_t \cdot (H - \mu_t)$, then there exists $a \in \mu_t$ and $x \in H - \mu_t$ such that $b \in a \cdot x$. Therefore, $\mu(x) < t \leq \mu(a)$ and since μ is right fuzzy closed we get $\min\{\mu(a), \mu(b)\} \leq \mu(x)$. Hence, $\mu(b) \leq \mu(x) < t$ which implies that $b \in H - \mu_t$. So, we have proved $\mu_t \cdot (H - \mu_t) \subseteq H - \mu_t$.

On the other hand, if $x \in H - \mu_t$, then for every $a \in \mu_t$ by the reproduction axiom there exists $y \in H$ such that $x \in a \cdot y$ and so it is enough to prove $y \in H - \mu_t$. Since μ is a fuzzy H_v-subgroup of H, by the definition we have $\min\{\mu(a), \mu(y)\} \leq \inf\limits_{\alpha \in a \cdot y} \{\mu(\alpha)\}$ which implies that

$$\min\{\mu(a), \mu(y)\} \leq \mu(x). \tag{3.1}$$

Since μ is right fuzzy closed so

$$\min\{\mu(x), \mu(a)\} \leq \mu(y). \tag{3.2}$$

Now, from $x \in H - \mu_t$ we get $a \in \mu_t$ and so $\mu(x) < t \leq \mu(a)$. Using 5.16 we obtain $\mu(x) \leq \mu(y)$. Therefore, $\mu(x) \leq \min\{\mu(a), \mu(y)\}$ and by 5.15 the relation $\min\{\mu(a), \mu(y)\} = \mu(x)$ is obtained. But $\mu(x) < \mu(a)$ and hence $\min\{\mu(a), \mu(y)\} = \mu(y)$. So, $\mu(x) = \mu(y)$. Since $x \in H - \mu_t$ we get $y \in H - \mu_t$ and the theorem is proved. ■

Now, we define anti fuzzy H_v-subgroup of an H_v-group and then we present some results in this connection.

Definition 3.1.12. Let $(H, \cdot)$ be an H_v-group and let μ be a fuzzy subset of H. Then, μ is said to be an *anti fuzzy H_v-subgroup* of H if the following axioms hold.

(1) $\sup\limits_{\alpha \in x \cdot y} \{\mu(\alpha)\} \leq \max\{\mu(x), \mu(y)\}$, for all $x, y \in H$.

(2) For all $x, a \in H$ there exists $y \in H$ such that $x \in a \cdot y$ and

$$\mu(y) \leq \max\{\mu(a), \mu(x)\}.$$

(3) For all $x, a \in H$ there exists $z \in H$ such that $x \in z \cdot a$ and

$$\mu(z) \leq \max\{\mu(a), \mu(x)\}.$$

Condition (2) is called the *left anti fuzzy reproduction axiom*, while (3) is called the *right anti fuzzy reproduction axiom.*

For the sake of similarity, only left reproduction axiom for the H_v-groups is verified throughout this section.

Lemma 3.1.13. *Let $(H, \cdot)$ be an H_v-group and μ be an anti fuzzy H_v-subgroup of H. Then,*

(1) $\sup\limits_{\alpha \in (\ldots((x_1 \cdot x_2) \cdot x_3) \ldots) \cdot x_n} \{\mu(\alpha)\} \leq \max\{\mu(x_1), \ldots, \mu(x_n)\}$, *for all* $x_1, x_2, \ldots, x_n \in H$,

(2) $\sup\{\mu(\alpha) \mid \alpha \in \bigcap\limits_{p(x_1,\ldots,x_n) \in P_n(x_1,\ldots,x_n)} p(x_1, \ldots, x_n)\} \leq \max\{\mu(x_1), \ldots, \mu(x_n)\}$.

Proof. The proof is similar to the proof of Lemma 3.1.2. ■

Proposition 3.1.14. *Let H be an H_v-group and μ be a fuzzy H_v-subgroup of H. Then, the set $\overline{\mu} = \{x \in H \mid \mu(x) = 1\}$ is empty or an H_v-subgroup of H.*

Proof. Let $\overline{\mu} \neq \emptyset$. Then, for all x, y in $\overline{\mu}$ we have $1 = \min\{\mu(x), \mu(y)\} \leq \inf\limits_{\alpha \in x \cdot y} \{\mu(\alpha)\}$. Therefore, for every $\alpha \in x \cdot y$ we have $\mu(\alpha) = 1$ which implies that $\alpha \in \overline{\mu}$. So, $x \cdot y \subseteq \overline{\mu}$ implying $x \cdot y \in \mathcal{P}^*(\overline{\mu})$. Hence, for every $a \in \overline{\mu}$, we have $a \cdot \overline{\mu} \subseteq \overline{\mu}$ and to prove the left reproduction axiom it is enough to prove $\overline{\mu} \subseteq a \cdot \overline{\mu}$.

Since μ is a fuzzy H_v-subgroup of H, for every $x \in \overline{\mu}$ there exists $y \in H$ such that $x \in a \cdot y$ and $\min\{\mu(a), \mu(x)\} \leq \mu(y)$. Since $x \in \overline{\mu}$ and $a \in \overline{\mu}$, we have $\min\{\mu(a), \mu(x)\} = 1$. Therefore, $\mu(y) = 1$ which implies that $y \in \overline{\mu}$ and the proposition is proved. ■

EXAMPLE 37 *Let $(G, \cdot)$ be a group and μ be a fuzzy subset of G. We define the hyperoperation $\circ : G \times G \to \mathcal{P}^*(G)$ as follows: $x \circ y = \{t \mid \mu(t) \leq \mu(x \cdot y)\}$. Then,*

(1) $(G, \circ)$ is an H_v-group.
(2) If μ is an anti fuzzy subgroup of $(G, \cdot)$, then μ is an anti fuzzy H_v-subgroup of $(G, \circ)$.

Theorem 3.1.15. *Let H be an H_v-group and μ be a fuzzy subset of H. Then μ is a fuzzy H_v-subgroup of H if and only if it's complement μ^c is an anti fuzzy H_v-subgroup of H.*

Proof. Let μ be a fuzzy H_v-subgroup of H, for every x, y in H, we have $\min\{\mu(x), \mu(y)\} \leq \inf_{\alpha \in x \cdot y} \{\mu(\alpha)\}$, or $\min\{1 - \mu^c(x), 1 - \mu^c(y)\} \leq \inf_{\alpha \in x \cdot y} \{1 - \mu^c(\alpha)\}$, or $\min\{1 - \mu^c(x), 1 - \mu^c(y)\} \leq 1 - \sup_{\alpha \in x \cdot y} \{\mu^c(\alpha)\}$, or $\sup_{\alpha \in x \cdot y} \{\mu^c(\alpha)\} \leq 1 - \min\{1 - \mu^c(x), 1 - \mu^c(y)\}$, or $\sup_{\alpha \in x \cdot y} \{\mu^c(\alpha)\} \leq \max\{\mu^c(x), \mu^c(y)\}$, and in this way the first condition is verified for μ^c.

Since μ is a fuzzy H_v-subgroup of H, for every a, x in H, there exists $y \in H$ such that $x \in a \cdot y$ and $\min\{\mu(a), \mu(x)\} \leq \mu(y)$, or $\min\{1 - \mu^c(a), 1 - \mu^c(x)\} \leq 1 - \mu^c(y)$, or $\mu^c(y) \leq 1 - \min\{1 - \mu^c(a), 1 - \mu^c(x)\}$, or $\mu^c(y) \leq \max\{\mu^c(a), \mu^c(x)\}$ and the second condition is satisfied. Thus, μ^c is an anti fuzzy H_v-subgroup. The converse also can be proved similarly. ■

Now, let H be a non-empty set and μ be a fuzzy subset of H. Then, for $0 \leq t \leq 1$, the lower level subset of μ is the set $\overline{\mu_t} = \{x \in H \mid \mu(x) \leq t\}$. Clearly $\overline{\mu_1} = H$ and if $t_1 < t_2$, then $\overline{\mu_{t_1}} \subseteq \overline{\mu_{t_2}}$.

Theorem 3.1.16. *Let H be an H_v-group and μ be a fuzzy subset of H. Then, μ is an anti fuzzy H_v-subgroup of H if and only if for every t, $0 \leq t \leq 1$, $\overline{\mu_t} \neq \emptyset$ is an H_v-subgroup of H.*

Proof. Let μ be an anti fuzzy H_v-subgroup of H. For every x, y in $\overline{\mu_t}$ we have $\mu(x) \leq t$ and $\mu(y) \leq t$. Hence, $\max\{\mu(x), \mu(y)\} \leq t$ and so $\sup_{\alpha \in x \cdot y} \{\mu(\alpha)\} \leq t$. Therefore, for every $\alpha \in x \cdot y$ we have $\mu(\alpha) \leq t$ which implies that $\alpha \in \overline{\mu_t}$, so $x \cdot y \subseteq \overline{\mu_t}$ implying $x \cdot y \in \mathcal{P}^*(\overline{\mu_t})$. Hence, for every $a \in \overline{\mu_t}$ we have $a \cdot \overline{\mu_t} \subseteq \overline{\mu_t}$ and to prove this part of the theorem it is enough to prove that $\overline{\mu_t} \subseteq a \cdot \overline{\mu_t}$.

Since μ is an anti fuzzy H_v-subgroup of H, for every $x \in \overline{\mu_t}$ there exists $y \in H$ such that $x \in a \cdot y$ and $\mu(y) \leq \max\{\mu(a), \mu(x)\}$. From $x \in \overline{\mu_t}$ and $a \in \overline{\mu_t}$ we get $\max\{\mu(x), \mu(a)\} \leq t$ and so $y \in \overline{\mu_t}$. Therefore, we have proved that for every $x \in \overline{\mu_t}$ there exists $y \in \overline{\mu_t}$ such that $x \in a \cdot y$ implying that $x \in a \cdot \overline{\mu_t}$ and this proves $\overline{\mu_t} \subseteq a \cdot \overline{\mu_t}$.

Conversely, assume that for every t, $0 \leq t \leq 1$, $\overline{\mu_t} \neq \emptyset$ is an H_v-subgroup of H. For every x, y in H we can write $\mu(x) \leq \max\{\mu(x), \mu(y)\}$ and $\mu(y) \leq \max\{\mu(x), \mu(y)\}$ and if we put $t_0 = \max\{\mu(x), \mu(y)\}$, then $x \in \overline{\mu_{t_0}}$ and $y \in \overline{\mu_{t_0}}$. Since $\overline{\mu_{t_0}}$ is an H_v-subgroup, so $x \cdot y \subseteq \overline{\mu_{t_o}}$. Therefore, for every $\alpha \in x \cdot y$ we have $\mu(\alpha) \leq t_o$ implying $\sup_{\alpha \in x \cdot y} \{\mu(\alpha)\} \leq t_0$ and so $\sup_{\alpha \in x \cdot y} \{\mu(\alpha)\} \leq$

$\max\{\mu(x), \mu(y)\}$ and in this way the first condition of Definition 3.1.12 is verified. To verify the second condition, if for every $a, x \in H$ we put $t_1 = \max\{\mu(a), \mu(x)\}$, then $x \in \overline{\mu_{t_1}}$ and $a \in \overline{\mu_{t_1}}$. Since $a \cdot \overline{\mu_{t_1}} = \overline{\mu_{t_1}}$, so there exists $y \in \overline{\mu_{t_1}}$ such that $x \in a \cdot y$. On the other hand, since $y \in \overline{\mu_{t_1}}$, then $\mu(y) \leq t_1$ and hence $\mu(y) \leq \max\{\mu(a), \mu(x)\}$ and the second condition of the Definition 3.1.12 is satisfied. ■

Definition 3.1.17. Let X be a non-empty set and x_t, with $t \in [0, 1]$, be a fuzzy point of X characterized by the fuzzy subset μ defined by $\mu(x) = t$ and $\mu(y) = 0$, for $y \in X - \{x\}$. We define the *hight* of x_t by $hgt(\mu) = t$. Moreover,

$$\widetilde{\mu} = \{x_t \mid \mu(x) \geq t, \ x \in X\}.$$

and $\widetilde{X}$ the family of all fuzzy points in X. The *support* of fuzzy subset μ of X is the set $supp(\mu) = \{x \in X \mid \mu(x) > 0\}$.

Proposition 3.1.18. *Let H be an H_v-group and μ be a fuzzy H_v-subgroup of H. Then, the set $supp(\mu)$ is an H_v-subgroup of H.*

Proof. For every x, y in $supp(\mu)$, we have $\mu(x) > 0, \mu(y) > 0$ and so $\min\{\mu(x), \mu(y)\} > 0$ which implies that $\inf\limits_{\alpha \in x \cdot y} > 0$. Therefore, for every $\alpha \in x \cdot y$ we have $\mu(\alpha) > 0$ which implies that $\alpha \in supp(\mu)$. Hence, for every $a \in supp(\mu)$, we have $a \cdot supp(\mu) \subseteq supp(\mu)$ and to prove the left reproduction axiom it is enough to prove $supp\mu \subseteq a \cdot supp(\mu)$.

Since μ is a fuzzy H_v-subgroup of H, for every $x \in supp(\mu)$ there exists $y \in H$ such that $x \in a \cdot y$ and $\min\{\mu(a), \mu(x)\} \leq \mu(y)$. Since $x \in supp(\mu)$ and $a \in supp(\mu)$ we have $\min\{\mu(a), \mu(x)\} > 0$. Therefore, $\mu(y) > 0$ which implies that $y \in supp(\mu)$ and the proposition is proved. ■

Definition 3.1.19. Let H be an H_v-group and μ be a fuzzy H_v-subgroup of H. We define the following hyperoperation on $\widetilde{\mu}$,

$$\begin{aligned} \circ : \widetilde{\mu} \times \widetilde{\mu} &\rightarrow \mathcal{P}^*(\widetilde{\mu}) \\ x_t \circ y_s &= \{\alpha_{t \wedge s} \mid \alpha \in x \cdot y\}, \end{aligned}$$

where $t \wedge s = \min\{t, s\}$.

Suppose that $x_t, y_s \in \widetilde{\mu}$. Then, $\mu(x) \geq t$, $\mu(y) \geq s$ and so $\min\{\mu(x), \mu(y)\} \geq t \wedge s$ which implies that $\inf\limits_{\alpha \in x \cdot y} \{\mu(\alpha)\} \geq t \wedge s$. Therefore, for every $\alpha \in x \cdot y$, we have $\alpha_{t \wedge s} \in \widetilde{\mu}$.

Lemma 3.1.20. *$(\widetilde{\mu}, \circ)$ is an H_v-semigroup.*

Proof. For every $x_t, y_s, z_r \in \widetilde{\mu}$, we have

$$\begin{aligned} (x_t \circ y_s) \circ z_r &= \{\alpha_{(t \wedge s) \wedge r} \mid \alpha \in (x \cdot y) \cdot z\}, \\ x_t \circ (y_s \circ z_r) &= \{\alpha_{t \wedge (s \wedge r)} \mid \alpha \in x \cdot (y \cdot z)\}. \end{aligned}$$

Since $(H, \cdot)$ is weak associative, $(H, \circ)$ is weak associative. ■

Definition 3.1.21. Let H_1, H_2 be two H_v-groups and μ_1, μ_2 be fuzzy H_v-subgroups of H_1, H_2, respectively. Let $\widetilde{f}$ be a mapping from $\widetilde{\mu}_1$ into $\widetilde{\mu}_2$ such that $supp\ \widetilde{f}\ (x_t) = supp\ \widetilde{f}\ (x_s)$, for all $x_t, x_s \in\widetilde{\mu}_1$. Then, f is called

(1) a *strong fuzzy homomorphism* if

$$\widetilde{f}\ (x_t \circ y_s) = \widetilde{f}\ (x_t)\circ \widetilde{f}\ (y_s), \text{ for all } x_t, y_s \in\widetilde{\mu}_1,$$

(2) an *inclusion fuzzy homomorphism* if

$$\widetilde{f}\ (x_t \circ y_s) \subseteq \widetilde{f}\ (x_t)\circ \widetilde{f}\ (y_s), \text{ for all } x_t, y_s \in\widetilde{\mu}_1,$$

(3) a *fuzzy H_v-homomorphism* if

$$\widetilde{f}\ (x_t \circ y_s)\cap \widetilde{f}\ (x_t)\circ \widetilde{f}\ (y_s) \neq \emptyset, \text{ for all } x_t, y_s \in\widetilde{\mu}_1 .$$

A mapping $\widetilde{f}:\widetilde{\mu}_1\to\widetilde{\mu}_2$ is called a *fuzzy isomorphism* if it is bijective and strong fuzzy homomorphism. Two fuzzy H_v-subgroups μ_1 and μ_2 are said to be *fuzzy isomorphic*, denoted by $\mu_1 \cong \mu_2$, if there exists a fuzzy isomorphism from $\widetilde{\mu}_1$ onto $\widetilde{\mu}_2$.

Theorem 3.1.22. *Let H_1, H_2 be two H_v-groups and μ_1, μ_2 be fuzzy H_v-subgroups of H_1, H_2, respectively. Let $\widetilde{f}:\widetilde{\mu}_1\to\widetilde{\mu}_2$ be a fuzzy inclusion homomorphism. Then,*

(1) *$hgt\ \widetilde{f}\ (x_t) = hgt\ \widetilde{f}\ (y_t)$.*
(2) *$hgt\ \widetilde{f}\ (x_t) \leq hgt\ \widetilde{f}\ (x_s)$, whenever $t \leq s$.*

Proof. (1) For every x_t and y_t in $\widetilde{\mu}_1$, there exists $z \in H_1$ such that $y \in x \cdot z$ and $\min\{\mu_1(x), \mu_1(y)\} \leq \mu_1(z)$. Since $\mu_1(x) \geq t$ and $\mu_1(y) \geq t$, we have $\mu_1(z) \geq t$ which implies that $z_t \in\widetilde{\mu}_1$. From $y_t \in x_t \circ z_t$ we get $\widetilde{f}\ (y_t) \in\widetilde{f}\ (x_t \circ z_t)$ or $\widetilde{f}\ (y_t) \in\widetilde{f}\ (x_t)\circ \widetilde{f}\ (z_t)$ and so

$$hgt\ \widetilde{f}\ (y_t) = \min\{hgt\ \widetilde{f}\ (x_t), hgt\ \widetilde{f}\ (z_t)\} \leq hgt\ \widetilde{f}\ (x_t).$$

Similarly, we obtain $hgt\ \widetilde{f}\ (x_t) \leq hgt\ \widetilde{f}\ (y_t)$. Therefore, $hgt\ \widetilde{f}\ (x_t) = hgt\ \widetilde{f}\ (y_t)$.

(2) Suppose that $t \leq s$. If $x_t \in\widetilde{\mu}_1$, then there exists $y \in H_1$ such that $x \in x \cdot y$ and $\mu_1(x) \leq \mu_1(y)$, so $y_t \in\widetilde{\mu}_1$. From $x_t \in x_t \circ y_t$ we have $\widetilde{f}\ (x_t) \in\widetilde{f}\ (x_t \circ y_t)$ which implies $\widetilde{f}\ (x_t) \in\widetilde{f}\ (x_s \circ y_t)$ or $\widetilde{f}\ (x_t) \in\widetilde{f}\ (x_s)\circ \widetilde{f}\ (y_t)$. Therefore,

$$hgt\ \widetilde{f}\ (x_t) = \min\{hgt\ \widetilde{f}\ (x_s), hgt\ \widetilde{f}\ (y_t)\} \leq hgt\ \widetilde{f}\ (x_s).$$

■

Theorem 3.1.23. *Let H_1, H_2 be two H_v-groups and μ_1, μ_2 be fuzzy H_v-subgroups of H_1, H_2, respectively. A mapping $\widetilde{f}:\widetilde{\mu_1}\to\widetilde{\mu_2}$ is a fuzzy strong homomorphism if and only if there exists an ordinary strong homomorphism of H_v-groups $f : supp\mu_1 \to supp\mu_2$ and an increasing function $\varphi : (0,1] \to (0,1]$ such that*

$$\widetilde{f}\,(x_t) = [f(x)]_{\varphi(t)}, \text{ for all } x_t \in\widetilde{\mu_1}\,.$$

Proof. Suppose that $\widetilde{f}:\widetilde{\mu_1}\to\widetilde{\mu_2}$ is a fuzzy strong homomorphism. We define a mapping $f : supp(\mu_1) \to supp(\mu_2)$ and a function $\varphi : (0,1] \to (0,1]$ as follows:

$$f(x) = supp\,\widetilde{f}\,(x_{\mu_1(x)}), \text{ for all } x \in supp\mu_1$$

and

$$\varphi(t) = hgt\,\widetilde{f}\,(x_t), \text{ for all } t \in (0,1].$$

Since $\widetilde{f}$ is a fuzzy strong homomorphism, then $supp\,\widetilde{f}\,(x_t) = supp\,\widetilde{f}\,(x_{\mu_1(x)})$ and so $supp\,\widetilde{f}\,(x_t) = f(x)$ which implies that $\widetilde{f}\,(x_t) = [f(x)]_{\varphi(t)}$. By definition of φ and Theorem 3.1.22, it is easy to see that φ is increasing. Therefore, it remains to show that f is a strong homomorphism from the H_v-group $supp\mu_1$ into the H_v-group $supp(\mu_2)$.

For every $x, y \in supp\mu_1$, we put $\mu_1(x) = t$ and $\mu_1(y) = s$. Then, we have

$$\begin{aligned}
[f(x\cdot y)]_{\varphi(t\wedge s)} &= \bigcup_{\alpha\in x\cdot y} [f(\alpha)]_{\varphi(t\wedge s)} \\
&= \bigcup_{\alpha\in x\cdot y} \widetilde{f}\,(\alpha_{t\wedge s}) \\
&=\widetilde{f}\,(\bigcup_{\alpha\in x\cdot y} \alpha_{t\wedge s}) \\
&=\widetilde{f}\,(x_t\circ y_s) =\widetilde{f}\,(x_t)\circ\widetilde{f}\,(y_s) \\
&= [f(x)]_{\varphi(t)}\circ[f(y)]_{\varphi(s)} \\
&= \bigcup_{z\in f(x)\cdot f(y)} z_{\varphi(t)\wedge\varphi(s)} \\
&= [f(x)\cdot f(y)]_{\varphi(t)\wedge\varphi(s)}.
\end{aligned}$$

Therefore, $f(x\cdot y) = f(x)\cdot f(y)$, i.e., f is a strong homomorphism.

Conversely, we consider a mapping $\widetilde{f}:\widetilde{\mu_1}\to\widetilde{\mu_2}$ such that $\widetilde{f}\,(x_t) = [f(x)]_{\varphi(t)}$. It is enough to show that $\widetilde{f}$ is a strong fuzzy homomorphism. For every $x_t, y_s \in\widetilde{\mu_1}$ $(t \le s)$, we have

$$
\begin{aligned}
\widetilde{f}\,(x_t \circ y_s) &= \bigcup_{\alpha \in x \cdot y} \widetilde{f}\,(\alpha_{t \wedge s}) \\
&= \bigcup_{\alpha \in x \cdot y} [f(\alpha)]_{\varphi(t \wedge s)} \\
&= \bigcup_{\alpha \in x \cdot y} [f(\alpha)]_{\varphi(t)} \\
&= [f(x \cdot y)]_{\varphi(t)} = [f(x) \cdot f(y)]_{\varphi(t) \wedge \varphi(s)} \\
&= [f(x)]_{\varphi(t)} \circ [f(y)]_{\varphi(s)} \\
&= \widetilde{f}\,(x_t) \circ \widetilde{f}\,(y_s). \ \blacksquare
\end{aligned}
$$

Let f be a strong homomorphism from H_1 into H_2. We can define a mapping $\widetilde{f}:\widetilde{H}_1 \to \widetilde{H}_2$ as follows: $\widetilde{f}\,(x_t) = [f(x)]_t$. Obviously, $\widetilde{f}$ is a strong fuzzy homomorphism from $\widetilde{H}_1$ into $\widetilde{H}_2$, where $\varphi(\lambda) = \lambda, \forall \lambda \in (0,1]$. Therefore, the concept of strong fuzzy homomorphism between two H_v- groups can be seen as an extension of the concept of strong homomorphism between two H_v-groups.

REMARK 2 Let $f : X \to Y$ and let $\varphi : (0,1] \to (0,1]$ be an increasing mapping. We define the mapping $f_\varphi : \widetilde{X} \to \widetilde{Y}$ by $f_\varphi(x_t) = [f(x)]_{\varphi(t)}$. Then, for every fuzzy subset μ of X we have

$$f_\varphi(\mu)(y) = \sup_{x \in f^{-1}(y)} \{\varphi(\mu(x))\}.$$

Theorem 3.1.24. *Let φ be bijective and $f : H_1 \to H_2$ be a surjective strong homomorphism and let μ be a fuzzy H_v-subgroup of H_1. Then, $f_\varphi(\mu)$ is a fuzzy H_v -subgroup of H_2.*

Proof. Let μ be a fuzzy H_v-subgroup of H_1. By Theorem 3.1.4, for every t, $0 \le t \le 1$, level subset $\mu_t (\mu_t \neq \emptyset)$ is an H_v-subgroup of H_1 and so $f(\mu_{\varphi^{-1}(t)})$ is an H_v-subgroup of H_2. Now, it is enough to show that

$$f(\mu_{\varphi^{-1}(t)}) = (f_\varphi(\mu))_t.$$

For every y in $(f_\varphi(\mu))_t$ we have $f_\varphi(\mu)(y) \ge t$ which implies that

$$\sup_{x \in f^{-1}(y)} \{\varphi(\mu(x))\} \ge t.$$

Therefore, there exists $x_0 \in f^{-1}(y)$ such that $\varphi(\mu(x_0)) \ge t$ which implies that $\mu(x_0) \ge \varphi^{-1}(t)$ or $x_0 \in \mu_{\varphi^{-1}(t)}$ and so $f(x_0) \in f(\mu_{\varphi^{-1}(t)})$ implying $y \in f(\mu_{\varphi^{-1}(t)})$. Now, for every y in $f(\mu_{\varphi^{-1}(t)})$, there exists $x \in \mu_{\varphi^{-1}(t)}$ such that $f(x) = y$. Since $x \in \mu_{\varphi^{-1}(t)}$, we have $\mu(x) \ge \varphi^{-1}(t)$ or $\varphi(\mu(x)) \ge t$ and so $\sup_{x \in f^{-1}(y)} \{\varphi(\mu(x))\} \ge t$ which implies that $f_\varphi(\mu)(y) \ge t$. Therefore, $y \in (f_\varphi(\mu))_t$. $\blacksquare$

3.2 Generalized Fuzzy H_v-Subgroups

Similar to Section 2.4, we extend the concept of $(\in, \in \vee q)$-fuzzy subgroups and $(\in, \in \vee q)$-fuzzy subpolygroups to $(\in, \in \vee q)$-fuzzy H_v-subgroups [80].

Definition 3.2.1. A fuzzy subset μ of an H_v-group H is said to be an $(\in, \in \vee q)$-fuzzy H_v-subgroup of H if for all $t, r \in (0, 1]$ and $x, y \in H$,

(1) $x_t, y_r \in \mu$ implies $z_{t\wedge r} \in \vee q\mu$ for all $z \in x \circ y$.
(2) $x_t, a_r \in \mu$ implies $y_{t\wedge r} \in \vee q\mu$ for some $y \in H$ with $x \in a \circ y$.
(3) $x_t, a_r \in \mu$ implies $z_{t\wedge r} \in \vee q\mu$ for some $z \in H$ with $x \in z \circ a$.

Proposition 3.2.2. *Conditions (1), (2) and (3) in Definition 3.2.1 are equivalent to the following conditions, respectively.*

(1′) $\mu(x) \wedge \mu(y) \wedge 0.5 \leq \bigwedge_{z \in x \circ y} \mu(z)$ *for all* $x, y \in H$.

(2′) *For all* $x, a \in H$ *there exists* $y \in H$ *such that* $x \in a \circ y$ *and*

$$\mu(a) \wedge \mu(x) \wedge 0.5 \leq \mu(y).$$

(3′) *For all* $x, a \in H$ *there exists* $z \in H$ *such that* $x \in z \circ a$ *and*

$$\mu(a) \wedge \mu(x) \wedge 0.5 \leq \mu(z).$$

Proof. $(1 \Rightarrow 1')$: Suppose that $x, y \in H$. We consider the following cases:

(a) $\mu(x) \wedge \mu(y) < 0.5$.
(b) $\mu(x) \wedge \mu(y) \geq 0.5$.

Case a: Assume that there exists $z \in x \circ y$ such that $\mu(z) < \mu(x) \wedge \mu(y) \wedge 0.5$, which implies that $\mu(z) < \mu(x) \wedge \mu(y)$. Choose t such that $\mu(z) < t < \mu(x) \wedge \mu(y)$. Then, $x_t, y_t \in \mu$, but $z_t \overline{\in \vee q\mu}$ which contradicts (1).

Case b: Assume that $\mu(z) < 0.5$ for some $z \in x \circ y$. Then, $x_{0.5}, y_{0.5} \in \mu$, but $z_{0.5} \overline{\in \vee q\mu}$, a contradiction.

Hence, (1′) holds.

$(2 \Rightarrow 2')$: Suppose that $x, a \in H$. We consider the following cases:

(a) $\mu(x) \wedge \mu(a) < 0.5$.
(b) $\mu(x) \wedge \mu(a) \geq 0.5$.

Case a: Assume that for all y with $x \in a \circ y$, we have $\mu(y) < \mu(x) \wedge \mu(a)$. Choose t such that $\mu(y) < t < \mu(x) \wedge \mu(a)$ and $t + \mu(y) < 1$. Then, $x_t, a_t \in \mu$, but $y_t \overline{\in \vee q\mu}$, which contradicts (2).

Case b: Assume that for all y with $x \in a \circ y$, we have

$$\mu(y) < \mu(x) \wedge \mu(a) \wedge 0.5.$$

Then, $x_{0.5}, a_{0.5} \in \mu$, but $y_{0.5} \overline{\in \vee q\mu}$, which contradicts (2).

Hence, (2′) holds.

$(3 \Rightarrow 3')$: The proof is similar to $(2 \Rightarrow 2')$.

($1' \Rightarrow 1$): Let $x_t, y_r \in \mu$, then $\mu(x) \geq t$ and $\mu(y) \geq r$. For every $z \in x \circ y$ we have

$$\mu(z) \geq \mu(x) \wedge \mu(y) \wedge 0.5 \geq t \wedge r \wedge 0.5.$$

If $t \wedge r > 0.5$, then $\mu(z) \geq 0.5$ which implies that $\mu(z) + t \wedge r > 1$.
If $t \wedge r \leq 0.5$, then $\mu(z) \geq t \wedge r$.
Therefore, $z_{t\wedge r} \in \vee q\mu$ for all $z \in x \circ y$.
($2' \Rightarrow 2$): Let $x_t, a_r \in \mu$. Then, $\mu(x) \geq t$ and $\mu(a) \geq r$. Now, for some y with $x \in a \circ y$ we have

$$\mu(y) \geq \mu(a) \wedge \mu(x) \wedge 0.5 \geq t \wedge r \wedge 0.5.$$

If $t \wedge r > 0.5$, then $\mu(y) \geq 0.5$ which implies $\mu(y) + t \wedge r > 1$.
If $t \wedge r \leq 0.5$, then $\mu(y) \geq t \wedge r$.
Therefore, $y_{t\wedge r} \in \vee q\mu$. Hence, (2) holds.
($3' \Rightarrow 3$): The proof is similar to ($2' \Rightarrow 2$). ■

By Definition 3.2.1 and Proposition 3.2.2, we immediately get:

Corollary 3.2.3. *A fuzzy subset μ of an H_v-group H is an $(\in, \in \vee q)$-fuzzy H_v-subgroup of H if and only if the conditions (1′), (2′) and (3′) in Proposition 3.2.2 hold.*

Now, we characterize $(\in, \in \vee q)$-fuzzy H_v-subgroups by their level H_v-subgroups.

Theorem 3.2.4. *Let μ be an $(\in, \in \vee q)$-fuzzy H_v-subgroup of H. Then, for all $0 < t \leq 0.5$, μ_t is an empty set or an H_v-subgroup of H. Conversely, if μ is a fuzzy subset of H such that μ_t $(\neq \emptyset)$ is an H_v-subgroup of H for all $0 < t \leq 0.5$, then μ is an $(\in, \in \vee q)$-fuzzy H_v-subgroup of H.*

Proof. Let μ be an $(\in, \in \vee q)$-fuzzy H_v-subgroup of H and $0 < t \leq 0.5$. Let $x, y \in \mu_t$. Then, $\mu(x) \geq t$ and $\mu(y) \geq t$. Now

$$\bigwedge_{z \in x \circ y} \mu(z) \geq \mu(x) \wedge \mu(y) \wedge 0.5 \geq t \wedge 0.5 = t.$$

Therefore, for every $z \in x \circ y$ we have $\mu(z) \geq t$ or $z \in \mu_t$, so $x \circ y \subseteq \mu_t$. Hence, for every $a \in \mu_t$ we have $a \circ \mu_t \subseteq \mu_t$. Now, let $x, a \in \mu_t$, then there exists $y \in H$ such that $x \in a \circ y$ and $\mu(a) \wedge \mu(x) \wedge 0.5 \leq \mu(y)$. From $x, a \in \mu_t$, we have $\mu(x) \geq t$ and $\mu(a) \geq t$, and so

$$t = t \wedge t \wedge 0.5 \leq \mu(a) \wedge \mu(x) \wedge 0.5 \leq \mu(y).$$

Hence, $y \in \mu_t$, and this prove that $\mu_t \subseteq a \circ \mu_t$.

Conversely, let μ be a fuzzy subset of H such that μ_t $(\neq \emptyset)$ is an H_v-subgroup of H for all $0 < t \leq 0.5$. For every $x, y \in H$, we can write

$$\mu(x) \geq \mu(x) \wedge \mu(y) \wedge 0.5 = t_0,$$
$$\mu(y) \geq \mu(x) \wedge \mu(y) \wedge 0.5 = t_0,$$

then $x \in \mu_{t_0}$ and $y \in \mu_{t_0}$, so $x \circ y \subseteq \mu_{t_0}$. Therefore, for every $z \in x \circ y$ we have $\mu(z) \geq t_0$ which implies

$$\bigwedge_{z \in x \circ y} \mu(z) \geq t_0,$$

and in this way the first condition of Proposition 3.2.1 is verified. To verify the second condition, if for every $a, x \in H$, we put $t_1 = \mu(a) \wedge \mu(x) \wedge 0.5$, then $x \in \mu_{t_1}$ and $a \in \mu_{t_1}$. So there exists $y \in \mu_{t_1}$ such that $x \in a \circ y$. Since $y \in \mu_{t_1}$, we have $\mu(y) \geq t_1$ or

$$\mu(y) \geq \mu(a) \wedge \mu(x) \wedge 0.5.$$

The third condition is verified similarly. ■

Naturally, a corresponding result should be considered when μ_t is an H_v-subgroup of H for all $t \in (0.5, 1]$.

Theorem 3.2.5. *Let μ be a fuzzy subset of an H_v-group H. Then, μ_t ($\neq \emptyset$) is an H_v-subgroup of H for all $t \in (0.5, 1]$ if and only if*

(1) $\mu(x) \wedge \mu(y) \leq \bigwedge_{z \in x \circ y} (\mu(z) \vee 0.5)$ *for all $x, y \in H$.*

(2) *For all $x, a \in H$ there exists $y \in H$ such that $x \in a \circ y$ and*

$$\mu(a) \wedge \mu(x) \leq \mu(y) \vee 0.5.$$

(3) *For all $x, a \in H$ there exists $z \in H$ such that $x \in z \circ a$ and*

$$\mu(a) \wedge \mu(x) \leq \mu(z) \vee 0.5.$$

Proof. If there exist $x, y, z \in H$ with $z \in x \circ y$ such that

$$\mu(z) \vee 0.5 < \mu(x) \wedge \mu(y) = t,$$

then $t \in (0.5, 1]$, $\mu(z) < t$, $x \in \mu_t$, and $y \in \mu_t$. Since $x, y \in \mu_t$ and μ_t is an H_v-subgroup, so $x \circ y \subseteq \mu_t$ and $\mu(z) \geq t$ for all $z \in x \circ y$, which is a contradiction with $\mu(z) < t$. Therefore,

$$\mu(x) \wedge \mu(y) \geq \mu(z) \vee 0.5 \text{ for all } x, y, z \in H \text{ with } z \in x \circ y,$$

which implies that

$$\mu(x) \wedge \mu(y) \geq \bigwedge_{z \in x \circ y} (\mu(z) \vee 0.5), \text{ for all } x, y \in H.$$

Hence, (1) holds.

Now, assume that there exist $x_0, a_0 \in H$ such that for all $y \in H$ with $x_0 \in a_0 \circ y$, the following inequality holds:

$$\mu(y) \vee 0.5 < \mu(a_0) \wedge \mu(x_0) = t.$$

Then, $t \in (0.5, 1]$, $x_0 \in \mu_t$, $a_0 \in \mu_t$ and $\mu(y) < t$. Since $x_0, a_0 \in \mu_t$ and μ_t is an H_v-subgroup, so there exists $y_0 \in \mu_t$ such that $x_0 \in a_0 \circ y_0$. From $y_0 \in \mu_t$, we get $\mu(y_0) \geq t$, which is a contradiction with $\mu(y_0) < t$. Therefore, for all $x, a \in H$ there exists $y \in H$ such that $x \in a \circ y$ and

$$\mu(a) \wedge \mu(x) \leq \mu(y) \vee 0.5.$$

Hence, (2) holds.

The proof of the third condition is similar to the proof of the second condition.

Conversely, assume that $t \in (0.5, 1]$ and $x, y \in \mu_t$. Then,

$$0.5 < t \leq \mu(x) \wedge \mu(y) \leq \bigwedge_{z \in x \circ y} (\mu(z) \vee 0.5).$$

It follows that for every $z \in x \circ y$, $0.5 < t \leq \mu(z) \vee 0.5$ and so $t \leq \mu(z)$, which implies $z \in \mu_t$. Hence, $x \circ y \subseteq \mu_t$.

Now, let $x, a \in \mu_t$. Then, by using the condition (2), there exists $y \in H$ such that $x \in a \circ y$ and

$$\mu(a) \wedge \mu(x) \leq \mu(y) \vee 0.5.$$

We show that $y \in \mu_t$. We have

$$0.5 < t \leq \mu(x) \leq \mu(a) \wedge \mu(x) \leq \mu(y) \vee 0.5.$$

It follows that $0.5 \leq \mu(y)$ and so $y \in \mu_t$. Therefore, μ_t is an H_v-subgroup of H for all $t \in (0.5, 1]$. ■

Let μ be a fuzzy subset of an H_v-group H and

$$J = \{t \mid t \in (0, 1] \text{ and } \mu_t \text{ is an empty} - \text{set or an } H_v\text{−subgroup of } H\}.$$

When $J = (0, 1]$, then μ is an ordinary fuzzy H_v-subgroup of the H_v-group H. When $J = (0, 0.5]$, μ is an $(\in, \in \vee q)$-fuzzy H_v-subgroup of the H_v-group H.

Now, we can define the concept of fuzzy H_v-subgroup with thresholds in the following way:

Definition 3.2.6. Let $\alpha, \beta \in [0, 1]$ and $\alpha < \beta$. Let μ be a fuzzy subset of an H_v-group H. Then, μ is called a *fuzzy H_v-subgroup with thresholds* (α, β) *of* H if for all $x, y \in H$,

(1) $\mu(x) \wedge \mu(y) \wedge \beta \leq \bigwedge_{z \in x \circ y} (\mu(z) \vee \alpha)$, for all $x, y \in H$.

(2) For all $x, a \in H$ there exists $y \in H$ such that $x \in a \circ y$ and

$$\mu(a) \wedge \mu(x) \wedge \beta \leq \mu(y) \vee \alpha.$$

(3) For all $x, a \in H$ there exists $z \in H$ such that $x \in z \circ a$ and

$$\mu(a) \wedge \mu(x) \wedge \beta \leq \mu(z) \vee \alpha.$$

If μ is a fuzzy H_v-subgroup with thresholds of H, then we can conclude that μ is an ordinary fuzzy H_v-subgroup when $\alpha = 0$, $\beta = 1$; and μ is an $(\in, \in \vee q)$-fuzzy H_v-subgroup when $\alpha = 0$, $\beta = 0.5$.

Now, we characterize fuzzy H_v-subgroups with thresholds by their level H_v-subgroups.

Theorem 3.2.7. *A fuzzy subset μ of an H_v-group H is a fuzzy H_v-subgroup with thresholds (α, β) of H if and only if μ_t $(\neq \emptyset)$ is an H_v-subgroup of H for all $t \in (\alpha, \beta]$.*

Proof. Let μ be a fuzzy H_v-subgroup with thresholds of H and $t \in (\alpha, \beta]$. Let $x, y \in \mu_t$. Then, $\mu(x) \geq t$ and $\mu(y) \geq t$. Now

$$\bigwedge_{z \in x \circ y} (\mu(z) \vee \alpha) \geq \mu(x) \wedge \mu(y) \wedge \beta \geq t \wedge \beta \geq t > \alpha.$$

So, for every $z \in x \circ y$ we have $\mu(z) \vee \alpha \geq t > \alpha$ which implies $\mu(z) \geq t$ and $z \in \mu_t$. Hence, $x \circ y \subseteq \mu_t$. Now, let $x, a \in \mu_t$, then there exists $y \in H$ such that $x \in a \circ y$ and $\mu(a) \wedge \mu(x) \wedge \beta \leq \mu(y) \vee \alpha$. From $x, a \in \mu_t$, we have $\mu(x) \geq t$ and $\mu(a) \geq t$, and so

$$\alpha < t \leq t \wedge \beta \leq \mu(a) \wedge \mu(x) \wedge \beta \leq \mu(y) \vee \alpha,$$

which implies that $\mu(y) \geq t$, and so $y \in \mu_t$. Therefore, we have $\mu_t = a \circ \mu_t$ for all $a \in \mu_t$. Similarly we get $\mu_t \circ a = \mu_t$ for all $a \in \mu_t$. Therefore, μ_t is an H_v-subgroup of H for all $t \in (\alpha, \beta]$.

Conversely, let μ be a fuzzy subset of H such that μ_t $(\neq \emptyset)$ is an H_v-subgroup of H for all $t \in (\alpha, \beta]$. If there exist $x, y, z \in H$ with $z \in x \circ y$ such that

$$\mu(z) \vee \alpha < \mu(x) \wedge \mu(y) \wedge \beta = t,$$

then $t \in (\alpha, \beta]$, $\mu(z) < t$, $x \in \mu_t$ and $y \in \mu_t$. Since μ_t is an H_v-subgroup of H and $x, y \in \mu_t$, so $x \circ y \subseteq \mu_t$. Hence, $\mu(z) \geq t$ for all $z \in x \circ y$. This is a contradiction with $\mu(z) < t$. Therefore,

$$\mu(x) \wedge \mu(y) \wedge \beta \leq \mu(z) \vee \alpha \text{ for all } x, y, z \in H \text{ with } z \in x \circ y,$$

which implies that

$$\mu(x) \wedge \mu(y) \wedge \beta \leq \bigwedge\nolimits_{z \in x \circ y} (\mu(z) \vee \alpha) \text{ for all } x, y \in H.$$

Hence, the condition (1) of Definition 3.2.6 holds.

Now, assume that there exist $x_0, a_0 \in H$ such that for all $y \in H$ which satisfies $x_0 \in a_0 \circ y$, the following inequality holds:

$$\mu(y) \vee \alpha < \mu(a_0) \wedge \mu(x_0) \wedge \beta = t.$$

Then, $t \in (\alpha, \beta]$, $x_0 \in \mu_t$, $a_0 \in \mu_t$ and $\mu(y) < t$. Since $x_0, a_0 \in \mu_t$ and μ_t is an H_v-subgroup, so there exists $y_0 \in \mu_t$ such that $x_0 \in a_0 \circ y_0$. From $y_0 \in \mu_t$, we get $\mu(y_0) \geq t$. This is a contradiction with $\mu(y_0) < t$. Therefore,

$$\mu(a) \wedge \mu(x) \wedge \beta \leq \mu(y) \vee \alpha.$$

Hence, the second condition of Definition 3.2.6 holds. The proof of third condition is similar to the proof of second condition. ■

3.3 Intuitionistic Fuzzy H_v-Subgroups

As an important generalization of the notion of fuzzy sets on X, Atanassov introduced in [7] the concept of *intuitionistic fuzzy sets* defined on a non-empty set X as objects having the form

$$A = \{(x, \mu_A(x), \lambda_A(x)) \mid x \in X\},$$

where the functions $\mu_A : X \to [0,1]$ and $\lambda_A : X \to [0,1]$ denote the *degree of membership* (namely $\mu_A(x)$) and the *degree of non-membership* (namely $\lambda_A(x)$) of each element $x \in X$ to the set A respectively, and $0 \leq \mu_A(x) + \lambda_A(x) \leq 1$ for all $x \in X$.

For every two intuitionistic fuzzy sets A and B on X we define (see [8]):

(1) $A \subseteq B$ if and only if $\mu_A(x) \leq \mu_B(x)$ and $\lambda_A(x) \geq \lambda_B(x)$, for all $x \in X$.
(2) $A^c = \{(x, \lambda_A(x), \mu_A(x)) \mid x \in X\}$.
(3) $A \cap B = \{(x, \min\{\mu_A(x), \mu_B(x)\}, \max\{\lambda_A(x), \lambda_B(x)\}) \mid x \in X\}$.
(4) $A \cup B = \{(x, \max\{\mu_A(x), \mu_B(x)\}, \min\{\lambda_A(x), \lambda_B(x)\}) \mid x \in X\}$.
(5) $\Box A = \{(x, \mu_A(x), \mu_A^c(x)) \mid x \in X\}$.
(6) $\Diamond A = \{(x, \lambda_A^c(x), \lambda_A(x)) \mid x \in X\}$.

EXAMPLE 38 Consider the universe $X = \{10, 100, 500, 1000, 1200\}$. An intutionistic fuzzy set "Large" of X denoted by L and may be defined by

$$L = \{< 10,\ 0.01,\ 0.9 >,\ < 100,\ 0.1,\ 0.88 >,\ < 500,\ 0.4,\ 0.5 >,\ < 1000,\ 0.8,\ 0.1 >,\ < 1200,\ 1,\ 0 >\}.$$

One may define an intuitionistic fuzzy set "Very Large" (denoted by VL) as follows:

$$\mu_{VL}(x) = (\mu_L(x))^2 \quad \text{and} \quad \nu_{VL}(x) = 1 - (1 - \nu_L(x))^2,$$

for all $x \in X$. Thus,

$$VL = \{< 10,\ 0.0001,\ 0.99 >,\ < 100,\ 0.01,\ 0.9856 >,\ < 500,\ 0.16,\ 0.75 >,\ < 1000,\ 0.64,\ 0.19 >,\ < 1200,\ 1,\ 0 >\}.$$

EXAMPLE 39 Consider the universe $\{a_1, a_2, a_3, a_4, a_5, a_6\}$. Let A and B be two intuitionistic fuzzy sets of X given by

$$A = \{< a_1,\ 0.2,\ 0.6 >,\ < a_2,\ 0.3,\ 0.7 >,\ < a_3,\ 1,\ 0 >,\\ < a_4,\ 0.8,\ 0.1 >,\ < a_5,\ 0.5,\ 0.4 >\}$$

and

$$B = \{< a_1,\ 0.4,\ 0.4 >,\ < a_2,\ 0.5,\ 0.2 >,\ < a_3,\ 0.6,\ 0.2,\\ < a_4,\ 0.1,\ 0.7 >,\ < a_5,\ 0,\ 1 >\}.$$

Then,

$$\begin{aligned} A^c &= \{< a_1,\ 0.6,\ 0.2 >,\ < a_2,\ 0.7,\ 0.3 >,\ < a_3,\ 0,\ 1 >,\\ &\quad < a_4,\ 0.1,\ 0.8 >,\ < a_5,\ 0.4,\ 0.5 >\},\\ A \cap B &= \{< a_1,\ 0.2,\ 0.6 >,\ < a_2,\ 0.3,\ 0.7 >,\ < a_3,\ 0.6,\ 0.2 >,\\ &\quad < a_4,\ 0.1,\ 0.7 >,\ < a_5,\ 0,\ 1 >\},\\ A \cup B &= \{< a_1,\ 0.4,\ 0.4 >,\ < a_2,\ 0.5,\ 0.2 >,\ < a_3,\ 1,\ 0 >,\\ &\quad < a_4,\ 0.8,\ 0.1 >,\ < a_5,\ 0.5,\ 0.4 >\},\\ \Box A &= \{< a_1,\ 0.2,\ 0.8 >,\ < a_2,\ 0.3,\ 0.7 >,\ < a_3,\ 1,\ 0 >,\\ &\quad < a_4,\ 0.8,\ 0.2 >,\ < a_5,\ 0.5,\ 0.5 >\},\\ \Diamond B &= \{< a_1,\ 0.6,\ 0.4 >,\ < a_2,\ 0.8,\ 0.2 >,\ < a_3,\ 0.8,\ 0.2 >,\\ &\quad < a_4,\ 0.3,\ 0.7 >,\ < a_5,\ 0,\ 1 >\}. \end{aligned}$$

For the sake of simplicity, we shall use the symbol $A = (\mu_A, \lambda_A)$ for the intuitionistic fuzzy set $A = \{(x, \mu_A(x), \lambda_A(x) \mid x \in X\}$.

In what follows, let H denote an H_v-group, and we start by defining the notion of intuitionistic fuzzy H_v-subgroup.

Definition 3.3.1. [103] An intuitionistic fuzzy set $A = (\mu_A, \lambda_A)$ in H is called an *intuitionistic fuzzy H_v-subgroup* of H (IFS of H for short) if

(1) $\min\{\mu_A(x), \mu_A(y)\} \leq \inf\{\mu_A(z) \mid z \in x \circ y\}$, for all $x, y \in H$.
(2) For all $x, a \in H$ there exist $y, z \in H$ such that $x \in (a \circ y) \cap (z \circ a)$ and

$$\min\{\mu_A(a), \mu_A(x)\} \leq \min\{\mu_A(y), \mu_A(z)\}.$$

(3) $\sup\{\lambda_A(z) \mid z \in x \circ y\} \leq \max\{\lambda_A(x), \lambda_A(y)\}$, for all $x, y \in H$.
(4) For all $x, a \in H$ there exist $y, z \in H$ such that $x \in (a \circ y) \cap (z \circ a)$ and

$$\max\{\lambda_A(y), \lambda_A(z)\} \leq \max\{\lambda_A(a), \lambda_A(x)\}.$$

Lemma 3.3.2. *If $A = (\mu_A, \lambda_A)$ is an IFS of H, then so is $\Box A = (\mu_A, \mu_A^c)$.*

Proof. It is sufficient to show that μ_A^c satisfies the third and fourth conditions of Definition 3.3.1. For $x, y \in H$ we have

$$\min\{\mu_A(x), \mu_A(y)\} \leq \inf\{\mu_A(z) \mid z \in x \circ y\}$$

and so

$$\min\{1-\mu_A^c(x), 1-\mu_A^c(y)\} \le \inf\{1-\mu_A^c(z) \mid z \in x \circ y\}.$$

Hence,

$$\min\{1-\mu_A^c(x), 1-\mu_A^c(y)\} \le 1-\sup\{\mu_A^c(z) \mid z \in x \circ y\}$$

which implies

$$\sup\{\mu_A^c(z) \mid z \in x \circ y\} \le 1-\min\{1-\mu_A^c(x), 1-\mu_A^c(y)\}.$$

Therefore,

$$\sup\{\mu_A^c(z) \mid z \in x \circ y\} \le \max\{\mu_A^c(x), \mu_A^c(y)\}$$

in this way the third condition of Definition 3.3.1 is verified.

Now, let $a, x \in H$ then there exist $y, z \in H$ such that $x \in a \circ y$, $x \in z \circ a$ and

$$\min\{\mu_A(a), \mu_A(x)\} \le \min\{\mu_A(y), \mu_A(z)\}.$$

So

$$\min\{1-\mu_A^c(a), 1-\mu_A^c(x)\} \le \min\{1-\mu_A^c(y), 1-\mu_A^c(z)\}.$$

Hence,

$$\max\{\mu_A^c(y), \mu_A^c(z)\} \le \max\{\mu_A^c(a), \mu_A^c(x)\},$$

and the fourth condition of Definition 3.3.1 is satisfied. ∎

Lemma 3.3.3. *If $A = (\mu_A, \lambda_A)$ is an IFS of H, then so is $\diamond A = (\lambda_A^c, \lambda_A)$.*

Proof. The proof is similar to the proof of Lemma 3.3.2. ∎

Combining the above two lemmas it is not difficult to see that the following theorem is valid.

Theorem 3.3.4. *$A = (\mu_A, \lambda_A)$ is an IFS of H if and only if $\square A$ and $\diamond A$ are $IFSs$ of H.*

Corollary 3.3.5. *$A = (\mu_A, \lambda_A)$ is an IFS of H if and only if μ_A and λ_A^c are fuzzy H_v-subgroups of H.*

In this section, we use the following notation instead of the notations in Definition 2.2.2. For any $t \in [0,1]$ and fuzzy set μ of H, the set

$$U(\mu; t) = \{x \in H \mid \mu(x) \ge t\} \text{ (respectively } L(\mu; t) = \{x \in H \mid \mu(x) \le t\})$$

is called an *upper* (respectively, *lower*) *t-level cut* of μ.

Theorem 3.3.6. *If $A = (\mu_A, \lambda_A)$ is an IFS of H, then the levels $U(\mu_A; t)$ and $L(\lambda_A; t)$ are H_v-subgroups of H for every $t \in Im(\mu_A) \cap Im(\lambda_A)$.*

Proof. Let $t \in Im(\mu_A) \cap Im(\lambda_A) \subseteq [0,1]$ and let $x, y \in U(\mu_A;t)$. Then, $\mu_A(x) \geq t$ and $\mu_A(y) \geq t$ and so $\min\{\mu_A(x), \mu_A(y)\} \geq t$. It follows from the first condition of Definition 3.3.1 that $\inf\{\mu_A(z) \mid z \in x \circ y\} \geq t$. Therefore, for all $z \in x \circ y$ we have $z \in U(\mu_A;t)$, so $x \circ y \subseteq U(\mu_A;t)$. Hence, for all $a \in U(\mu_A;t)$ we have $a \circ U(\mu_A;t) \subseteq U(\mu_A;t)$ and $U(\mu_A;t) \circ a \subseteq U(\mu_A;t)$. Now, let $x \in U(\mu_A;t)$. Then, there exist $y, z \in H$ such that $x \in a \circ y$, $x \in z \circ a$ and $\min\{\mu_A(x), \mu_A(a)\} \leq \min\{\mu(y), \mu(z)\}$. Since $x, a \in U(\mu_A;t)$, we have $t \leq \min\{\mu_A(x), \mu_A(a)\}$ and so $t \leq \min\{\mu_A(y), \mu_A(z)\}$ which implies that $y \in U(\mu_A;t)$, $z \in U(\mu_A;t)$ and these prove that $U(\mu_A;t) \subseteq a \circ U(\mu_A;t)$ and $U(\mu_A;t) \subseteq U(\mu_A;t) \circ a$. Hence, $a \circ U(\mu_A;t) = U(\mu_A;t) = U(\mu_A;t) \circ a$.

Now let $x, y \in L(\lambda_A;t)$. Then, $\lambda_A(x) \leq t$, $\lambda_A(y) \leq t$ and, in the consequence, $\max\{\lambda_A(x), \lambda_A(y)\} \leq t$. It follows from the third condition of Definition 3.3.1 that $\sup\{\lambda_A(z) \mid z \in x \circ y\} \leq t$. Therefore, for all $z \in x \circ y$ we have $z \in L(\lambda_A;t)$, so $x \circ y \subseteq L(\lambda_A;t)$. Hence, for all $a \in L(\lambda_A;t)$ we have $a \circ L(\lambda_A;t) \subseteq L(\lambda_A;t)$ and $L(\lambda_A;t) \circ a \subseteq L(\lambda_A;t)$. Now, let $x \in L(\lambda_A;t)$. Then, there exist $y, z \in H$ such that $x \in a \circ y$, $x \in z \circ a$ and $\max\{\lambda_A(y), \lambda_A(z)\} \leq \max\{\lambda_A(a), \lambda_A(x)\}$. Since $x, a \in L(\lambda_A;t)$, we have $\max\{\lambda_A(a), \lambda_A(x)\} \leq t$ and so $\max\{\lambda_A(y), \lambda_A(z)\} \leq t$ which implies $y \in L(\lambda_A;t)$, $z \in L(\lambda_A;t)$ and these prove that $L(\lambda_A;t) \subseteq a \circ L(\lambda_A;t)$ and $L(\lambda_A;t) \subseteq L(\lambda_A;t) \circ a$. Thus, $a \circ L(\lambda_A;t) = L(\lambda_A;t) = L(\lambda_A;t) \circ a$. ■

Theorem 3.3.7. *If $A = (\mu_A, \lambda_A)$ is an intuitionistic fuzzy set of H such that the non-empty sets $U(\mu_A;t)$ and $L(\lambda_A;t)$ are H_v-subgroups of H for all $t \in [0,1]$, then, $A = (\mu_A, \lambda_A)$ is an IFS of H.*

Proof. For $t \in [0,1]$, assume that $U(\mu_A;t) \neq \emptyset$ and $L(\lambda_A;t) \neq \emptyset$ are subhyperquasigroups of H. We must show that $A = (\mu_A, \lambda_A)$ satisfies the all conditions in Definition 3.3.1. Let $x, y \in H$, we put $t_0 = \min\{\mu_A(x), \mu_A(y)\}$ and $t_1 = \max\{\lambda_A(x), \lambda_A(y)\}$. Then, $x, y \in U(\mu_A;t_0)$ and $x, y \in L(\lambda_A;t_1)$. So $x \circ y \subseteq U(\mu_A;t_0)$ and $x \circ y \subseteq L(\lambda_A;t_1)$. Therefore, for all $z \in x \circ y$ we have $\mu_A(z) \geq t_0$ and $\lambda_A(z) \leq t_1$ which imply

$$\inf\{\mu_A(z) \mid z \in x \circ y\} \geq \min\{\mu_A(x), \mu_A(y)\}$$

and

$$\sup\{\lambda_A(z) \mid z \in x \circ y\} \leq \max\{\lambda_A(x), \lambda_A(y)\}$$

The conditions (1) and (3) of Definition 3.3.1 are verified.

Now, let $x, a \in H$. If $t_2 = \min\{\mu_A(a), \mu_A(x)\}$, then $a, x \in U(\mu_A;t_2)$. So there exist $y_1, z_1 \in U(\mu_A;t_2)$ such that $x \in a \circ y_1$ and $x \in z_1 \circ a$. Also we have $t_2 \leq \min\{\mu_A(y_1), \mu_A(z_1)\}$. Therefore, the condition (2) of Definition 3.3.1 is verified. If we put $t_3 = \max\{\lambda_A(a), \lambda_A(x)\}$, then $a, x \in L(\lambda_A;t_3)$. So there exist $y_2, z_2 \in L(\lambda_A;t_3)$ such that $x \in a \circ y_2$ and $x \in z_2 \circ a$ and we have $\max\{\lambda_A(y_2), \lambda_A(y_2)\} \leq t_3$, and so the condition (4) of Definition 3.3.1 is verified. This completes the proof. ■

Corollary 3.3.8. *Let K be an H_v-subgroup of an H_v-group $(H, \circ)$. If fuzzy sets μ and λ are defined on H by*

$$\mu(x) = \begin{cases} \alpha_0 & \text{if } x \in K, \\ \alpha_1 & \text{if } x \in H \setminus K, \end{cases} \qquad \lambda(x) = \begin{cases} \beta_0 & \text{if } x \in K, \\ \beta_1 & \text{if } x \in H \setminus K, \end{cases}$$

where $0 \leq \alpha_1 < \alpha_0$, $0 \leq \beta_0 < \beta_1$ and $\alpha_i + \beta_i \leq 1$ for $i = 0, 1$, then $A = (\mu, \lambda)$ is an IFS of H and $U(\mu; \alpha_0) = K = L(\lambda; \beta_0)$. □

Corollary 3.3.9. *Let χ_K be the characteristic function of a H_v-subgroup K of $(H, \circ)$. Then, $K = (\chi_K, \chi_K^c)$ is an IFS of H.* □

Theorem 3.3.10. *If $A = (\mu_A, \lambda_A)$ is an IFS of H, then for all $x \in H$ we have*

$$\mu_A(x) = \sup\{\alpha \in [0, 1] \mid x \in U(\mu_A; \alpha)\}$$

and

$$\lambda_A(x) = \inf\{\alpha \in [0, 1] \mid x \in L(\lambda_A; \alpha)\}.$$

Proof. Let $\delta = \sup\{\alpha \in [0, 1] \mid x \in U(\mu_A; \alpha)\}$ and let $\varepsilon > 0$ be given. Then, $\delta - \varepsilon < \alpha$ for some $\alpha \in [0, 1]$ such that $x \in U(\mu_A; \alpha)$. This means that $\delta - \varepsilon < \mu_A(x)$ so that $\delta \leq \mu_A(x)$ since ε is arbitrary.

We now show that $\mu_A(x) \leq \delta$. If $\mu_A(x) = \beta$, then $x \in U(\mu_A; \beta)$ and so

$$\beta \in \{\alpha \in [0, 1] \mid x \in U(\mu_A; \alpha)\}.$$

Hence,

$$\mu_A(x) = \beta \leq \sup\{\alpha \in [0, 1] \mid x \in U(\mu_A; \alpha)\} = \delta.$$

Therefore,

$$\mu_A(x) = \delta = \sup\{\alpha \in [0, 1] \mid x \in U(\mu_A; \alpha)\}.$$

Now let $\eta = \inf\{\alpha \in [0, 1] \mid x \in L(\lambda_A; \alpha)\}$. Then,

$$\inf\{\alpha \in [0, 1] \mid x \in L(\lambda_A; \alpha)\} < \eta + \varepsilon$$

for any $\varepsilon > 0$, and so $\alpha < \eta + \varepsilon$ for some $\alpha \in [0, 1]$ with $x \in L(\lambda_A; \alpha)$. Since $\lambda_A(x) \leq \alpha$ and ε is arbitrary, it follows that $\lambda_A(x) \leq \eta$.

To prove $\lambda_A(x) \geq \eta$, let $\lambda_A(x) = \zeta$. Then, $x \in L(\lambda_A; \zeta)$ and thus $\zeta \in \{\alpha \in [0, 1] \mid x \in L(\lambda_A; \alpha)\}$. Hence,

$$\inf\{\alpha \in [0, 1] \mid x \in L(\lambda_A; \alpha)\} \leq \zeta,$$

i.e. $\eta \leq \zeta = \lambda_A(x)$. Consequently

$$\lambda_A(x) = \eta = \inf\{\alpha \in [0, 1] \mid x \in L(\lambda_A; \alpha)\},$$

which completes the proof. ■

Theorem 3.3.11. *Let Ω be a non-empty finite subset of $[0,1]$. If $\{K_\alpha \mid \alpha \in \Omega\}$ is a collection of H_v-subgroups of H such that*

(1) $H = \bigcup_{\alpha \in \Omega} K_\alpha$,

(2) $\alpha > \beta \leftrightarrow K_\alpha \subset K_\beta$, *for all* $\alpha, \beta \in \Omega$,

then an intuitionistic fuzzy set $A = (\mu_A, \lambda_A)$ defined on H by

$$\mu_A(x) = \sup\{\alpha \in \Omega \mid x \in K_\alpha\} \quad \text{and} \quad \lambda_A(x) = \inf\{\alpha \in \Omega \mid x \in K_\alpha\}$$

is an IFS of H.

Proof. According to Theorem 3.3.7, it is sufficient to show that the non-empty sets $U(\mu_A;\alpha)$ and $L(\lambda_A;\beta)$ are H_v-subgroups of H. We show that $U(\mu_A;\alpha) = K_\alpha$. This holds, since

$$\begin{aligned} x \in U(\mu_A;\alpha) &\leftrightarrow \mu_A(x) \geq \alpha \\ &\leftrightarrow \sup\{\gamma \in \Omega \mid x \in K_\gamma\} \geq \alpha \\ &\leftrightarrow \exists \gamma_0 \in \Omega,\ x \in K_{\gamma_0},\ \gamma_0 \geq \alpha \\ &\leftrightarrow x \in K_\alpha \quad (\text{since } K_{\gamma_0} \subseteq K_\alpha). \end{aligned}$$

Now, we prove that $L(\lambda;\beta) \neq \emptyset$ is an H_v-subgroup of H. We have

$$\begin{aligned} x \in L(\lambda_A;\beta) &\leftrightarrow \lambda_A(x) \leq \beta \\ &\leftrightarrow \inf\{\gamma \in \Omega \mid x \in K_\gamma\} \leq \beta \\ &\leftrightarrow \exists \gamma_0 \in \Omega,\ x \in K_{\gamma_0},\ \gamma_0 \leq \beta \\ &\leftrightarrow x \in \bigcup_{\gamma \leq \beta} K_\gamma \end{aligned}$$

and hence $L(\lambda_A;\beta) = \bigcup_{\gamma \leq \beta} K_\gamma$. It is not difficult to see that the union of any family of increasing H_v-subgroups of a given hyperquasigroup is an H_v-subgroup. This completes the proof. ∎

Let $\alpha \in [0,1]$ be fixed and let $IFS(H)$ be the family of all intuitionistic fuzzy H_v-subgroups of an H_v-group H. For any $A = (\mu_A, \lambda_A)$ and $B = (\mu_B, \lambda_B)$ from $IFS(H)$ we define two binary relations $\mathfrak{U}^\alpha$ and $\mathfrak{L}^\alpha$ on $IFS(H)$ as follows:

$$(A,B) \in \mathfrak{U}^\alpha \leftrightarrow U(\mu_A;\alpha) = U(\mu_B;\alpha)$$

and

$$(A,B) \in \mathfrak{L}^\alpha \leftrightarrow L(\lambda_A;\alpha) = L(\lambda_B;\alpha).$$

These two relations $\mathfrak{U}^\alpha$ and $\mathfrak{L}^\alpha$ are equivalence relations. Hence, $IFS(H)$ can be divided into the equivalence classes of $\mathfrak{U}^\alpha$ and $\mathfrak{L}^\alpha$, denoted by $[A]_{\mathfrak{U}^\alpha}$ and $[A]_{\mathfrak{L}^\alpha}$ for any $A = (\mu_A, \lambda_A) \in IFS(H)$, respectively. The corresponding the quotient sets will be denoted as $IFS(H)/\mathfrak{U}^\alpha$ and $IFS(H)/\mathfrak{L}^\alpha$, respectively.

For the family $S(H)$ of all subhyperquasigroups of H we define two maps U_α and L_α from $IFS(H)$ to $S(H) \cup \{\emptyset\}$ putting

$$U_\alpha(A) = U(\mu_A;\alpha) \quad \text{and} \quad L_\alpha(A) = L(\lambda_A;\alpha)$$

for each $A = (\mu_A, \lambda_A) \in IFS(H)$.

It is not difficult to see that these maps are well-defined.

Lemma 3.3.12. *For any $\alpha \in (0,1)$ the maps U_α and L_α are surjective.*

Proof. Let $\mathbf{0}$ and $\mathbf{1}$ be fuzzy sets on H defined by $\mathbf{0}(x) = 0$ and $\mathbf{1}(x) = 1$ for all $x \in H$. Then, $\mathbf{0}_\sim = (\mathbf{0}, \mathbf{1}) \in IFS(H)$ and $U_\alpha(\mathbf{0}_\sim) = L_\alpha(\mathbf{0}_\sim) = \emptyset$ for any $\alpha \in (0,1)$. Moreover for any $K \in S(H)$ we have $K_\sim = (\chi_K, \chi_K^c) \in IFS(H)$, $U_\alpha(K_\sim) = U(\chi_K; \alpha) = K$ and $L_\alpha(K_\sim) = L(\chi_K^c; \alpha) = K$. Hence, U_α and L_α are surjective. ■

Theorem 3.3.13. *For any $\alpha \in (0,1)$ the sets $IFS(H)/\mathfrak{U}^\alpha$ and $IFS(H)/\mathfrak{L}^\alpha$ are equipotent to $S(H) \cup \{\emptyset\}$.*

Proof. Let $\alpha \in (0,1)$. Putting $U_\alpha^*([A]_{\mathfrak{U}^\alpha}) = U_\alpha(A)$ and $L_\alpha^*([A]_{\mathfrak{L}^\alpha}) = L_\alpha(A)$ for any $A = (\mu_A, \lambda_A) \in IFS(H)$, we obtain two maps

$$U_\alpha^* : IFS(H)/\mathfrak{U}^\alpha \to S(H) \cup \{\emptyset\} \quad \text{and} \quad L_\alpha^* : IFS(H)/\mathfrak{L}^\alpha \to S(H) \cup \{\emptyset\}.$$

If $U(\mu_A; \alpha) = U(\mu_B; \alpha)$ and $L(\lambda_A; \alpha) = L(\lambda_B; \alpha)$ for some $A = (\mu_A, \lambda_A)$ and $B = (\mu_B, \lambda_B)$ from $IFS(H)$, then $(A, B) \in \mathfrak{U}^\alpha$ and $(A, B) \in \mathfrak{L}^\alpha$, whence $[A]_{\mathfrak{U}^\alpha} = [B]_{\mathfrak{U}^\alpha}$ and $[A]_{\mathfrak{L}^\alpha} = [B]_{\mathfrak{L}^\alpha}$, which means that U_α^* and L_α^* are injective.

To show that the maps U_α^* and L_α^* are surjective, let $K \in S(H)$. Then, for $K_\sim = (\chi_K, \chi_K^c) \in IFS(H)$ we have $U_\alpha^*([K_\sim]_{\mathfrak{U}^\alpha}) = U(\chi_K; \alpha) = K$ and $L_\alpha^*([K_\sim]_{\mathfrak{L}^\alpha}) = L(\chi_K^c; \alpha) = K$. Also $\mathbf{0}_\sim = (\mathbf{0}, \mathbf{1}) \in IFS(H)$. Moreover $U_\alpha^*([\mathbf{0}_\sim]_{\mathfrak{U}^\alpha}) = U(\mathbf{0}; \alpha) = \emptyset$ and $L_\alpha^*([\mathbf{0}_\sim]_{\mathfrak{L}^\alpha}) = L(\mathbf{1}; \alpha) = \emptyset$. Hence, U_α^* and L_α^* are surjective. ■

Now for any $\alpha \in [0,1]$ we obtain the new relation $\mathfrak{R}^\alpha$ on $IFS(H)$ putting:

$$(A, B) \in \mathfrak{R}^\alpha \leftrightarrow U(\mu_A; \alpha) \cap L(\lambda_A; \alpha) = U(\mu_B; \alpha) \cap L(\lambda_B; \alpha),$$

where $A = (\mu_A, \lambda_A)$ and $B = (\mu_B, \lambda_B)$. Obviously $\mathfrak{R}^\alpha$ is an equivalence relation.

Lemma 3.3.14. *The map $I_\alpha : IFS(H) \to S(H) \cup \{\emptyset\}$ defined by*

$$I_\alpha(A) = U(\mu_A; \alpha) \cap L(\lambda_A; \alpha),$$

where $A = (\mu_A, \lambda_A)$, is surjective for any $\alpha \in (0,1)$.

Proof. Indeed, if $\alpha \in (0,1)$ is fixed, then for $\mathbf{0}_\sim = (\mathbf{0}, \mathbf{1}) \in IFS(H)$ we have

$$I_\alpha(\mathbf{0}_\sim) = U(\mathbf{0}; \alpha) \cap L(\mathbf{1}; \alpha) = \emptyset,$$

and for any $K \in S(H)$ there exists $K_\sim = (\chi_K, \chi_K^c) \in IFS(H)$ such that $I_\alpha(K_\sim) = U(\chi_K; \alpha) \cap L(\chi_K^c; \alpha) = K$. ■

Theorem 3.3.15. *For any $\alpha \in (0,1)$ the quotient set $IFS(H)/\mathfrak{R}^\alpha$ is equipotent to $S(H) \cup \{\emptyset\}$.*

Proof. Let $I_\alpha^* : IFS(H)/\mathfrak{R}^\alpha \to S(H) \cup \{\emptyset\}$, where $\alpha \in (0,1)$, be defined by the formula:

$$I_\alpha^*([A]_{\mathfrak{R}^\alpha}) = I_\alpha(A) \quad \text{for each} \quad [A]_{\mathfrak{R}^\alpha} \in IFS(H)/\mathfrak{R}^\alpha.$$

If $I_\alpha^*([A]_{\mathfrak{R}^\alpha}) = I_\alpha^*([B]_{\mathfrak{R}^\alpha})$ for some $[A]_{\mathfrak{R}^\alpha}$, $[B]_{\mathfrak{R}^\alpha} \in IFS(H)/\mathfrak{R}^\alpha$, then

$$U(\mu_A;\alpha) \cap L(\lambda_A;\alpha) = U(\mu_B;\alpha) \cap L(\lambda_B;\alpha),$$

which implies $(A,B) \in \mathfrak{R}^\alpha$ and, in the consequence, $[A]_{\mathfrak{R}^\alpha} = [B]_{\mathfrak{R}^\alpha}$. Thus, I_α^* is injective.

It is also onto because $I_\alpha^*(\mathbf{0}_\sim) = I_\alpha(\mathbf{0}_\sim) = \emptyset$ for $\mathbf{0}_\sim = (\mathbf{0},\mathbf{1}) \in IFS(H)$, and $I_\alpha^*(K_\sim) = I_\alpha(K) = K$ for $K \in S(H)$ and $K_\sim = (\chi_K, \chi_K^c) \in IFS(H)$. ∎

3.4 T-Product of Fuzzy H_v-Subgroups

In this section, by using the definition of t-norm (Definition 2.3.20), we study the notion of fuzzy H_v-subgroup of an H_v-group under a t-norm T. The main references for this section are [59, 62, 63].

Definition 3.4.1. Let $(H,\cdot)$ be an H_v-group and let μ be a fuzzy subset of H. Then μ is said to be a *fuzzy H_v-subgroup of H under a t-norm T*, if the following axioms hold.

(1) $T(\mu(x),\mu(y)) \leq \bigwedge_{\alpha \in x \circ y} \{\mu(\alpha)\}$, for all $x,y \in H$.

(2) For all $x,a \in H$ there exists $y \in H$ such that $x \in a \circ y$ and

$$T(\mu(a),\mu(x)) \leq \mu(y).$$

(3) For all $x,a \in H$ there exists $z \in H$ such that $x \in z \circ a$ and

$$T(\mu(a),\mu(x)) \leq \mu(z).$$

Let H_1 and H_2 be two H_v-groups. Then, in $H_1 \times H_2$ we can define a *hyperproduct* as follows:

$$(x_1,y_1) * (x_2,y_2) = \{(a,b) \mid a \in x_1 \circ x_2, b \in y_1 \circ y_2\}$$

and we call this the *direct H_v-product.* It is easy to see that $H_1 \times H_2$ equipped with the direct H_v-product becomes an H_v-group.

Suppose that T_1 and T_2 are two t-norms. Then, T_2 is said to *dominate* T_1 and write $T_1 << T_2$ if for all $a,b,c,d \in [0,1]$,

$$T_1(T_2(a,c),T_2(b,d)) \leq T_2(T_1(a,b),T_1(c,d))$$

and T_1 is said *weaker than* T_2 or T_2 is *stronger than* T_1 and write $T_1 \leq T_2$ if for all $x,y \in [0,1]$,

$$T_1(x,y) \leq T_2(x,y).$$

Definition 3.4.2. Let H_1, H_2 be two H_v-groups and μ, λ be fuzzy H_v-subgroups of H_1, H_2 under t-norm T, respectively. The T-*product* of μ, λ is defined to be the fuzzy subset $\mu \times \lambda$ of $H_1 \times H_2$ with

$$(\mu \times \lambda)(x, y) = T(\mu(x), \lambda(y)), \text{ for all } (x, y) \in H_1 \times H_2.$$

Proposition 3.4.3. *Let H_1, H_2 be H_v-groups and μ, λ be fuzzy H_v-subgroups of H_i under t-norms $T_i, i = 1, 2$ respectively and T' be a t-norm such that $T' \leq T_1, T_2$ and let T be a t-norm such that $T' << T$. Then, T-product $\mu \times \lambda$ is a fuzzy H_v-subgroup of $H_1 \times H_2$ under t-norm T'.*

Proof. Let $x, y \in H_1 \times H_2$ such that $x = (x_1, x_2), y = (y_1, y_2)$. For every $\alpha = (\alpha_1, \alpha_2) \in x * y = (x_1, x_2) * (y_1, y_2)$ we have

$$\begin{aligned}(\mu \times \lambda)(\alpha) &= (\mu \times \lambda)(\alpha_1, \alpha_2) = T(\mu(\alpha_1), \lambda(\alpha_2)) \\ &\geq T(T_1(\mu(x_1), \mu(y_1)), T_2(\lambda(x_2), \lambda(y_2))) \\ &\geq T(T'(\mu(x_1), \mu(y_1)), T'(\lambda(x_2), \lambda(y_2))) \\ &\geq T'(T(\mu(x_1), \lambda(x_2)), T(\mu(y_1), \lambda(y_2))) \quad \text{since } T >> T' \\ &= T'((\mu \times \lambda)(x_1, x_2), (\mu \times \lambda)(y_1, y_2)).\end{aligned}$$

Therefore, the first condition of Definition 3.4.1 is satisfied. Now, we prove the second condition of Definition 3.4.1 as follows. For every (x_1, x_2) and (a_1, a_2) in $H_1 \times H_2$ there exists (y_1, y_2) in $H_1 \times H_2$ such that

$$T_1(\mu(x_1), \mu(a_1)) \leq \mu(y_1) \ , \ T_2(\lambda(x_2), \lambda(a_2)) \leq \lambda(y_2).$$

Therefore, we have $(x_1, x_2) \in (a_1, a_2) * (y_1, y_2)$ and

$$\begin{aligned}(\mu \times \lambda)(y_1, y_2) &= T(\mu(y_1), \lambda(y_2)) \\ &\geq T(T_1(\mu(x_1), \mu(a_1)), T_2(\lambda(x_2), \lambda(a_2))) \\ &\geq T(T'(\mu(x_1), \mu(a_1)), T'(\lambda(x_2), \lambda(a_2))) \\ &\geq T'(T(\mu(x_1), \lambda(x_2)), T(\mu(a_1), \lambda(a_2))) \\ &= T'((\mu \times \lambda)(x_1, x_2), (\mu \times \lambda)(a_1, a_2)).\end{aligned}$$

The proof of the third condition of Definition 3.4.1 is similar to the proof of second condition. ■

Corollary 3.4.4. *Let H_1, H_2 be two H_v-groups and let μ, λ be fuzzy H_v-subgroups of H_1, H_2 under t-norm T, respectively. Then, $\mu \times \lambda$ is a fuzzy H_v-subgroup of $H_1 \times H_2$ under t-norm T.*

In the following result the T-product is considered only for the t-norm minimum.

Corollary 3.4.5. *Let μ and λ be fuzzy H_v-subgroups of H_1 and H_2 respectively, then $(\mu \times \lambda)_t = \mu_t \times \lambda_t$.*

Definition 3.4.6. Let μ be a fuzzy H_v-subgroup of $H_1 \times H_2$ and let (x_1, x_2), $(a_1, a_2) \in H_1 \times H_2$. Then, by the second condition of Definition 3.4.1, there exists $(y_1, y_2) \in H_1 \times H_2$ such that $(x_1, x_2) \in (a_1, a_2) \circ (y_1, y_2)$ and

$$T(\mu(x_1, x_2), \mu(a_1, a_2)) \leq \mu(y_1, y_2).$$

Now, if for every $r_1, s_1 \in H_1$ there exists $t_1 \in H_1$ such that $(r_1, x_2) \in (s_1, a_2) * (t_1, y_2)$ and

$$T(\mu(r_1, x_2), \mu(s_1, a_2)) \leq \mu(t_1, y_2),$$

and for every $r_2, s_2 \in H_2$ there exists $t_2 \in H_2$ such that $(x_1, r_2) \in (a_1, s_2) * (y_1, t_2)$ and

$$T(\mu(x_1, r_2), \mu(a_1, s_2)) \leq \mu(y_1, t_2),$$

then we say that μ satisfies the *strong left fuzzy reproduction axiom*. Similarly, we can define the *strong right fuzzy reproduction axiom*. μ is called a *strong fuzzy H_v-subgroup* of H if μ satisfies the strong left (and right) reproduction axiom.

Each subset A of H may be regarded as a fuzzy subset by identifying it with its characteristic function χ_A. Let H_1, H_2 be H_v-groups and A be an H_v-subgroup of $H_1 \times H_2$ then χ_A is a strong fuzzy H_v-subgroup of $H_1 \times H_2$ under t-norm min.

Lemma 3.4.7. [141] *If T is the t-norm min and X, Y are any index sets, then*

$$\bigvee_{i \in X, j \in Y} \{T(x_i, y_j)\} = T\Big(\bigvee_{i \in X} \{x_i\}, \bigvee_{j \in Y} \{y_j\}\Big).$$

Proposition 3.4.8. *Let H_1, H_2 be two H_v-groups and μ be a strong fuzzy H_v-subgroup of $H_1 \times H_2$ under t-norm T=min. Then, $\mu_i, i = 1, 2$ is a fuzzy H_v-subgroup of H_i, $i = 1, 2$ respectively, where*

$$\mu_1(x) = \bigvee_{a \in H_2} \{\mu(x, a)\} \text{ and } \mu_2(y) = \bigvee_{b \in H_1} \{\mu(b, y)\}.$$

Proof. We show that μ_1 is a fuzzy H_v-subgroup of H_1. Suppose that $x, y \in H_1$. Then, for every $\alpha \in x \circ y$ we have $\mu_1(\alpha) = \bigvee_{a \in H_2} \{\mu(\alpha, a)\}$. For every $a \in H_2$ there exist $r_a, s_a \in H_2$ such that $a \in a \circ r_a$ and $a \in s_a \circ a$. Now we have

$$\begin{aligned}
\mu_1(\alpha) &\geq \bigvee_{a \in H_2} \Big\{T\left(\mu(x, a), \mu(y, r_a)\right)\Big\} \\
&\geq \bigvee_{a \in H_2} \Big\{T(\mu(x, a), \bigvee_{b \in H_2} \{\mu(y, b)\})\Big\} \\
&= T\Big(\bigvee_{a \in H_2} \{\mu(x, a)\}, \bigwedge_{b \in H_2} \{\mu(y, b)\}\Big) \\
&= T(\mu_1(x), \bigvee_{b \in H_2} \{\mu(y, b)\})
\end{aligned}$$

and also we have

$$\begin{aligned}\mu_1(\alpha) &\geq \bigvee_{a\in H_2} \{T(\mu(x,s_a),\mu(y,a))\}\\ &\geq \bigvee_{a\in H_2}\Big\{T(\bigwedge_{c\in H_2}\{\mu(x,c)\},\mu(y,a))\Big\}\\ &= T\Big(\bigwedge_{c\in H_2}\{\mu(x,c)\},\ \bigvee_{a\in H_2}\{\mu(y,a)\}\Big)\\ &= T\Big(\bigwedge_{c\in H_2}\{\mu(x,c)\},\mu_1(y)\Big).\end{aligned}$$

Therefore,

$$\begin{aligned}\mu_1(\alpha) &\geq T\Big(\mu_1(x),\ \bigwedge_{b\in H_2}\{\mu(y,b)\}\Big)\vee T\Big(\bigvee_{c\in H_2}\{\mu(x,c)\},\mu_1(y)\Big)\\ &= T\Big(\mu_1(x)\vee(\bigvee_{c\in H_2}\{\mu(x,c)\}),(\bigwedge_{b\in H_2}\{\mu(y,b)\})\vee\mu_1(y)\Big)\\ &\geq T(\mu_1(x),\mu_1(y)).\end{aligned}$$

Now, let $x,a\in H_1$. Then, for every $r,s\in H_2$ there exists $(y,\gamma_{r,s})\in H_1\times H_2$ such that $(x,r)\in(a,s)\circ(y,\gamma_{r,s})$ and $T(\mu(x,r),\mu(a,s))\leq\mu(y,\gamma_{r,s})$. Therefore,

$$\begin{aligned}T(\mu_1(x),\mu_1(a)) &= T(\bigvee_{r\in H_2}\{\mu(x,r)\},\ \bigvee_{s\in H_2}\{\mu(a,s)\})\\ &= \bigvee_{r\in H_2,s\in H_2}\{T(\mu(x,r),\mu(a,s))\}\\ &\leq \bigvee_{r\in H_2,s\in H_2}\{\mu(y,\gamma_{r,s})\}\\ &\leq \bigvee_{\gamma\in H_2}\{\mu(y,\gamma)\}\\ &= \mu_1(y).\end{aligned}$$

The proof of condition (3) of Definition 3.4.1 is similar to condition (2). Therefore, μ_1 is a fuzzy H_v-subgroup of H_1. Similarly, we can prove that μ_2 is a fuzzy H_v-subgroup of H_2. ∎

Corollary 3.4.9. *Let H_1,H_2 be two H_v-groups and μ,λ be fuzzy subsets of H_1, H_2, respectively. If $\bigvee_{x\in H_1}\{\mu(x)\}=\bigvee_{y\in H_2}\{\lambda(y)\}=1$ and $\mu\times\lambda$ is a strong fuzzy H_v-subgroup of $H_1\times H_2$, then μ and λ are fuzzy H_v-subgroups of H_1, H_2, respectively.*

Theorem 3.4.10. *Let H_1, H_2 be two H_v-groups. Let β_1^*, β_2^* and β^* be fundamental equivalence relations on H_1, H_2 and $H_1\times H_2$, respectively. Then,*

$$(H_1\times H_2)/\beta^*\cong H_1/\beta_1^*\times H_2/\beta_2^*.$$

Corollary 3.4.11. *Let β_1^*, β_2^* and β^* be fundamental equivalence relations on H_1, H_2 and $H_1\times H_2$, respectively. Then,*

$$(x_1,y_1)\beta^*(x_2,y_2)\ \Leftrightarrow\ x_1\beta_1^*x_2,\ \ y_1\beta_2^*y_2,$$

for all $(x_i,y_i)\in H_1\times H_2$, $i=1,2$.

Definition 3.4.12. Let H be an H_v-group and μ be a fuzzy subset of H. The fuzzy subset μ_{β^*} on H/β^* is defined as follows:

$$\mu_{\beta^*} : H/\beta^* \to [0,1]$$
$$\mu_{\beta^*}(\beta^*(x)) = \bigvee_{a\in\beta^*(x)} \{\mu(a)\}.$$

Theorem 3.4.13. *Let T be the t-norm min, H be an H_v-group and μ be a fuzzy H_v-subgroup of H under a t-norm T. Then, μ_{β^*} is a fuzzy subgroup of H/β^* under the t-norm T.*

Proof. First, we considering H/β^* as a hypergroup and we show that μ_{β^*} is a fuzzy H_v-subgroup of H/β^* under T. Let $\beta^*(x)$ and $\beta^*(y)$ be two elements of H/β^*. We can write:

$$T(\mu_{\beta^*}(\beta^*(x)), \mu_{\beta^*}(\beta^*(y))) = T\left(\bigvee_{a\in\beta^*(x)} \{\mu(a)\}, \bigvee_{b\in\beta^*(y)} \{\mu(b)\}\right)$$
$$= \bigvee_{\substack{a\in\beta^*(x)\\ b\in\beta^*(y)}} \{T(\mu(a),\mu(b))\} \leq \bigvee_{\substack{a\in\beta^*(x)\\ b\in\beta^*(y)}} \left\{\bigwedge_{\alpha\in a\circ b} \{\mu(\alpha)\}\right\}$$
$$\leq \bigvee_{\substack{a\in\beta^*(x)\\ b\in\beta^*(y)}} \left\{\bigvee_{\alpha\in a\circ b} \{\mu(\alpha)\}\right\} \leq \bigvee_{\substack{a\in\beta^*(x)\\ b\in\beta^*(y)}} \left\{\bigvee_{\alpha\in\beta^*(a\circ b)} \{\mu(\alpha)\}\right\}$$
$$= \bigvee_{\substack{a\in\beta^*(x)\\ b\in\beta^*(y)}} \{\mu_{\beta^*}(\beta^*(a\circ b))\} = \mu_{\beta^*}(\beta^*(a\circ b)) = \mu_{\beta^*}(\beta^*(a)\odot\beta^*(b)).$$

Therefore, the first condition of Definition 3.4.1 is satisfied.

Now, suppose that $\beta^*(x)$ and $\beta^*(a)$ are two arbitrary elements of H/β^*. Since μ is a T-fuzzy H_v-subgroup of H, it follows that for all $r \in \beta^*(a), s \in \beta^*(x)$ there exists $y_{r,s} \in H$ such that $r \in s \circ y_{r,s}$ and $T(\mu(r),\mu(s)) \leq \mu(y_{r,s})$. From $r \in s\circ y_{r,s}$ it follows that $\beta^*(s)\odot\beta^*(y_{r,s}) = \{\beta^*(r)\}$ which implies that $\beta^*(x)\odot\beta^*(y_{r,s}) = \{\beta^*(a)\}$.

Now, if $r_1 \in \beta^*(a)$ and $s_1 \in \beta^*(x)$, then there exists $y_{r_1,s_1} \in H$ such that $\beta^*(s_1)\odot\beta^*(y_{r_1,s_1}) = \{\beta^*(r_1)\}$ and since $\beta^*(r_1) = \beta^*(r)$ we get $\beta^*(s_1)\odot\beta^*(y_{r_1,s_1}) = \beta^*(s)\odot\beta^*(y_{r,s})$ and therefore $\beta^*(y_{r,s}) = \beta^*(y_{r_1,s_1})$. So, all the $y_{r,s}$ satisfying $T(\mu(r),\mu(s)) \leq \mu(y_{r,s})$ belong to the same equivalence class. Now we have :

$$T(\mu_{\beta^*}(\beta^*(a)), \mu_{\beta^*}(\beta^*(x))) = T(\textstyle\bigvee_{r\in\beta^*(a)}\{\mu(r)\}, \bigvee_{s\in\beta^*(x)}\{\mu(s)\})$$
$$= \bigvee_{\substack{r\in\beta^*(a)\\ s\in\beta^*(x)}} \{T(\mu(r),\mu(s))\} \leq \bigvee_{\substack{r\in\beta^*(a)\\ s\in\beta^*(x)}} \{\mu(y_{r,s})\}$$
$$\leq \bigvee_{y\in\beta^*(y_{r,s})} \{\mu(y)\} = \mu_{\beta^*}(\beta^*(y_{r,s})),$$

in this way the second condition of Definition 3.4.1 is satisfied. Now, we have

(1) $\mu_{\beta^*}(\beta^*(x)) \wedge \mu_{\beta^*}(\beta^*(y)) \leq \bigwedge_{\beta^*(\alpha)\in\beta^*(x)\odot\beta^*(y)} \{\mu_{\beta^*}(\beta^*(\alpha))\},$

$$\forall \beta^*(x), \beta^*(y) \in H/\beta^*$$

(2) for all $\beta^*(x), \beta^*(a) \in H/\beta^*$ there exists $\beta^*(y) \in H/\beta^*$ such that $\beta^*(x) = \beta^*(a)\odot\beta^*(y)$ and $\mu_{\beta^*}(\beta^*(x)) \wedge \mu_{\beta^*}(\beta^*(a)) \leq \mu_{\beta^*}(\beta^*(y))$.

Now, for all $\beta^*(x)$ in H/β^* we prove that $\mu_{\beta^*}(\beta^*(x)) \leq \mu_{\beta^*}(\beta^*(x)^{-1})$. Since $\beta^*(x) \in H/\beta^*$, by considering $\beta^*(a) = \beta^*(x)$ which is obtained from the second condition there exists $\beta^*(y_1)$ in H/β^* such that $\beta^*(x) = \beta^*(x) \odot \beta^*(y_1)$ and $\mu_{\beta^*}(\beta^*(x)) \wedge \mu_{\beta^*}(\beta^*(x)) \leq \mu_{\beta^*}(\beta^*(y_1))$. From $\beta^*(x) = \beta^*(x) \odot \beta^*(y_1)$ we obtain $\omega_H = \beta^*(y_1)$, where ω_H denotes the unit of the group H/β^*. Therefore,

$$\mu_{\beta^*}(\beta^*(x)) \leq \mu_{\beta^*}(\omega_H). \tag{3.3}$$

Now, considering $\beta^*(x), \omega_H$ in H/β^*, by the condition (2) above there exists $\beta^*(y_2)$ in H/β^* such that $\omega_H = \beta^*(x) \odot \beta^*(y_2)$ and $\mu_{\beta^*}(\omega_H) \wedge \mu_{\beta^*}(\beta^*(x)) \leq \mu_{\beta^*}(\beta^*(y_2))$. From $\omega_H = \beta^*(x) \odot \beta^*(y_2)$ we obtain $\beta^*(y_2) = \beta^*(x)^{-1}$. Thus,

$$\mu_{\beta^*}(\omega_H) \wedge \mu_{\beta^*}(\beta^*(x)) \leq \mu_{\beta^*}(\beta^*(x)^{-1}). \tag{3.4}$$

By (5.17) and (5.18) the inequality $\mu_{\beta^*}(\beta^*(x)) \leq \mu_{\beta^*}(\beta^*(x)^{-1})$ is obtained. ■

Theorem 3.4.14. *Suppose that*

(1) H_1, H_2 *are* H_v*-groups.*
(2) β_1^*, β_2^* *and* β^* *are fundamental equivalence relations on* H_1, H_2 *and* $H_1 \times H_2$, *respectively.*
(3) T *is the t-norm min.*
(4) μ *is a fuzzy* H_v*-subgroup of* H_1 *under t-norm* T.
(5) λ *is a fuzzy* H_v*-subgroup of* H_2 *under t-norm* T.

Then, we have

$$(\mu \times \lambda)_{\beta^*} = \mu_{\beta_1^*} \times \lambda_{\beta_2^*}.$$

Proof. Clearly, by the conditions (4) and (5) we get $\mu \times \lambda$ is a fuzzy H_v-subgroup of $H_1 \times H_2$ under t-norm T. Then, by Theorem 3.4.13 we have $(\mu \times \lambda)_{\beta^*}$ is a fuzzy subgroup of the group $(H_1 \times H_2)/\beta^*$ under t-norm T.

Now, assume that $x \in H_1$ and $y \in H_2$. Then,

$$\begin{aligned}
(\mu \times \lambda)_{\beta^*}(\beta^*(x, y)) &= \bigvee_{(a,b) \in \beta^*(x,y)} \{(\mu \times \lambda)(a, b)\} \\
&= \bigvee_{(a,b) \in \beta^*(x,y)} \{T(\mu(a), \lambda(b))\} \\
&= \bigvee_{\substack{a \in \beta_1^*(x) \\ b \in \beta_2^*(y)}} \{T(\mu(a), \lambda(b))\} \\
&= T\Big(\bigvee_{a \in \beta_1^*(x)} \{\mu(a)\}, \bigvee_{b \in \beta_2^*(y)} \{\lambda(b)\}\Big) \\
&= T(\mu_{\beta_1^*}(\beta_1^*(x)), \lambda_{\beta_2^*}(\beta_2^*(y))) \\
&= (\mu_{\beta_1^*} \times \lambda_{\beta_2^*})(\beta_1^*(x), \beta_2^*(y)).
\end{aligned}$$

■

Theorem 3.4.15. *Suppose that*

(1) H_1, H_2 *are* H_v*-groups.*

(2) β_1^*, β_2^* *and* β^* *are fundamental equivalence relations on* H_1, H_2 *and* $H_1 \times H_2$ *respectively.*
(3) T *is the t-norm min.*
(4) μ *is a strong fuzzy* H_v*-subgroup of* $H_1 \times H_2$ *under t-norm* T.
(5) $\mu_1(x) = \bigvee_{a\in H_2}\{\mu(x,a)\}$, $\mu_2(y) = \bigvee_{b\in H_1}\{\mu(b,y)\}$.

Then, $(\mu_1)_{\beta_1^*}$ *is a fuzzy subgroup of* H_1/β_1^* *under t-norm* T *and* $(\mu_2)_{\beta_2^*}$ *is a fuzzy subgroup of* H_2/β_2^* *under t-norm* T*. Moreover, we have*

$$\mu_{\beta^*} \subseteq (\mu_1)_{\beta_1^*} \times (\mu_2)_{\beta_2^*}.$$

Proof. By Proposition 3.4.8 and Theorem 3.4.13 we get $(\mu_i)_{\beta_i^*}$ is a fuzzy subgroup of H_i/β_i^* under t-norm T, i=1,2. Also, we have

$$\begin{aligned}((\mu_1)_{\beta_1^*} \times (\mu_2)_{\beta_2^*})(\beta_1^*(x), \beta_2^*(y)) &= T((\mu_1)_{\beta_1^*}(\beta_1^*(x)), (\mu_2)_{\beta_2^*}(\beta_2^*(y)))\\ &= T(\bigvee_{a\in\beta_1^*(x)}\{\mu_1(a)\}, \bigvee_{b\in\beta_2^*(y)}\{\mu_2(b)\})\\ &= \bigvee_{\substack{a\in\beta_1^*(x)\\ b\in\beta_2^*(y)}}\{T(\mu_1(a), \mu_2(b))\}\\ &\geq \bigvee_{\substack{a\in\beta_1^*(x)\\ b\in\beta_2^*(y)}}\{T(\mu(a,b), \mu(a,b))\}\\ &\geq \bigvee_{(a,b)\in\beta^*(x,y)}\{\mu(a,b)\}\\ &= \mu_{\beta^*}(\beta^*(x,y)).\end{aligned}$$

and by Corollary 3.4.11 the proof of theorem is completed. ■

The kernel of the canonical map $\varphi : H \to H/\beta^*$ is called the core of H and is denoted by ω_H. Here we also denote by ω_H the unit element of H/β^*.

Theorem 3.4.16. *Let* H_1, H_2 *be two* H_v*-groups and* T *be the t-norm min. Suppose that* μ *is a strong fuzzy* H_v*-subgroup of* $H_1 \times H_2$ *under t-norm* T*. Then,* λ_i *is a fuzzy subgroup of* H_i/β_i^* *under t-norm* T, $i = 1, 2$, *where*

$$\begin{aligned}\lambda_1(\beta_1^*(x)) &= \mu_{\beta^*}(\beta_1^*(x), \omega_{H_2}), \text{ for all } \beta_1^*(x) \in H_1/\beta_1^*,\\ \lambda_2(\beta_2^*(y)) &= \mu_{\beta^*}(\omega_{H_1}, \beta_2^*(y)), \text{ for all } \beta_2^*(y) \in H_2/\beta_2^*.\end{aligned}$$

Moreover, we have

$$\lambda_1 \times \lambda_2 \subseteq \mu_{\beta^*}.$$

Proof. The proof is straightforward and omitted. ■

Corollary 3.4.17. *We consider the same hypothesis as in the above theorem.*

(1) *If* $\mu_1(x) = \bigvee_{a\in H_2}\{\mu(x,a)\} = \bigvee_{a\in\omega_{H_2}}\{\mu(x,a)\}$, *then* $\lambda_1 = (\mu_1)_{\beta_1^*}$.
(2) *If* $\mu_2(y) = \bigvee_{b\in H_1}\{\mu(b,y)\} = \bigvee_{b\in\omega_{H_1}}\{\mu(b,y)\}$, *then* $\lambda_2 = (\mu_2)_{\beta_2^*}$.

From (1) and (2) we get $\mu_{\beta^*} = (\mu_1)_{\beta_1^*} \times (\mu_2)_{\beta_2^*}$.

In the following theorems, assume that β_A^*, β_B^* are fundamental equivalence relations on A, B, respectively.

Theorem 3.4.18. *Let μ, λ be fuzzy subsets of H_v-groups A, B, respectively. If $\mu \times \lambda$ is a fuzzy H_v-subgroup of $A \times B$, then at least one of the following two statements must be hold.*

(1) $\lambda_{\beta^*_B}(\omega_B) \geq \mu_{\beta^*_A}(\beta^*_A(a))$, *for all* $a \in A$.
(2) $\mu_{\beta^*_A}(\omega_A) \geq \lambda_{\beta^*_B}(\beta^*_B(b))$, *for all* $b \in B$.

Proof. Suppose that $\mu \times \lambda$ is a fuzzy H_v-subgroup of $A \times B$. Then, by Theorem 3.4.13 $(\mu \times \lambda)_{\beta^*_{A\times B}}$ is a fuzzy subgroup of $(A \times B)/\beta^*_{A\times B}$. Now, we have $(\mu \times \lambda)_{\beta^*_{A\times B}} = \mu_{\beta^*_A} \times \lambda_{\beta^*_B}$. By contraposition, suppose that none of the statements (1) and (2) hold. Then, we can find $a_0 \in A$ and $b_0 \in B$ such that

$$\mu_{\beta^*_A}(\beta^*_A(a_0)) > \lambda_{\beta^*_B}(\omega_B) \text{ and } \lambda_{\beta^*_B}(\beta^*_B(b_0)) > \mu_{\beta^*_A}(\omega_A).$$

Now, we have

$$\begin{aligned}(\mu_{\beta^*_A} \times \lambda_{\beta^*_B})(\beta^*_A(a_0),\ \beta^*_B(b_0)) &= \mu_{\beta^*_A}(\beta^*_A(a_0)) \wedge \lambda_{\beta^*_B}(\beta^*_B(b_0)) \\ &> \mu_{\beta^*_A}(\omega_A) \wedge \lambda_{\beta^*_B}(\omega_B) \\ &= (\mu_{\beta^*_A} \times \lambda_{\beta^*_B})(\omega_A,\ \omega_B).\end{aligned}$$

On the other hand, always we have

$$(\mu_{\beta^*_A} \times \lambda_{\beta^*_B})(\omega_A,\ \omega_B) \geq (\mu_{\beta^*_A} \times \lambda_{\beta^*_B})(\beta^*_A(a_0),\ \beta^*_B(b_0)).$$

Thus, $\mu_{\beta^*_A} \times \lambda_{\beta^*_B}$ is not a fuzzy subgroup of $A/\beta^*_A \times B/\beta^*_B$. Therefore, either $\lambda_{\beta^*_B}(\omega_B) \geq \mu_{\beta^*_A}(\beta^*_A(a))$ for all $a \in A$ or $\mu_{\beta^*_A}(\omega_A) \geq \lambda_{\beta^*_B}(\beta^*_B(b))$ for all $b \in B$. ∎

Theorem 3.4.19. *Let μ, λ be fuzzy subsets of H_v-groups A, B respectively, such that $\mu \times \lambda$ is a fuzzy H_v-subgroup of $A \times B$. If $\mu_{\beta^*_A}(\beta^*_A(a)) \leq \lambda_{\beta^*_B}(\omega_B)$ for all $a \in A$, then $\mu_{\beta^*_A}$ is a fuzzy subgroup of A/β^*_A.*

Proof. Suppose that $x, y \in A$. Then, we have

$$\begin{aligned}\mu_{\beta^*_A}(\beta^*_A(x) \odot \beta^*_A(y)) &= \mu_{\beta^*_A}(\beta^*_A(x) \odot \beta^*_A(y)) \wedge \lambda_{\beta^*_B}(\omega_B \odot \omega_B) \\ &= (\mu_{\beta^*_A} \times \lambda_{\beta^*_B})\left((\beta^*_A(x), \omega_B) \odot (\beta^*_A(y), \omega_B)\right) \\ &\geq (\mu_{\beta^*_A} \times \lambda_{\beta^*_B})\left(\beta^*_A(x), \omega_B\right) \wedge (\mu_{\beta^*_A} \times \lambda_{\beta^*_B})\left(\beta^*_A(y), \omega_B\right) \\ &= \left(\mu_{\beta^*_A}(\beta^*_A(x)) \wedge \lambda_{\beta^*_B}(\omega_B)\right) \wedge \left(\mu_{\beta^*_A}(\beta^*_A(y)) \wedge \lambda_{\beta^*_B}(\omega_B)\right) \\ &= \mu_{\beta^*_A}(\beta^*_A(x)) \wedge \mu_{\beta^*_A}(\beta^*_A(y)).\end{aligned}$$

Also, we have

$$\begin{aligned}\mu_{\beta^*_A}(\beta^*_A(x)^{-1}) &= \mu_{\beta^*_A}(\beta^*_A(x)^{-1}) \wedge \lambda_{\beta^*_B}\left((\omega_B)^{-1}\right) \\ &= (\mu_{\beta^*_A} \times \lambda_{\beta^*_B})\left(\beta^*(x)^{-1},\ \omega_B^{-1}\right) \\ &= (\mu_{\beta^*_A} \times \lambda_{\beta^*_B})(\beta^*_A(x),\ \omega_B)^{-1} \\ &\geq (\mu_{\beta^*_A} \times \lambda_{\beta^*_B})(\beta^*_A(x),\ \omega_B) \\ &= \mu_{\beta^*_A}(\beta^*_A(x)) \wedge \lambda_{\beta^*_B}(\omega_B) \\ &= \mu_{\beta^*_A}(\beta^*_A(x)).\end{aligned}$$

Therefore, $\mu_{\beta^*_A}$ is a fuzzy subgroup of A/β^*_A. ∎

Theorem 3.4.20. *Let μ, λ be fuzzy subsets of H_v-groups A, B respectively, such that $\mu \times \lambda$ is a fuzzy H_v-subgroup of $A \times B$. If $\lambda_{\beta_B^*}(\beta_B^*(b)) \leq \mu_{\beta_A^*}(\omega_A)$ for all $b \in B$, then $\lambda_{\beta_B^*}$ is a fuzzy subgroup of B/β_B^*.*

Proof. The proof is similar to the proof of Theorem 3.4.19. ■

Corollary 3.4.21. *Let μ, λ be fuzzy subsets of H_v-groups A, B respectively. If $\mu \times \lambda$ is a fuzzy H_v-subgroup of $A \times B$, then either $\mu_{\beta_A^*}$ is a fuzzy subgroup of A/β_A^* or $\lambda_{\beta_B^*}$ is a fuzzy subgroup of B/β_B^*.*

Proof. The proof follows from Theorems 3.4.18, 3.4.19 and 3.4.20. ■

3.5 Probabilistic Fuzzy Semihypergroups

In this section, we study the connection between TL-sub-semihypergroups and the probability space. The main references for this section are [64, 65].

Definition 3.5.1. Let T be a t-norm on a complete lattice $(L, \leq, \vee, \wedge)$. An L-subset $\mu \in F^L(H)$ of the semihypergroup H is a *TL-sub-semihypergroup* of H if the following axioms hold.

(1) $Im(\mu) \subseteq I_T$.
(2) $T(\mu(x), \mu(y)) \leq \bigwedge\limits_{\alpha \in x \circ y} \mu(\alpha)$, for all $x, y \in H$.

Theorem 3.5.2. *Let T be a t-norm on the complete lattice $(L, \leq, \vee, \wedge)$ and let μ be an L-subset of H such that $Im(\mu) \subseteq I_T$ and $b = \bigvee Im(\mu)$. Then, the following two statements are equivalent:*

(1) *μ is a TL-sub-semihypergroup of H.*
(2) *$\mu^{-1}[a, b]$ is a sub-semihypergroup of H whenever $a \in I_T$ and $0 < a \leq b$.*

Corollary 3.5.3. *Let $A \subseteq H$. Then, the characteristic function χ_A is a TL-sub-semihypergroup of H if and only if A is a sub-semihypergroup of H.*

Corollary 3.5.4. *Let T be a t-norm on the complete lattice $(L, \leq, \vee, \wedge)$, and let $\{\mu_i\}_{i \in I}$ be a family of TL-sub-semihypergroups of H. Then, $\bigcap_{i \in I} \mu_i$ is a TL-sub-semihypergroup of H.*

Corollary 3.5.5. *Let $f : H_1 \rightarrow H_2$ be a strong homomorphism, μ any TL-sub-semihypergroup of H_1 and λ any TL-sub-semihypergroup of H_2. Then, $f(\mu)$ and $f^{-1}(\lambda)$ are TL-sub-semihypergroup of H_2 and H_1, respectively.*

Definition 3.5.6. Let H_1, H_2 be two semihypergroups and let μ, λ be TL-sub- semihypergroups of H_1, H_2, respectively. The *product* of μ, λ is defined to be the TL-subset $\mu \times \lambda$ of $H_1 \times H_2$ with $(\mu \times \lambda)(x, y) = T(\mu(x), \lambda(x))$ for all $(x, y) \in H_1 \times H_2$.

Corollary 3.5.7. *In the above definition, $\mu \times \lambda$ is a TL-subsemihypergroup of $H_1 \times H_2$.*

Definition 3.5.8. Let H be a semihypergroup and μ a L-subset of H. The L-subset μ_{β^*} on H/β^* is defined as follows:

$$\mu_{\beta^*} : H/\beta^* \to L,$$
$$\mu_{\beta^*}(\beta^*(x)) = \bigvee_{\alpha \in \beta^*(x)} \{\mu(a)\}.$$

Theorem 3.5.9. *Let H be a semihypergroup and μ a TL-subsemihypergroup of H. Then μ_{β^*} is a TL-subsemigroup of H/β^*.*

Proof. The proof is similar to the proof of Theorem 3.4.13. ■

Theorem 3.5.10. *Let H_1 and H_2 be two semihypergroups, β_1^*, β_2^* and β^* fundamental equivalence relations on H_1, H_2 and $H_1 \times H_2$, respectively. Let T be a t-norm on the complete lattice $(L, \leq, \vee, \wedge)$. If μ, λ are TL-subsemihypergroups of H_1, H_2, respectively, then $(\mu \times \lambda)_{\beta^*} = \mu_{\beta_1^*} \times \lambda_{\beta_2^*}$.*

Proof. The proof is similar to the proof of Theorem 3.4.14. ■

In the following, we consider $L = [0, 1]$.

In the theory of probability we start with $(\Omega, \mathbb{A}, P)$, where Ω is the set of elementary events and $\mathbb{A}$ a σ-algebra of subsets of Ω called events. A probability on $\mathbb{A}$ is defined as a countable additive and nonnegative function P such that $P(\Omega) = 1$.

Definition 3.5.11. Let H be a semihypergroup, $(\Omega, \mathbb{A}, P)$ be a probability space, and let $R : \Omega \to P(H)$ be a random set. If for any $\omega \in \Omega$, $R(\omega)$ is a subsemihypergroup of H, then the falling shadow S of the random set R, i.e., $S(x) = P(\omega | x \in R(\omega))$, is called a *$\pi$-fuzzy subsemihypergroup* of H.

Proposition 3.5.12. *Let S be a π-fuzzy subsemihypergroup of semihypergroup $(H, \circ)$. Then, $\bigwedge\{S(z) \mid z \in x \circ y\} \geq T^m(S(x), S(y))$ for all $x, y \in H$.*

Proof. We know $R(\omega)$ is a subsemihypergroup. Now, let $x \in R(\omega)$ and $y \in R(\omega)$. Then, $x \circ y \subseteq R(\omega)$. So, for every $z \in x \circ y$ we have $\{\omega \mid z \in R(\omega)\} \supseteq \{\omega | \, x \in R(\omega)\} \cap \{\omega | \, y \in R(\omega)\}$. Thus,

$$\begin{aligned} S(z) &= P(\omega \mid z \in R(\omega)) \\ &\geq P(\{\omega \mid x \in R(\omega)\} \cap \{\omega \mid y \in R(\omega)\}) \\ &\geq P(\omega \mid x \in R(\omega)) + P(\omega | \, y \in R(\omega)) - P(\omega | \, x \in R(\omega) \text{ or } y \in R(\omega)) \\ &\geq S(x) + S(y) - 1. \end{aligned}$$

Hence, $\bigwedge\{S(z) \mid z \in g(x, y)\} \geq T^m(S(x), S(y))$. ■

Theorem 3.5.13. (1) *Let $\mathbb{H}$ denote the set of all subsemihypergroups of H. For each $x \in H$, write $H_x = \{A \mid A \in \mathbb{H}, \ x \in A\}$. Let $(\mathbb{H}, \sigma)$ be a measurable space where σ is a σ-algebra that contains $\{H_x \mid x \in H\}$, and P a probability measure on $(\mathbb{H}, \sigma)$. We define $\mu : H \to [0, 1]$ as follows: $\mu(x) = P(H_x)$ for $x \in H$. Then, μ is a T^m-fuzzy subsemihypergroup of H.*

(2) *Suppose that there exists* $\mathbb{A} \in \sigma$ *such that* $\mathbb{A}$ *is a chain with respect to the set inclusion and* $P(\mathbb{A}) = 1$. *Then,* μ *is a fuzzy subsemihypergroup of* H.

Proof. (1) Suppose that $x, y \in H$. Then, $H_z \supseteq H_x \cup H_y$ for all $z \in x \circ y$, and so

$$\mu(z) = P(H_z) \geq P(H_x \cap H_y) \geq P(H_x) + P(H_y) - 1 \vee 0 = T^m(\mu(x), \mu(y)).$$

Therefore, $\inf\{\mu(z) |\ z \in x \circ y \geq T^m(\mu(x), \mu(y))$.

(2) Since P is a probability measure and $P(\mathbb{A}) = 1$ we have $P(H_x \cap \mathbb{A}) = P(H_x)$ for all $x \in H$. Therefore, for every $z \in x \circ y$ we have

$$\mu(z) = P(H_z) \geq P(H_x \cap H_y) = P(H_x \cap \mathbb{A}) \cap (H_y \cap \mathbb{A}).$$

Since $\mathbb{A}$ with the set inclusion forms a chain, it follows that either $H_x \cap \mathbb{A} \subseteq H_y \cap \mathbb{A}$ or $H_y \cap \mathbb{A} \subseteq H_x \cap \mathbb{A}$. Therefore

$$\mu(z) \geq P(H_x \cap \mathbb{A}) \wedge P(H_y \cap \mathbb{A}) = \mu(x) \wedge \mu(y)$$

and so $\wedge\{\mu(z) \mid z \in x \circ y)\} \geq \mu(x) \wedge \mu(y)$. ■

Theorem 3.5.14. *Let* H *be a semihypergroup and* μ *a fuzzy subsemihypergroup of* H. *Then, there exists a probability space* $(\Omega, \mathbb{A}, P)$ *such that for some* $A \in \mathbb{A}$, $\mu(x) = P(A)$.

3.6 F-Hypergroups

Motivated by the definition of F-polygroup given by Zahedi and Hasankani in Definition 2.6.1 and using four types of reproducibility, Corsini and Tofan [45] introduced four distinct types of fuzzy hypergroups (shortly F-hypergroups).

Let H be a non-empty set and I_*^H denote the set of all nonzero fuzzy subsets of H. Any F-hyperoperation $*$ on H leads to

- a hyperoperation on H, defined by

$$a \odot b = \{x \in H \mid (a * b)(x) \neq \emptyset\} = supp(a * b),$$

  called the *hyperoperation extracted from* $*$, as we have already seen in Theorem 2.6.5, and
- two partial hyperoperations on H
  (1) $a \otimes b = \{x \in H \mid (a * b)(x) = 1\}$
  (2) $a *_\epsilon b = \{x \in H \mid (a * b)(x) \geq \epsilon\}$, for any $\epsilon \in (0, 1]$.
  It is clear that $\otimes$ and $*_\epsilon$ become hyperoperations on H whenever the F-hyperoperation $*$ satisfies a condition of quasi-normality, that is

$$\forall (a, b) \in H^2, \exists x \in H : (a * b)(x) = 1,$$

  or a condition of normality, that is

$$\forall (a, b) \in H^2, \exists! x \in H : (a * b)(x) = 1.$$

We say that $(a*b)(x)$ represents the *degree of membership* of the element x in the set $a \odot b$ and $a *_\epsilon b$ represents the ϵ-*cut* of the fuzzy set $a*b$.

Besides, a fuzzy hyperoperation $*$ on H induces also a binary operation on I_*^H by taking

$$(\mu * \nu) = \bigvee_{(a,b)\in H^2} \{(a*b)(x) \mid \mu(a) \neq 0, \nu(b) \neq 0\}.$$

If $*$ is normal, then one may define a binary operation on H, taking

$$a \nabla b = x, \text{ where } x \text{ is the unique element such that } (a*b)(x) = 1.$$

In the following, we introduce some notations. For any $a \in H, \emptyset \neq K \subseteq H$, $\mu \in I_*^H$, $\epsilon \in (0,1]$, set

$$a * K : H \to [0,1],\ (a*K)(x) = \bigvee_{k\in K}\{(a*k)(x)\},$$

$$a * \mu : H \to [0,1],\ \ a*\mu = a*K, \text{ where } K = \{k \in H \mid \mu(k) \neq 0\},$$

$$a *_\epsilon \mu : H \to [0,1],\ a *_\epsilon \mu = a*K, \text{ where } K = \{k \in H \mid \mu(k) \geq \epsilon\}.$$

Similarly one defines $K*a$, $\mu*a$ and $\mu *_\epsilon a$.

It is clear that the associativity of the fuzzy hyperoperation $*$ implies the associativity of the hyperoperation $\odot$ extracted from $*$, and of the binary operation $*$ on I_*^H. But what about the reproducibility? Having defined on H one fuzzy hyperoperation $*$ and three hyperoperations $\odot, \otimes$ and $*_\epsilon$, it is possible to define four types of reproduction axioms, that lead to four different types of fuzzy hypergroups.

Consider the following definitions, for any $a \in H$:

$(R_1) a * H = \chi_H = H * a$, where χ_H denotes the characteristic function of H.
$(R_2) a \odot H = H = H \odot a$.
$(R_3) a \otimes H = H = H \otimes a$.
(R_4)For any $\epsilon \in (0,1]$, $a *_\epsilon H = H = H *_\epsilon$.

The next result provides the connections between the above conditions of reproducibility.

Theorem 3.6.1. *The following implications are valid:*

$$\begin{array}{ccc} (R_3) & \Longleftrightarrow & (R_4) \\ \Downarrow & & \Downarrow \\ (R_1) & \Longrightarrow & (R_2) \end{array}$$

Proof. Since $a \otimes H \subseteq a *_\epsilon H \subseteq a \odot H$, for any $a \in H$, the implication $(R_3) \Rightarrow (R_4) \Rightarrow (R_2)$ clearly follows. Further suppose that $a \otimes H = H$, so the reproducibility (R_3) is valid. Then, for any $x \in H$, there exists $y_0 \in H$ such that $x \in a \otimes y_0$, that is $(a \otimes y_0)(x) = 1$. Thereby, $\sup\limits_{y\in H}\{(a*y)(x)\} = 1$, whence $a*H = \chi_H$ and similarly, $H*a = \chi_H$, therefore $(R_3) \Rightarrow (R_1)$.

Suppose now that H satisfies the reproducibility (R_4). In particular, for $\epsilon = 1$, we have $a *_\epsilon b = a \otimes b$, which is equivalent with the implication $(R_4) \Rightarrow (R_3)$.

It remains to prove the implication $(R_1) \Rightarrow (R_2)$. Suppose that, for any $a \in H$, $a * H = \chi_H$ is valid. Then, for any $x \in H$, we have $(a * H)(x) = 1$ and $\sup_{y \in H}\{(a * y)(x)\} = 1$. Thus, there exists $y_0 \in H$ such that $(a * y_0)(x) \neq 0$, for any $x \in H$. It follows that $H \subseteq \bigcup_{y \in H} a \odot y = a \odot H \subseteq H$, whence $a \odot H = H$ and similarly, $H \odot a = H$. So the reproducibility (R_2) is valid. ∎

Under supplementary conditions, the reproduction axioms (R_1) and (R_3) are equivalent, as one can see below.

Corollary 3.6.2. *If, for any $a, b \in H$, the families of fuzzy sets $(H, a*h)_{h \in H}$ and $(H, h * b)_{h \in H}$ have the supremum property, then the implication $(R_1) \Rightarrow (R_3)$ holds.*

Proof. Let a be an arbitrary element of H. Because $(a * H)(x) = 1$, for any $x \in H$, we have $\sup_{h \in H}\{(a * h)(x)\} = 1$. Accordingly with the hypothesis, there exists $h_0 \in H$ such that $(a * h_0)(x) = 1$, hence $x \in a \otimes h_0 \subset a \otimes H$. We obtain $H \subseteq a \otimes H \subseteq H$, i.e. $a \otimes H = H$. Similarly, $H \otimes a = H$, which completes the proof. ∎

Next we introduce the following four types of hypergroups.

Definition 3.6.3. A non-empty set H endowed with a F-hyperoperation $* : H \times H \to I_*^H$ satisfying the associativity and the reproducibility (R_i), $1 \leq i \leq 4$, is called an *F_i-hypergroup*.

Accordingly with Theorem 3.6.1, there exist the following relations between the above four kinds of hypergroups:

$$\begin{array}{ccc} F_3 & \Longleftrightarrow & F_4 \\ \Downarrow & & \Downarrow \\ F_1 & \Longrightarrow & F_2 \end{array}$$

Moreover, if the fuzzy hyperoperation $*$ satisfies the conditions of Corollary 3.6.2, then F_1-hypergroups and F_2-hypergroups represent the same class of hyperstructures.

But, how to obtain F-hypergroups? The following result gives a simple method to construct an F_3-hypergroup starting from a hypergroup, and conversely, to obtain a hypergroup from an F_i-hypergroup. We stated a similar result for the F-polygroups (see Theorem 2.6.5), thus we omit here the proof.

Theorem 3.6.4. *Let $(H, \circ)$ be a hypergroup. Define on H the fuzzy hyperoperation $* : H \times H \to I_*^H$ by $a * b = \chi_{a \circ b}$. Then $(H, *)$ is an F_3-hypergroup, and in particular, it is an F_i-hypergroup, $i = 1, 2, 3, 4$.*

Definition 3.6.5. A hypergroup $(H, \circ)$ endowed with a family of fuzzy sets $F = \{\mu_{ab} \mid \mu_{ab} \in I_*^H, (a, b) \in H^2\}$ satisfying the condition

$$\mu_{ab}(x) \neq 0 \Leftrightarrow x \in a \circ b \tag{3.5}$$

is called an *F-hypergroup*. Denote by $\mathcal{F}_{(H,\circ)}$ the class (clearly not empty) of the above families.

The following result provides necessary and sufficient conditions to endowed a non-empty set with an F-hypergroup structure.

Theorem 3.6.6. *Let H be a non-empty set. The family $F = \{\mu_{ab} \mid \mu_{ab} \in I_*^H, (a, b) \in H^2\}$ gives an F-hypergroup structure on H, setting $a \circ b = \{x \in H \mid \mu_{ab}(x) \neq 0\}$, if and only if the following conditions are verified:*

(1) *For any $(a, b, c, x) \in H^4$, there exists $y \in H$, such that $\mu_{ay}(x)\mu_{bc}(y) \neq 0$ if and only if there exists $z \in H$ such that $\mu_{ab}(z)\mu_{zc}(x) \neq 0$.*
(2) *For any $(a, b) \in H^2$, there exists $(x, y) \in H^2$ such that $\mu_{ax}(b)\mu_{ya}(b) \neq 0$.*

Proof. It is straightforward. ■

Theorem 3.6.7. *Let $(H, \circ_1)$ and $(H, \circ_2)$ be hypergroups with the same support H. Defining on H the hyperoperation*

$$a \circ b = a \circ_1 b \cup a \circ_2 b,$$

the following statements are equivalent:

(1) *The hyperoperation $\circ$ is associative.*
(2) *$(H, \circ)$ is a hypergroup.*
(3) *For every $F_1 \in \mathcal{F}_{(H,\circ_1)}$ and $F_2 \in \mathcal{F}_{(H,\circ_2)}$, the union $F_1 \cup F_2$ satisfies relation (5.15).*
(4) *There exist $F_1 \in \mathcal{F}_{(H,\circ_1)}$ and $F_2 \in \mathcal{F}_{(H,\circ_2)}$ such that the union $F_1 \cup F_2$ satisfies relation (5.15).*

Proof. $(1 \Rightarrow 2)$: It is clear that the reproducibility of the hyperoperations $\circ_1$ and $\circ_2$ implies the reproducibility of the hyperoperation $\circ$.

$(2 \Rightarrow 3)$: For any $(a, b) \in H^2$, and for any $F_1 = \{\mu^1_{ab} \mid (a, b) \in H^2\} \in \mathcal{F}_{(H,\circ_1)}$, $F_2 = \{\mu^2_{ab} \mid (a, b) \in H^2\} \in \mathcal{F}_{(H,\circ_2)}$, we have $(\mu^1_{ab} \cup \mu^2_{ab})(x) = max\{\mu^1_{ab}(x), \mu^2_{ab}(x)\}$. It follows that, $(\mu^1_{ab} \cup \mu^2_{ab})(x) \neq 0$ if and only if $x \in a \circ_1 b \cup a \circ_2 b = a \circ b$, which means that $F_1 \cup F_2$ satisfies relation (5.15).

$(3 \Rightarrow 4)$: It is obvious.

$(4 \Rightarrow 1)$: It is straightforward. ■

Chapter 4
H_v-Rings (Hyperrings) and H_v-Ideals

4.1 Fuzzy H_v-Ideals

The concept of a fuzzy ideal of a ring is introduced by W.J. Liu in [137]. Then, in [58], Davvaz introduced the concept of fuzzy H_v-ideal of an H_v-ring.

Definition 4.1.1. If R is a ring and $\mu : R \to [0,1]$ is a fuzzy subset of R, then μ is called a *left* (respectively, *right*) *fuzzy ideal* if it satisfies the following conditions:

(1) $\min\{\mu(x), \mu(y)\} \leq \mu(x-y)$, for all $x, y \in R$.
(2) $\mu(y) \leq \mu(x \cdot y)$ (respectively, $\mu(x) \leq \mu(x \cdot y)$), for all $x, y \in R$.

The fuzzy subset μ of R is called a *fuzzy ideal* if it is a left and right fuzzy ideal.

In this section, we define the concept of fuzzy H_v-ideal of an H_v-ring and present some of their properties.

Definition 4.1.2. [58] Let R be an H_v-ring and μ be a fuzzy subset of R. Then, μ is said to be a *left* (respectively, *right*) *fuzzy H_v-ideal* of R if the following axioms hold.

(1) $\min\{\mu(x), \mu(y)\} \leq \inf\limits_{\alpha \in x+y} \{\mu(\alpha)\}$, for all $x, y \in R$.
(2) For all $x, a \in R$ there exists $y \in R$ such that $x \in a + y$ and

$$\min\{\mu(a), \mu(x)\} \leq \mu(y).$$

(3) For all $x, a \in R$ there exists $z \in R$ such that $x \in z + a$ and

$$\min\{\mu(a), \mu(x)\} \leq \mu(z).$$

(4) $\mu(y) \leq \inf\limits_{\alpha \in x \cdot y} \{\mu(\alpha)\}$ (respectively, $\mu(x) \leq \inf\limits_{\alpha \in x \cdot y} \{\mu(\alpha)\}$), for all $x, y \in R$.

Here, we present all the proofs for the left H_v-ideals. For the right H_v-ideals similar results hold as well.

EXAMPLE 40 Let $(R, +, \cdot)$ be an H_v-ring and μ be a fuzzy H_v-ideal of R. The set $I = \{x \in R \mid \mu(x) = 1\}$ is either the empty-set or an H_v-ideal of R.

B. Davvaz and I. Cristea, *Fuzzy Algebraic Hyperstructures*,
Studies in Fuzziness and Soft Computing 321, DOI: 10.1007/978-3-319-14762-8_4

EXAMPLE 41 Let $(R, +, \cdot)$ be a ring and μ be a fuzzy subset of R. We define hyperoperations $\oplus, \odot, *$ on R as follows:

$$x \oplus y = \{t \mid \mu(t) = \mu(x + y)\},$$
$$x \odot y = \{t \mid \mu(t) = \mu(x \cdot y)\},$$
$$x * y = y * x = \{t \mid \mu(x) \leq \mu(t) \leq \mu(y)\} \text{ (if } \mu(x) \leq \mu(y)).$$

Then, $(R, *, *), (R, *, \odot), (R, *, \oplus), (R, \oplus, *), (R, \oplus, \odot), (R, \odot, *)$ are H_v-rings. Because for all x, y, z in R we have $[(x+y)+z \in (x\oplus y)\oplus z,\ x+(y+z) \in x\oplus(y\oplus z)]$; $[(x\cdot y)\cdot z \in (x\odot y)\odot z,\ x\cdot(y\cdot z) \in x\odot(y\odot z)]$ and $[x \in (x*y)*z,\ x \in x*(y*z)]$. Thus, $\oplus, \odot, *$ are weak associative. Moreover, it is clear that reproduction axioms are valid, i.e., $a\oplus R = R\oplus a = R$, $a\odot R = R\odot a = R$ and $a*R = R*a = R$. We also have $[x\cdot(y+z) \in x\odot(y\oplus z),\ (x\cdot y)+(x\cdot z) \in (x\odot y)\oplus(x\odot z)]$; $[y+z \in x*(y\oplus z),\ y+z \in (x*y)\oplus(x*z)]$; $[y\cdot z \in x*(y\odot z),\ y\cdot z \in (x*y)\odot(x*z)]$; $[x+y \in x\oplus(y*z),\ x+y \in (x\oplus y)*(x\oplus z)]$; $[x\cdot y \in x\odot(y*z),\ x\cdot y \in (x\odot y)*(x\odot z)]$ and $[x \in x*(y*z),\ x \in (x*y)*(x*z)]$. Therefore, $(R, *, *)$, $(R, *, \odot)$, $(R, *, \oplus)$, $(R, \oplus, *)$, $(R, \oplus, \odot)$ and $(R, \odot, *)$ are H_v-rings.

EXAMPLE 42 In Example 41, if μ is a fuzzy ideal of R, then μ is a fuzzy H_v-ideal of $(R, \oplus, \odot)$.

EXAMPLE 43 In Example 41, if μ is a fuzzy ideal of R, then μ is a fuzzy H_v-ideal of $(R, *, \odot)$.

Now, let μ_t be the level set of μ.

Theorem 4.1.3. *Let R be an H_v-ring and μ be a fuzzy subset of R. Then, μ is a fuzzy H_v-ideal of R if and only if for every $0 \leq t \leq 1$, $\mu_t \neq \emptyset$ is an H_v-ideal of R.*

Proof. Let μ be a fuzzy H_v-ideal of R. For every x, y in μ_t we have $\mu(x) \geq t$, $\mu(y) \geq t$. Hence, $\min\{\mu(x), \mu(y)\} \geq t$ and so $\inf_{\alpha \in x+y}\{\mu(\alpha)\} \geq t$. Therefore, for every $\alpha \in x + y$ we get $\mu(\alpha) \geq t$ which implies that $\alpha \in \mu_t$, so $x + y \subseteq \mu_t$. Thus, for every $a \in \mu_t$ we have $a + \mu_t \subseteq \mu_t$and to prove this part of the theorem it is enough to prove that $\mu_t \subseteq a + \mu_t$.

Since μ is a fuzzy H_v-ideal of R, it follows that for every $x \in \mu_t$ there exists $y \in R$ such that $x \in a + y$ and $min\{\mu(a), \mu(x)\} \leq \mu(y)$. From $x \in \mu_t$ and $a \in \mu_t$ we obtain $\min\{\mu(x), \mu(a)\} \geq t$ and so $y \in \mu_t$. Therefore, we have proved that, for every $x \in \mu_t$ there exists $y \in \mu_t$ such that $x \in a+y$ implying that $x \in a+\mu_t$ and this proves $\mu_t \subseteq a+\mu_t$. Therefore, $\mu_t = a+\mu_t$, i.e., the left reproduction axiom is valid for $(\mu_t, +)$. The proof of the right reproduction axiom is similar.

Now, we prove that $R \cdot \mu_t \subseteq \mu_t$. For every $x \in \mu_t$ and $r \in R$ we show that $r \cdot x \subseteq \mu_t$. Since μ is a left fuzzy H_v-ideal we have $t \leq \mu(x) \leq \inf_{\alpha \in r\cdot x}\{\mu(\alpha)\}$. Therefore, for every $\alpha \in r \cdot x$ we obtain $\mu(\alpha) \geq t$ which implies that $\alpha \in \mu_t$. So, $r \cdot x \subseteq \mu_t$.

Conversely, assume that for every t, $0 \leq t \leq 1$, $\mu_t \neq \emptyset$ is an H_v-ideal of R. For every x, y in R we can write $\mu(x) \geq t_0$, $\mu(y) \geq t_0$ where $t_0 = \min\{\mu(x), \mu(y)\}$. Then, $x \in \mu_{t_0}$, $y \in \mu_{t_0}$. Since μ_{t_0} is an H_v-ideal, $x+y \subseteq \mu_{t_0}$. Therefore, for every $\alpha \in x+y$ we have $\mu(\alpha) \geq t_0$ implying that $\inf\limits_{\alpha \in x+y} \{\mu(\alpha)\} \geq t_0$ and so

$$\min\{\mu(x), \mu(y)\} \leq \inf_{\alpha \in x+y} \{\mu(\alpha)\},$$

and in this way the first condition of Definition 4.1.2 is verified. To verify the second condition, if for every $a, x \in R$ we put $t_1 = min\{\mu(a), \mu(x)\}$ then $x \in \mu_{t_1}$ and $a \in \mu_{t_1}$. Since we have $a + \mu_{t_1} = \mu_{t_1}$, so there exists $y \in \mu_{t_1}$ such that $x \in a + y$. On the other hand, since $y \in \mu_{t_1}$, then $\mu(y) \geq t_1$. Hence, $\min\{\mu(a), \mu(x)\} \leq \mu(y)$ and the second condition of Definition 4.1.2 is satisfied. In the similar way, the third condition of Definition 4.1.2 is valid.

Now, we prove the fourth condition of Definition 4.1.2. For every $x, y \in R$, we put $t_2 = \mu(y)$. Then, $y \in \mu_{t_2}$. Since μ_{t_2} is an H_v-ideal of R, $x \cdot y \subseteq \mu_{t_2}$. Therefore, for every $\alpha \in x \cdot y$ we have $\alpha \in \mu_{t_2}$ which implies that $\mu(\alpha) \geq t_2$. Hence, $\inf\limits_{\alpha \in x \cdot y} \{\mu(\alpha)\} \geq t_2$ implying that $\mu(y) \leq \inf\limits_{\alpha \in x \cdot y} \{\mu(\alpha)\}$. ■

The following two corollaries are exactly obtained from Theorem 4.1.3.

Corollary 4.1.4. *Let $(R, +, \cdot)$ be an H_v-ring and μ be a fuzzy H_v-ideal of R. Let $0 \leq t_1 < t_2 \leq 1$. Then, $\mu_{t_1} = \mu_{t_2}$ if and only if there is no x in R such that $t_1 \leq \mu(x) < t_2$.*

Corollary 4.1.5. *Let $(R, +, \cdot)$ be an H_v-ring and μ be a fuzzy H_v-ideal of R. If the range of μ is the finite set $\{t_1, t_2, \ldots, t_n\}$, then the set $\{\mu_{t_i} \mid 1 \leq i \leq n\}$ contains all the level H_v-ideals μ. Moreover if $t_1 < t_2 < \ldots < t_n$, then all the level H_v-ideals μ_{t_i} form the following chain $\mu_{t_1} \subseteq \mu_{t_2} \subseteq \ldots \subseteq \mu_{t_n}$.*

Let μ be a fuzzy subset of the set R. Consider the relation $\sim$ in R defined by $x \sim y$ if and only if $\mu(x) = \mu(y)$. Then, $\sim$ is an equivalence relation in R.

Definition 4.1.6. Suppose that $\overline{x}$ is the equivalence class containing x. The fuzzy subset $\mu^*_{\overline{x}}$ of R defined by

$$\mu^*_{\overline{x}}(r) = \max\{\mu(x), \mu(r)\}, \text{ for all } r \in R$$

is called the *fuzzy max determined by $\overline{x}$ and μ*, and we denote $\mathcal{M}$, the set of all fuzzy max of μ in R.

Now let $(R, +, \cdot)$ be an H_v-ring and μ be a fuzzy H_v-ideal of R. We define two hyperoperations on $\mathcal{M}$ as follows:

$$\mu^*_{\overline{x}} \uplus \mu^*_{\overline{y}} = \mu^*_{\overline{x+y}}\,, \quad \mu^*_{\overline{x}} \otimes \mu^*_{\overline{y}} = \mu^*_{\overline{x \cdot y}}, \text{ for all } x, y \in R,$$

where $\mu^*_A = \{\mu^*_{\overline{\alpha}} \mid \alpha \in A\}$, for all $A \subseteq R$ and $\overline{x+y} = \overline{x} + \overline{y}$, $\overline{x \cdot y} = \overline{x} \cdot \overline{y}$.

Lemma 4.1.7. *If $\mu^*_{\overline{x}} = \mu^*_{\overline{y}}$, then $\overline{x} = \overline{y}$.*

Proof. We have $\mu^*_{\overline{x}}(x) = \mu^*_{\overline{y}}(x)$. Thus, $\mu(x) = \max\{\mu(x), \mu(y)\}$ which implies that $\mu(y) \leq \mu(x)$. Similarly, we get $\mu(x) \leq \mu(y)$. So, $\mu(x) = \mu(y)$. Therefore, $\overline{x} = \overline{y}$. ■

Corollary 4.1.8. *The hyperoperations $\uplus$ and $\otimes$ are well defined.*

Theorem 4.1.9. *$(\mathcal{M}, \uplus, \otimes)$ is an H_v-ring.*

Proof. It is straightforward. ■

Theorem 4.1.10. *If μ is any fuzzy H_v-ideal of an H_v-ring R, then the map $\psi : R \to \mathcal{M}$ defined by $\psi(x) = \mu^*_{\overline{x}}$, for all $x \in R$ is an inclusion homomorphism.*

Proof. It is straightforward. ■

Now, let R be an H_v-ring and γ^* be the fundamental relation on R. Suppose $\gamma^*(a)$ is the equivalence class containing $a \in R$. Then both the sum $\oplus$ and the product $\odot$ on R/γ^*, the set of all equivalence classes, are defined as follows:

$$\gamma^*(a) \oplus \gamma^*(b) = \gamma^*(c), \text{ for all } c \in \gamma^*(a) + \gamma^*(b),$$
$$\gamma^*(a) \odot \gamma^*(b) = \gamma^*(d), \text{ for all } d \in \gamma^*(a) \cdot \gamma^*(b).$$

If $\varphi : R \to R/\gamma^*$ is the canonical map and μ is a fuzzy subset of R, then $\varphi(\mu)$ is fuzzy subset of R/γ^*. In this case, we use μ_{γ^*} instead of $\varphi(\mu)$.

Theorem 4.1.11. *Let R be an H_v-ring and μ be a fuzzy H_v-ideal of R. Then, μ_{γ^*} is a fuzzy ideal of the ring R/γ^*.*

Proof. Since μ is a fuzzy H_v-ideal of R, we have

(1) $\min\{\mu_{\gamma^*}(\gamma^*(x)), \mu_{\gamma^*}(\gamma^*(y))\} \leq \inf\limits_{\gamma^*(\alpha) \in \gamma^*(x) \oplus \gamma^*(y)} \{\mu_{\gamma^*}(\gamma^*(\alpha))\}$,
for all $\gamma^*(x), \gamma^*(y) \in R/\gamma^*$.

(2) For all $\gamma^*(x), \gamma^*(a) \in R/\gamma^*$, there exists $\gamma^*(y) \in R/\gamma^*$ such that

$$\gamma^*(x) = \gamma^*(a) \oplus \gamma^*(y) \text{ and } \min\{\mu_{\gamma^*}(\gamma^*(x)), \mu_{\gamma^*}(\gamma^*(a))\} \leq \mu_{\gamma^*}(\gamma^*(y)).$$

From (1), we obtain $\min\{\mu_{\gamma^*}(\gamma^*(x)), \mu_{\gamma^*}(\gamma^*(y))\} \leq \mu_{\gamma^*}(\gamma^*(x) \oplus \gamma^*(y))$. Now, for all $\gamma^*(x)$ in R/γ^* we prove that $\mu_{\gamma^*}(\gamma^*(x)) \leq \mu_{\gamma^*}(-\gamma^*(x))$. Since $\gamma^*(x) \in R/\gamma^*$, by considering $\gamma^*(a) = \gamma^*(x)$ which is obtained from the second condition above there exists $\gamma^*(y_1)$ in R/γ^* such that $\gamma^*(x) = \gamma^*(x) \odot \gamma^*(y_1)$ and $\min\{\mu_{\gamma^*}(\gamma^*(x)), \mu_{\gamma^*}(\gamma^*(x))\} \leq \mu_{\gamma^*}(\gamma^*(y_1))$. From $\gamma^*(x) = \gamma^*(x) \odot \gamma^*(y_1)$ we obtain $\omega_H = \gamma^*(y_1)$, where ω_H denotes the unit of the group $(R/\gamma^*, \oplus)$. Therefore,

$$\mu_{\gamma^*}(\gamma^*(x)) \leq \mu_{\gamma^*}(\omega_H). \tag{4.1}$$

Now, by considering $\gamma^*(x), \omega_H$ in R/γ^*, by (2) above there exists $\gamma^*(y_2)$ in R/γ^* such that $\omega_H = \gamma^*(x) \oplus \gamma^*(y_2)$ and $\min\{\mu_{\gamma^*}(\omega_H), \mu_{\gamma^*}(\gamma^*(x))\} \leq \mu_{\gamma^*}(\gamma^*(y_2))$. From $\omega_H = \gamma^*(x) \oplus \gamma^*(y_2)$ we obtain $\gamma^*(y_2) = -\gamma^*(x)$, so

$$\min\{\mu_{\gamma^*}(\omega_H), \mu_{\gamma^*}(\gamma^*(x))\} \leq \mu_{\gamma^*}(-\gamma^*(x)). \tag{4.2}$$

By (4.1) and (4.2) the inequality $\mu_{\gamma^*}(\gamma^*(x)) \leq \mu_{\gamma^*}(-\gamma^*(x))$ is obtained. Now, from $\min\{\mu_{\gamma^*}(\gamma^*(x)), \mu_{\gamma^*}(\gamma^*(y))\} \leq \mu_{\gamma^*}(\gamma^*(x) \oplus \gamma^*(y))$ and $\mu_{\gamma^*}(\gamma^*(x)) \leq \mu_{\gamma^*}(-\gamma^*(x))$ we get

$$\{\mu_{\gamma^*}(\gamma^*(x)), \mu_{\gamma^*}(\gamma^*(y))\} \leq \mu_{\gamma^*}(\gamma^*(x) - \gamma^*(y)).$$

Now, we prove the second condition of the definition of fuzzy ideal. For all $\gamma^*(x), \gamma^*(y) \in R/\gamma^*$, we have $\mu_{\gamma^*}(\gamma^*(x) \odot \gamma^*(y)) = \mu_{\gamma^*}(\gamma^*(x) \odot \gamma^*(a))$, for all $a \in \gamma^*(y)$ and so

$$\begin{aligned}
\mu_{\gamma^*}(\gamma^*(x) \odot \gamma^*(y)) &= \mu_{\gamma^*}(\gamma^*(x \cdot a)) \\
&= \sup_{\alpha \in \gamma^*(x \cdot a)} \{\mu(\alpha)\} \\
&\geq \sup_{\alpha \in x \cdot a} \{\mu(\alpha)\} \\
&\geq \inf_{\alpha \in x \cdot a} \{\mu(\alpha)\} \\
&\geq \mu(a).
\end{aligned}$$

Therefore, $\mu_{\gamma^*}(\gamma^*(x) \odot \gamma^*(y)) \geq \sup_{a \in \gamma^*(y)} \{\mu(a)\}$ which implies that $\mu_{\gamma^*}(\gamma^*(x) \odot \gamma^*(y)) \geq \gamma^*(y)$, and the theorem is proved. ■

Let γ_μ^* be the fundamental equivalence relation on $\mathcal{M}$, the H_v-ring defined in Theorem 4.1.9, and $\mathcal{U}_\mu$ denotes the set of finite polynomials of elements of $\mathcal{M}$ over $\mathbb{N}$. We denote sum and product on $\mathcal{M}/\gamma_\mu^*$ by $\oplus, \odot$.

Proposition 4.1.12. *Let $(R, +, \cdot)$ be an H_v-ring. Then, there exists a strong homomorphism $f : R/\gamma^* \to \mathcal{M}/\gamma_\mu^*$.*

Proof. We define f as follows: $f(\gamma^*(r)) = \gamma_\mu^*(\mu_{\overline{r}}^*)$, for all $r \in R$. We prove that f is well defined. Suppose that $\gamma^*(r) = \gamma^*(r_1)$. Then, $r\gamma^* r_1$. We have $r\gamma^* r_1$ if and only if there exist $x_1, \ldots, x_{m+1}$, where $x_1 = r,\ x_{m+1} = r_1,\ u_1, \ldots, u_m \in \mathcal{U}$ such that $\{x_i, x_{i+1}\} \subseteq u_i,\ i = 1, \ldots, m$. Therefore, $\{\mu_{\overline{x_i}}^*,\ \mu_{\overline{x_{i+1}}}^*\} \subseteq \mu_{\overline{u_i}}^*$, where $\mu_{\overline{u_i}}^* \in \mathcal{U}_\mu,\ i = 1, \ldots, m$, which implies that $\mu_{\overline{r}}^*\ \gamma_\mu^*\ \mu_{\overline{r_1}}^*$ implying $\gamma_\mu^*(\mu_{\overline{r}}^*) = \gamma_\mu^*(\mu_{\overline{r_1}}^*)$.

Now, we show that f is strong homomorphism. For every $\gamma^*(a), \gamma^*(b)$ in R/γ^*, we can write:

$$\begin{aligned}
&f(\gamma^*(a) \oplus \gamma^*(b)) = f(\gamma^*(c)) = \gamma_\mu^*(\mu_{\overline{c}}^*),\ \text{for all } c \in \gamma^*(a) + \gamma^*(b), \\
&f(\gamma^*(a) \oplus \gamma^*(b)) = \gamma_\mu^*(\mu_{\overline{a}}^*) \oplus \gamma_\mu^*(\mu_{\overline{b}}^*) = \gamma_\mu^*(\mu_{\overline{d}}^*),\ \text{for all } \mu_{\overline{d}}^* \in \gamma_\mu^*(\mu_{\overline{a}}^*) \uplus \gamma_\mu^*(\mu_{\overline{b}}^*).
\end{aligned}$$

If we put $c \in a+b$ and $\mu_{\overline{d}}^* \in \mu_{\overline{a}}^* \uplus \mu_{\overline{b}}^*$, then $f(\gamma^*(a) \oplus \gamma^*(b)) = f(\gamma^*(a)) \oplus f(\gamma^*(b))$. In the similar way, we have $f(\gamma^*(a) \odot \gamma^*(b)) = f(\gamma^*(a)) \odot f(\gamma^*(b))$. ■

Corollary 4.1.13. *The following diagram is commutative:*

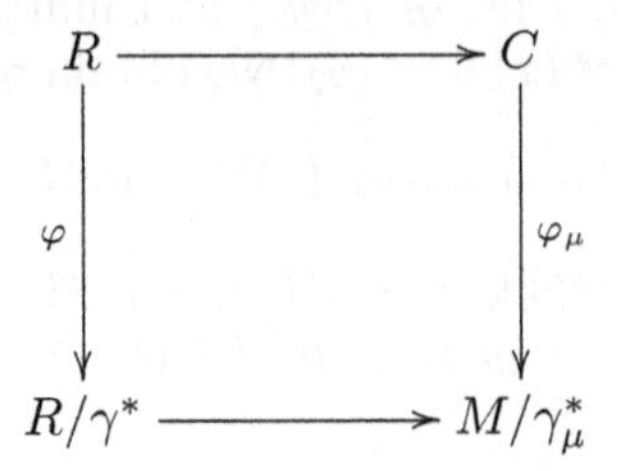

Let $\{R_\alpha \mid \alpha \in \Gamma\}$ be a collection of H_v-rings and

$$\prod_{\alpha \in \Gamma} R_\alpha = \{< x_\alpha > \mid x_\alpha \in R_\alpha\},$$

be the Cartesian product of R_α $(\alpha \in \Gamma)$. We can define two hyperoperations as follows:

$$< x_\alpha > \oplus < y_\alpha >= \{< z_\alpha > \mid z_\alpha \in x_\alpha + y_\alpha,\ \alpha \in \Gamma\},$$
$$< x_\alpha > \odot < y_\alpha >= \{< z_\alpha > \mid z_\alpha \in x_\alpha \cdot y_\alpha,\ \alpha \in \Gamma\}.$$

It follows that $\prod\limits_{\alpha \in \Gamma} R_\alpha$ is an H_v-ring. We call $\prod\limits_{\alpha \in \Gamma} R_\alpha$ the *external direct product* of R_α $(\alpha \in \Gamma)$. Let μ and λ be fuzzy subsets of a non-empty set X. The *Cartesian cross-product* $\mu \times \lambda$ is usually defined by:

$$(\mu \times \lambda)(x, y) = \min\{\mu(x), \lambda(y)\}, \text{ for all } x, y \in X.$$

Let $\{X_\alpha \mid \alpha \in \Gamma\}$ be a collection of non-empty sets and let μ_α be a fuzzy subset of X_α for all $\alpha \in \Gamma$. Define the Cartesian cross-product of the μ_α by $(\prod\limits_{\alpha \in \Gamma} \mu_\alpha)(x) = \inf\limits_{\alpha \in \Gamma}\{\mu_\alpha(x_\alpha)\}$ where $x =< x_\alpha >$ and $< x_\alpha >$ denotes an element of the Cartesian cross-product $\prod\limits_{\alpha \in \Gamma} X_\alpha$.

Proposition 4.1.14. *Let $\{R_\alpha \mid \alpha \in \Gamma\}$ be a collection of H_v-rings and let μ_α be a fuzzy H_v-ideal of R_α. Then $\prod\limits_{\alpha \in \Gamma} \mu_\alpha$ is a fuzzy H_v-ideal of $\prod\limits_{\alpha \in \Gamma} R_\alpha$.*

Proof. Let $x =< x_\alpha >,\ y =< y_\alpha >\in \prod\limits_{\alpha \in \Gamma} R_\alpha$. Then, for every $z =< z_\alpha >\in x + y =< x_\alpha > \oplus < y_\alpha >$ we have

$$\begin{aligned}
(\prod_{\alpha \in \Gamma} \mu_\alpha)(z) &= \inf_{\alpha \in \Gamma}\{\mu_\alpha(z_\alpha)\} \\
&\geq \inf_{\alpha \in \Gamma}\{\min\{\mu_\alpha(x_\alpha), \mu_\alpha(y_\alpha)\}\} \\
&= \min\left\{\inf_{\alpha \in \Gamma}\{\mu_\alpha(x_\alpha)\}, \inf_{\alpha \in \Gamma}\{\mu_\alpha(y_\alpha)\}\right\} \\
&= \min\left\{(\prod_{\alpha \in \Gamma} \mu_\alpha)(x), (\prod_{\alpha \in \Gamma} \mu_\alpha)(y)\right\}.
\end{aligned}$$

Therefore, the first condition of the definition of an H_v-ideal is satisfied. Now, we prove the second condition as follows. For every $x =< x_\alpha >$ and $a =<$

$a_\alpha >$ in $\prod_{\alpha\in\Gamma} R_\alpha$ there exists $y =< y_\alpha >$ in $\prod_{\alpha\in\Gamma} R_\alpha$ such that $\min\{\mu_\alpha(x_\alpha), \mu_\alpha(a_\alpha)\} \leq \mu_\alpha(y_\alpha)$. Therefore, we have $< x_\alpha >\in< a_\alpha > \oplus < y_\alpha >$ and

$$\begin{aligned}(\prod_{\alpha\in\Gamma} \mu_\alpha)(y) &= \inf_{\alpha\in\Gamma}\{\mu_\alpha(y_\alpha)\} \geq \inf_{\alpha\in\Gamma}\{\min\{\mu_\alpha(x_\alpha), \mu_\alpha(a_\alpha)\}\} \\ &= \min\Big\{\inf_{\alpha\in\Gamma}\{\mu_\alpha(x_\alpha)\}, \inf_{\alpha\in\Gamma}\{\mu_\alpha(a_\alpha)\}\Big\} \\ &= \min\Big\{(\prod_{\alpha\in\Gamma} \mu_\alpha)(x), (\prod_{\alpha\in\Gamma} \mu_\alpha)(a)\Big\}.\end{aligned}$$

The proof of third condition is similar to that of second condition. To verify the fourth condition, for every $z =< z_\alpha >\in x \odot y =< x_\alpha > \odot < y_\alpha >$ we have

$$(\prod_{\alpha\in\Gamma} \mu_\alpha)(z) = \inf_{\alpha\in\Gamma}\{\mu_\alpha(z_\alpha)\} \geq \inf_{\alpha\in\Gamma}\{\mu_\alpha(y_\alpha)\} = (\prod_{\alpha\in\Gamma} \mu_\alpha)(y).$$

Hence, $(\prod_{\alpha\in\Gamma} \mu_\alpha)(y) \leq \inf_{z\in x\cdot y}\Big\{(\prod_{\alpha\in\Gamma} \mu_\alpha)(z)\Big\}$. ■

The following corollary is exactly obtained from the above proposition.

Corollary 4.1.15. *Let μ be a fuzzy subset of an H_v-ring R. Then, $\mu \times \mu$ is a fuzzy left (right) H_v-ideal of $R \times R$ if and only if μ is a fuzzy left (right) H_v-ideal of R.*

Definition 4.1.16. Let μ be a fuzzy H_v-ideal of $R_1 \times R_2$ and let (x_1, x_2), $(a_1, a_2) \in R_1 \times R_2$. Then, there exists $(y_1, y_2) \in R_1 \times R_2$ such that $(x_1, x_2) \in (a_1, a_2) \oplus (y_1, y_2)$ and

$$\min\{\mu(x_1, x_2), \mu(a_1, a_2)\} \leq \mu(y_1, y_2).$$

Now, if for every $r_1, s_1 \in R_1$ there exists $t_1 \in R_1$ such that $(r_1, x_2) \in (s_1, a_2)\oplus (t_1, y_2)$ and

$$\min\{\mu(r_1, x_2), \mu(s_1, a_2)\} \leq \mu(t_1, y_2),$$

and for every $r_2, s_2 \in R_2$ there exists $t_2 \in R_2$ such that $(x_1, r_2) \in (a_1, s_2) \oplus (y_1, t_2)$ and

$$\min\{\mu(x_1, r_2), \mu(a_1, s_2)\} \leq \mu(y_1, t_2),$$

then we say that μ satisfies in the *regular left fuzzy reproduction axiom*. Similarly, we can define the *regular right fuzzy reproduction axiom*. μ is called the *regular fuzzy H_v-ideal* of R if μ satisfies the regular left and right fuzzy reproduction axioms.

Theorem 4.1.17. *Let R_1, R_2 be H_v-rings with scalar units and μ be a regular fuzzy H_v-ideal of $R_1 \times R_2$. Then, μ_i, $i = 1, 2$ is a fuzzy H_v-ideal of R_i, $i = 1, 2$, respectively, where $\mu_1(x) = \sup_{a\in R_2}\{\mu(x, a)\}$ and $\mu_2(y) = \sup_{b\in R_1}\{\mu(b, y)\}$.*

Proof. We show that μ_1 is a fuzzy H_v-ideal of R_1. Suppose that $x, y \in R_1$. Then, for every $\alpha \in x + y$ we have $\mu_1(\alpha) = \sup\limits_{a\in R_2} \{\mu(\alpha, a)\}$. For every $a \in R_2$ there exist $r_a, s_a \in R_2$ such that $a \in a + r_a$ and $a \in s_a + a$. Now, we have

$$\begin{aligned}
\mu_1(\alpha) &\geq \sup_{a\in R_2} \{\min\{\mu(x,a), \mu(y, r_a)\}\} \\
&\geq \sup_{a\in R_2} \{\min\{\mu(x,a), \inf_{b\in R_2} \{\mu(y,b)\}\}\} \\
&= \min\left\{ \sup_{a\in R_2} \{\mu(x,a)\}, \inf_{b\in R_2} \{\mu(y,b)\}\right\} \\
&= \min\left\{\mu_1(x), \inf_{b\in R_2} \{\mu(y,b)\}\right\}
\end{aligned}$$

and also we have

$$\begin{aligned}
\mu_1(\alpha) &\geq \sup_{a\in R_2} \{\min\{\mu(x, s_a), \mu(y,a)\}\} \\
&\geq \sup_{a\in R_2} \left\{ \min\{ \inf_{c\in R_2} \{\mu(x,c)\}, \mu(y,a)\}\right\} \\
&= \min\left\{ \inf_{c\in R_2} \{\mu(x,c)\}, \sup_{a\in R_2} \{\mu(y,a)\right\} \\
&= \min\left\{ \inf_{c\in R_2} \{\mu(x,c)\}, \mu_1(y)\right\}.
\end{aligned}$$

Therefore,

$$\begin{aligned}
\mu_1(\alpha) \geq \max &\left\{ \min\left\{\mu_1(x), \inf_{b\in R_2} \{\mu(y,b)\}\right\}, \min\left\{ \inf_{c\in R_2} \{\mu(x,c)\}, \mu_1(y)\right\}\right\} \\
= \min &\left\{ \max\left\{\mu_1(x), \inf_{c\in R_2} \{\mu(x,c)\right\}, \max\left\{ \inf_{b\in R_2} \{\mu(y,b)\}, \mu_1(y)\right\}\right\} \\
\geq \min &\{\mu_1(x), \mu_1(y)\},
\end{aligned}$$

Now, if $x, a \in R_1$, then for every $r, s \in R_2$ there exists $(y, y_{r,s}) \in R_1 \times R_2$ such that $(x, r) \in (a, s) \oplus (y, y_{r,s})$ and $\min\{\mu(x,r), \mu(a,s)\} \leq \mu(y, y_{r,s})$. Thus,

$$\begin{aligned}
\min\{\mu_1(x), \mu_1(a)\} &= \min\left\{ \sup_{r\in R_2} \{\mu(x,r)\}, \sup_{s\in R_2} \{\mu(a,s)\}\right\} \\
&= \sup_{\substack{r\in R_2 \\ s\in R_2}} \{\min\{\mu(x,r), \mu(a,s))\}\} \\
&\leq \sup_{\substack{r\in R_2 \\ s\in R_2}} \{\mu(y, y_{r,s})\} \\
&\leq \sup_{z\in R_2} \{\mu(y,z)\} = \mu_1(y).
\end{aligned}$$

The proof of the third condition is similar to the second condition. Now, we verify the fourth condition of the definition. Suppose that 1 be the unit scalar of R_2. For every $\alpha \in x \cdot y$, we have $(\alpha, a) \in (x, 1) \odot (y, a)$. Since μ is a fuzzy H_v-ideal of $R_1 \times R_2$, we obtain $\mu(\alpha, a) \geq \mu(y, a)$ which implies that $\sup\limits_{a\in R_2} \{\mu(\alpha, a)\} \geq \sup\limits_{a\in R_2} \{\mu(y,a)\}$. Therefore $\mu_1(\alpha) \geq \mu_1(y)$ for every $\alpha \in x \cdot y$, and so $\inf\limits_{\alpha\in x\cdot y} \{\mu_1(\alpha)\} \geq \mu_1(y)$. Hence, μ_1 is a fuzzy H_v-ideal of R_1. Similarly, we can prove that μ_2 is a fuzzy H_v-ideal of R_2. ■

Corollary 4.1.18. *Let R_1, R_2 be two H_v-rings with scalar units and μ, λ be fuzzy subsets of R_1, R_2, respectively. If $\sup_{x\in R_1}\{\mu(x)\} = \sup_{y\in R_2}\{\lambda(y)\} = 1$ and $\mu \times \lambda$ is a strong fuzzy H_v-ideal of $R_1 \times R_2$, then μ, λ are fuzzy H_v-ideals of R_1, R_2, respectively.*

Proof. Suppose that

$$\mu_1(x) = \sup_{a\in R_2}\{(\mu \times \lambda)(x, a)\} \text{ and } \mu_2(y) = \sup_{b\in R_1}\{(\mu \times \lambda)(b, y)\},$$

then it is enough to show that $\mu(x) = \mu_1(x)$ and $\lambda(y) = \lambda_2(y)$. Let $x \in R_1$ then

$$\begin{aligned}\mu(x) &= \min\{\mu(x), 1\}\\ &= \min\left\{\mu(x), \sup_{y\in R_2}\{\lambda(y)\}\right\}\\ &= \sup_{y\in R_2}\{\min\{\mu(x), \lambda(y)\}\}\\ &= \sup_{y\in R_2}\{(\mu \times \lambda)(x, y)\}\\ &= \mu_1(x).\end{aligned}$$

■

4.2 Generalized Fuzzy H_v-Ideals

In this section, we consider a special case of (α, β)-fuzzy H_v-ideals. The notion of $(\in, \in\vee q)$-fuzzy H_v-ideal is an important and useful generalization of ordinary fuzzy H_v-ideal.

Definition 4.2.1. Let R be an H_v-ring. A fuzzy subset μ of R is said to be an *$(\in, \in\vee q)$-fuzzy left (right) H_v-ideal* of R if for all $t, r \in (0, 1]$,

(1) $x_t \in \mu, y_r \in \mu$ implies $z_{t\wedge r} \in\vee q\mu$, for all $z \in x + y$.
(2) $x_t \in \mu, a_r \in \mu$ implies $y_{t\wedge r} \in\vee q\mu$, for some $y \in R$ with $x \in a + y$.
(3) $x_t \in \mu, a_r \in \mu$ implies $z_{t\wedge r} \in\vee q\mu$, for some $z \in R$ with $x \in z + a$.
(4) $y_t \in \mu$ and $x \in R$ imply $z_t \in\vee q\mu$ for all $z \in x \cdot y$
 ($x_t \in \mu$ and $y \in R$ imply $z_t \in\vee q\mu$, for all $z \in x \cdot y$).

It is easy to see that for any subset μ of R, χ_μ is an $(\in, \in\vee q)$-fuzzy left (right) H_v-ideal of R if and only if μ is a left (right) H_v-ideal of R.

EXAMPLE 44 Let $R = \{a, b, c, d\}$ be a set, and consider addition and multiplication tables below:

$+$	a	b	c	d
a	a	b	c	d
b	b	$\{a, b\}$	d	c
c	c	d	$\{a, c\}$	b
d	d	c	b	$\{a, d\}$

$\cdot$	a	b	c	d
a	a	a	a	a
b	a	b	b	b
c	a	c	c	c
d	a	d	d	d

Then, we can easily see that $(R, +, \cdot)$ is an H_v-ring. Let $\mu : R \to [0, 1]$ be defined by

$$\mu(a) = 0.6, \; \mu(b) = \mu(c) = \mu(d) = 0.8.$$

Then, μ is an $(\in, \in \vee q)$-fuzzy H_v-ideal of R but not an ordinary fuzzy H_v-ideal.

Proposition 4.2.2. *Conditions (1)-(4) in Definition 4.2.1 are equivalent to the following conditions respectively.*

(1′) $\mu(x) \wedge \mu(y) \wedge 0.5 \leq \bigwedge\limits_{z \in x+y} \mu(z)$, *for all* $x, y \in R$.

(2′) For all $x, a \in R$ *there exists* $y \in R$ *such that* $x \in a + y$ *and*

$$\mu(a) \wedge \mu(x) \wedge 0.5 \leq \mu(y).$$

(3′) For all $x, a \in R$ *there exists* $z \in R$ *such that* $x \in z + a$ *and*

$$\mu(a) \wedge \mu(x) \wedge 0.5 \leq \mu(z).$$

(4′) $\mu(y) \wedge 0.5 \leq \bigwedge\limits_{z \in x \cdot y} \mu(z)$, *for all* $x, y \in R$

$(\mu(x) \wedge 0.5 \leq \bigwedge\limits_{z \in x \cdot y} \mu(z)$, *for all* $x, y \in R)$.

Proof. $(1 \Rightarrow 1')$: Suppose that $x, y \in R$. We consider the following cases:

(a) $\mu(x) \wedge \mu(y) < 0.5$.
(b) $\mu(x) \wedge \mu(y) \geq 0.5$.

Case a: Assume that there exists $z \in x + y$ such that $\mu(z) < \mu(x) \wedge \mu(y) \wedge 0.5$, which implies $\mu(z) < \mu(x) \wedge \mu(y)$. Choose t such that $\mu(z) < t < \mu(x) \wedge \mu(y)$. Then $x_t, y_t \in \mu$, but $z_t \overline{\in \vee q} \mu$ which contradicts (1).

Case b: Assume that $\mu(z) < 0.5$ for some $z \in x + y$. Then $x_{0.5}, y_{0.5} \in \mu$, but $z_{0.5} \overline{\in \vee q} \mu$, a contradiction.

Hence $(1')$ holds.

$(2 \Rightarrow 2')$: Suppose that $x, a \in R$. We consider the following cases:

(a) $\mu(x) \wedge \mu(a) < 0.5$.
(b) $\mu(x) \wedge \mu(a) \geq 0.5$.

Case a: Assume that for all y with $x \in a + y$, we have $\mu(y) < \mu(x) \wedge \mu(a)$. Choose t such that $\mu(y) < t < \mu(x) \wedge \mu(a)$ and $t + \mu(y) < 1$. Then $x_t, a_t \in \mu$, but $y_t \overline{\in \vee q} \mu$, which contradicts (2).

Case b: Assume that for all y with $x \in a + y$, we have

$$\mu(y) < \mu(x) \wedge \mu(a) \wedge 0.5.$$

Then $x_{0.5}, a_{0.5} \in \mu$, but $y_{0.5} \overline{\in \vee q} \mu$, which contradicts (2).

Hence (2′) holds.

(3 ⇒ 3′): The proof is similar to (2 ⇒ 2′).

(4 ⇒ 4′): Suppose that $x, y \in R$. We consider the following cases:

(a) $\mu(y) < 0.5$.
(b) $\mu(y) \geq 0.5$.

Case a: Assume that there exists $z \in x \cdot y$ such that $\mu(z) < \mu(y) \wedge 0.5$, which implies $\mu(z) < \mu(y)$. Choose t such that $\mu(z) < t < \mu(y)$. Then $y_t \in \mu$, but $z_t \overline{\in \vee q} \mu$, which contradicts (4).

Case b: Assume that $\mu(z) < 0.5$ for some $z \in x \cdot y$. Then, $y_{0.5} \in \mu$, but $z_{0.5} \overline{\in \vee q} \mu$, a contradiction. Hence (4′) holds.

(1′ ⇒ 1): Let $x_t, y_r \in \mu$. Then, $\mu(x) \geq t$ and $\mu(y) \geq r$. For every $z \in x + y$ we have

$$\mu(z) \geq \mu(x) \wedge \mu(y) \wedge 0.5 \geq t \wedge r \wedge 0.5.$$

If $t \wedge r > 0.5$, then $\mu(z) \geq 0.5$ which implies that $\mu(z) + t \wedge r > 1$.

If $t \wedge r \leq 0.5$, then $\mu(z) \geq t \wedge r$.

Therefore, $z_{t \wedge r} \in \vee q \mu$ for all $z \in x + y$.

(2′ ⇒ 2): Let $x_t, a_r \in \mu$. Then, $\mu(x) \geq t$ and $\mu(a) \geq r$. Now, for some y with $x \in a + y$ we have

$$\mu(y) \geq \mu(a) \wedge \mu(x) \wedge 0.5 \geq t \wedge r \wedge 0.5.$$

If $t \wedge r > 0.5$, then $\mu(y) \geq 0.5$ which implies $\mu(y) + t \wedge r > 1$.

If $t \wedge r \leq 0.5$, then $\mu(y) \geq t \wedge r$.

Therefore $y_{t \wedge r} \in \vee q \mu$. Hence (2) holds.

(3′ ⇒ 3): The proof is similar to (2′ ⇒ 2).

(4′ ⇒ 4): Let $y_t \in \mu$ and $x \in R$. Then, $\mu(y) \geq t$. For every $z \in x \cdot y$ we have

$$\mu(z) \geq \mu(y) \wedge 0.5 \geq t \wedge 0.5.$$

If $t > 0.5$, then $\mu(z) \geq 0.5$ which implies $\mu(z) + t > 1$.

If $t \leq 0.5$, then $\mu(z) \geq t$.

Therefore, $z_t \in \vee q \mu$ for all $z \in x \cdot y$. ■

By Definition 4.2.1 and Proposition 4.2.2, we immediately obtain:

Corollary 4.2.3. *A fuzzy subset μ of an H_v-ring R is an $(\in, \in \vee q)$-fuzzy left (right) H_v-ideal of R if and only if the conditions (1′)-(4′) in Proposition 4.2.2 hold.*

Now, we characterize $(\in, \in \vee q)$-fuzzy left (right) H_v-ideals by their level H_v-ideals.

Theorem 4.2.4. *Let R be an H_v-ring and μ a fuzzy subset of R. If μ is an $(\in, \in \vee q)$-fuzzy left (right) H_v-ideal of R, then for all $0 < t \leq 0.5$, μ_t is the empty set or a left (right) H_v-ideal of R. Conversely, if μ_t $(\neq \emptyset)$ is a left (right) H_v-ideal of R for all $0 < t \leq 0.5$, then μ is an $(\in, \in \vee q)$-fuzzy left (right) H_v-ideal of R.*

Proof. Let μ be an $(\in, \in \vee q)$-fuzzy left H_v-ideal of R and $0 < t \leq 0.5$. Let $x, y \in \mu_t$. Then, $\mu(x) \geq t$ and $\mu(y) \geq t$. Now

$$\bigwedge_{z \in x+y} \mu(z) \geq \mu(x) \wedge \mu(y) \wedge 0.5 \geq t \wedge 0.5 = t.$$

Therefore, for every $z \in x + y$ we have $\mu(z) \geq t$ or $z \in \mu_t$, so $x + y \subseteq \mu_t$. Hence, for every $a \in \mu_t$ we have $a + \mu_t \subseteq \mu_t$. Now, let $x, a \in \mu_t$. Then, there exists $y \in R$ such that $x \in a+y$ and $\mu(a) \wedge \mu(x) \wedge 0.5 \leq \mu(y)$. From $x, a \in \mu_t$, we have $\mu(x) \geq t$ and $\mu(a) \geq t$, and so

$$t = t \wedge t \wedge 0.5 \leq \mu(a) \wedge \mu(x) \wedge 0.5 \leq \mu(y).$$

Hence, $y \in \mu_t$, and this proves that $\mu_t \subseteq a + \mu_t$.

Now, let $y \in \mu_t$ and $x \in R$. Then, $\mu(y) \geq t$ and so

$$\bigwedge_{z \in x \cdot y} \mu(z) \geq \mu(y) \wedge 0.5 \geq t \wedge 0.5 = t.$$

Thus, for every $z \in x \cdot y$ we have $\mu(z) \geq t$ or $z \in \mu_t$, so $x \cdot y \subseteq \mu_t$.

Conversely, let μ be a fuzzy subset of R such that μ_t $(\neq \emptyset)$ is a left H_v-ideal of R for all $0 < t \leq 0.5$. For every $x, y \in R$, we can write

$$\mu(x) \geq \mu(x) \wedge \mu(y) \wedge 0.5 = t_0,$$
$$\mu(y) \geq \mu(x) \wedge \mu(y) \wedge 0.5 = t_0,$$

then $x \in \mu_{t_0}$ and $y \in \mu_{t_0}$, so $x + y \subseteq \mu_{t_0}$. Thus, for every $z \in x + y$ we have $\mu(z) \geq t_0$ which implies that

$$\bigwedge_{z \in x+y} \mu(z) \geq t_0,$$

and in this way the condition $(1')$ of Proposition 4.2.2 is verified. To verify the second condition, for every $a, x \in R$, we put $t_1 = \mu(a) \wedge \mu(x) \wedge 0.5$. Then $x \in \mu_{t_1}$ and $a \in \mu_{t_1}$. So there exists $y \in \mu_{t_1}$ such that $x \in a+y$. Since $y \in \mu_{t_1}$, we have $\mu(y) \geq t_1$ or

$$\mu(y) \geq \mu(a) \wedge \mu(x) \wedge 0.5.$$

The third condition is verified similarly.

Now, let $x, y \in R$. We can write

$$\mu(y) \geq \mu(y) \wedge 0.5 = t_0.$$

Then, $y \in \mu_{t_0}$ and so $x \cdot y \subseteq \mu_{t_0}$. Therefore, for every $z \in x \cdot y$ we have $\mu(z) \geq t_0$ which implies that

$$\bigwedge_{z \in x \cdot y} \mu(z) \geq t_0,$$

and in this way the condition $(4')$ of Proposition 4.2.2 is verified. ■

Naturally, a corresponding result should be considered when μ_t is a left H_v-ideal of R for all $t \in (0.5, 1]$.

Theorem 4.2.5. *Let R be an H_v-ring and μ a fuzzy subset of R. Then, μ_t $(\neq \emptyset)$ is a left (right) H_v-ideal of R for all $t \in (0.5, 1]$ if and only if*

(1) $\mu(x) \wedge \mu(y) \leq \bigwedge_{z \in x+y} (\mu(z) \vee 0.5)$, *for all $x, y \in R$.*

(2) *For all $x, a \in R$ there exists $y \in R$ such that $x \in a + y$ and*

$$\mu(a) \wedge \mu(x) \leq \mu(y) \vee 0.5.$$

(3) *For all $x, a \in R$ there exists $z \in R$ such that $x \in z + a$ and*

$$\mu(a) \wedge \mu(x) \leq \mu(z) \vee 0.5.$$

(4) $\mu(y) \leq \bigwedge_{z \in x \cdot y} (\mu(z) \vee 0.5)$, *for all $x, y \in R$.*

Proof. If there exist $x, y, z \in R$ with $z \in x + y$ such that

$$\mu(z) \vee 0.5 < \mu(x) \wedge \mu(y) = t,$$

then $t \in (0.5, 1]$, $\mu(z) < t$, $x \in \mu_t$, and $y \in \mu_t$. Since $x, y \in \mu_t$ and μ_t is a left H_v-ideal, so $x + y \subseteq \mu_t$ and $\mu(z) \geq t$, for all $z \in x + y$, which is a contradiction with $\mu(z) < t$. Thus,

$$\mu(x) \wedge \mu(y) \leq \mu(z) \vee 0.5, \text{ for all } x, y, z \in R \text{ with } z \in x + y,$$

which implies that

$$\mu(x) \wedge \mu(y) \leq \bigwedge_{z \in x+y} (\mu(z) \vee 0.5), \text{ for all } x, y \in R.$$

Hence, (1) holds.

Now, assume that there exist $x_0, a_0 \in R$ such that for all $y \in R$ with $x_0 \in a_0 + y$, the following inequality holds:

$$\mu(y) \vee 0.5 < \mu(a_0) \wedge \mu(x_0) = t.$$

Then, $t \in (0.5, 1]$, $x_0 \in \mu_t$, $a_0 \in \mu_t$ and $\mu(y) < t$. Since $x_0, a_0 \in \mu_t$ and μ_t is a left H_v-ideal, so there exists $y_0 \in \mu_t$ such that $x_0 \in a_0 + y_0$. From $y_0 \in \mu_t$, we get $\mu(y_0) \geq t$, which is a contradiction with $\mu(y_0) < t$. Therefore, for all $x, a \in R$ there exists $y \in R$ such that $x \in a + y$ and

$$\mu(a) \wedge \mu(x) \leq \mu(y) \vee 0.5.$$

Hence, (2) holds.

The proof of third condition is similar to the proof of second condition.

Now, if there exist $x, y \in R$ with $z \in x \cdot y$ such that

$$\mu(z) \vee 0.5 < \mu(y) = t,$$

then $t \in (0.5, 1]$, $\mu(z) < t$, $y \in \mu_t$. Since $y \in \mu_t$ and μ_t is a left H_v-ideal, so $x \cdot y \subseteq \mu_t$ and $\mu(z) \geq t$ for all $z \in x \cdot y$, which is a contradiction with $\mu(z) < t$. Therefore

$$\mu(y) \leq \mu(z) \vee 0.5, \text{ for all } y \in R \text{ with } z \in x \cdot y,$$

which implies that

$$\mu(y) \leq \bigwedge_{z \in x \cdot y} (\mu(z) \vee 0.5), \text{ for all } x, y \in R.$$

Hence, (4) holds.

Conversely, assume that $t \in (0.5, 1]$ and $x, y \in \mu_t$. Then,

$$0.5 < t \leq \mu(x) \wedge \mu(y) \leq \bigwedge_{z \in x+y} (\mu(z) \vee 0.5).$$

It follows that for every $z \in x+y$, $0.5 < t \leq \mu(z) \vee 0.5$ and so $t \leq \mu(z)$, which implies $z \in \mu_t$. Hence $x + y \subseteq \mu_t$.

Now, we prove the reproducibility rule. Let $x, a \in \mu_t$. Then using condition (2), there exists $y \in R$ such that $x \in a + y$ and

$$\mu(a) \wedge \mu(x) \leq \mu(y) \vee 0.5.$$

We show that $y \in \mu_t$. We have

$$0.5 < t \leq \mu(x) \leq \mu(a) \wedge \mu(x) \leq \mu(y) \vee 0.5.$$

It follows that $0.5 \leq \mu(y)$ and so $y \in \mu_t$. Therefore, $\mu_t = a + \mu_t$, for all $a \in \mu_t$. Similarly, we have $\mu_t = \mu_t + a$, for all $a \in \mu_t$.

Now, assume that $t \in (0.5, 1]$, $y \in \mu_t$ and $x \in R$. Then

$$0.5 < t \leq \mu(y) \leq \bigwedge_{z \in x \cdot y} (\mu(z) \vee 0.5).$$

It follows that for every $z \in x \cdot y$, $0.5 < t \leq \mu(z) \vee 0.5$ and so $t \leq \mu(z)$, which implies that $z \in \mu_t$. It follows that $x \cdot y \subseteq \mu_t$. Therefore, μ_t is a left H_v-ideal of R for all $t \in (0.5, 1]$. ■

Let μ be a fuzzy subset of an H_v-ring R and

$$J = \{t \mid t \in (0, 1] \text{ and } \mu_t \text{ is the empty set or a left (right) } H_v\text{−ideal of } R\}.$$

When $J = (0, 1]$, then μ is an ordinary fuzzy left (right) H_v-ideal of the H_v-ring R. When $J = (0, 0.5]$, μ is an $(\in, \in \vee q)$-fuzzy left (right) H_v-ideal of R.

Now, we define the concept of fuzzy H_v-ideal with thresholds in the following way.

Definition 4.2.6. Let $r, s \in [0,1]$ and $r < s$. Let μ be a fuzzy subset of an H_v-ring R. Then μ is called a *fuzzy left (right) H_v-ideal with thresholds* (r, s) *of* R if

(1) $\mu(x) \wedge \mu(y) \wedge s \leq \bigwedge_{z \in x+y} (\mu(z) \vee r)$, for all $x, y \in R$.

(2) For all $x, a \in R$ there exists $y \in R$ such that $x \in a + y$ and

$$\mu(a) \wedge \mu(x) \wedge s \leq \mu(y) \vee r.$$

(3) For all $x, a \in R$ there exists $z \in R$ such that $x \in z + a$ and

$$\mu(a) \wedge \mu(x) \wedge s \leq \mu(z) \vee r.$$

(4) $\mu(y) \wedge s \leq \bigwedge_{z \in x \cdot y} (\mu(z) \vee r)$, for all $x, y \in R$
$(\mu(x) \wedge s \leq \bigwedge_{z \in x \cdot y} (\mu(z) \vee r)$, for all $x, y \in R)$.

If μ is a fuzzy left (right) H_v-ideal with thresholds of R, then we can conclude that μ is an ordinary fuzzy left (right) H_v-ideal when $r = 0$, $s = 1$; and μ is an $(\in, \in \vee q)$-fuzzy left (right) H_v-ring when $r = 0$, $s = 0.5$.

Now, we characterize fuzzy left (right) H_v-ideals with thresholds by their level left (right) H_v-ideals.

Theorem 4.2.7. *A fuzzy subset μ of an H_v-ring R is a fuzzy left (right) H_v-ideal with thresholds (r, s) of R if and only if μ_t $(\neq \emptyset)$ is a left (right) H_v-ideal of R for all $t \in (r, s]$.*

Proof. Let μ be a fuzzy left H_v-ideal with thresholds of R and $t \in (r, s]$. Let $x, y \in \mu_t$. Then, $\mu(x) \geq t$ and $\mu(y) \geq t$. Now,

$$\bigwedge_{z \in x+y} (\mu(z) \vee r) \geq \mu(x) \wedge \mu(y) \wedge s \geq t \wedge s \geq t > r.$$

So for every $z \in x + y$ we have $\mu(z) \vee r \geq t > r$ which implies $\mu(z) \geq t$ and $z \in \mu_t$. Hence $x + y \subseteq \mu_t$. Now, let $x, a \in \mu_t$, then there exists $y \in R$ such that $x \in a + y$ and $\mu(a) \wedge \mu(x) \wedge s \leq \mu(y) \vee r$. From $x, a \in \mu_t$, we have $\mu(x) \geq t$ and $\mu(a) \geq t$, and so

$$r < t \leq t \wedge s \leq \mu(a) \wedge \mu(x) \wedge s \leq \mu(y) \vee r,$$

which implies that $\mu(y) \geq t$, and so $y \in \mu_t$. Therefore, we have $\mu_t = a + \mu_t$ for all $a \in \mu_t$. Similarly, we get $\mu_t + a = \mu_t$ for all $a \in \mu_t$.

Now, let $y \in \mu_t$ and $x \in R$. Then, $\mu(x) \geq t$, and so

$$\bigwedge_{z \in x \cdot y} (\mu(z) \vee r) \geq \mu(x) \wedge s \geq t \wedge s \geq t > r.$$

So for every $z \in x \cdot y$ we have $\mu(z) \vee r \geq t > r$ which implies $\mu(z) \geq t$ and $z \in \mu_t$. Hence $x \cdot y \subseteq \mu_t$.

Therefore μ_t is a left H_v-ideal of R, for all $t \in (r, s]$.

Conversely, let μ be a fuzzy subset of R such that μ_t $(\neq \emptyset)$ is a left H_v-ideal of R for all $t \in (r, s]$. If there exist $x, y, z \in R$ with $z \in x + y$ such that

$$\mu(z) \vee r < \mu(x) \wedge \mu(y) \wedge s = t.$$

Then, $t \in (r, s]$, $\mu(z) < t$, $x \in \mu_t$ and $y \in \mu_t$. Since μ_t is a left H_v-ideal of R and $x, y \in \mu_t$, it follows that $x + y \subseteq \mu_t$. Hence $\mu(z) \geq t$ for all $z \in x + y$. This is a contradiction with $\mu(z) < t$. Therefore,

$$\mu(x) \wedge \mu(y) \wedge s \leq \mu(z) \vee r, \quad \text{for all } x, y, z \in R \text{ with } z \in x + y,$$

which implies that

$$\mu(x) \wedge \mu(y) \wedge s \leq \bigwedge_{z \in x+y} (\mu(z) \vee r), \text{ for all } x, y \in R.$$

Hence, condition (1) of Definition 4.2.6 holds.

Now, assume that there exist $x_0, a_0 \in R$ such that for all $y \in R$ which satisfies $x_0 \in a_0 + y$, the following inequality holds:

$$\mu(y) \vee r < \mu(a_0) \wedge \mu(x_0) \wedge s = t.$$

Then, $t \in (r, s]$, $x_0 \in \mu_t$, $a_0 \in \mu_t$ and $\mu(y) < t$. Since $x_0, a_0 \in \mu_t$ and μ_t is a left H_v-ideal, it follows that there exists $y_0 \in \mu_t$ such that $x_0 \in a_0 + y_0$. From $y_0 \in \mu_t$, we get $\mu(y_0) \geq t$. This is a contradiction with $\mu(y_0) < t$. Therefore

$$\mu(a) \wedge \mu(x) \wedge s \leq \mu(y) \vee r.$$

Hence, the second condition of Definition 4.2.6 holds. The proof of third condition is similar to the proof of second condition.

If there exist $x, z \in R$ with $z \in x \cdot y$ such that

$$\mu(z) \vee r < \mu(x) \wedge \mu(y) \wedge s = t,$$

then $t \in (r, s]$, $\mu(z) < t$, $y \in \mu_t$. Since μ_t is a left H_v-ideal of R and $x \in \mu_t$, so $x \cdot y \subseteq \mu_t$. Hence $\mu(z) \geq t$, for all $z \in x \cdot y$. This is a contradiction with $\mu(z) < t$. Therefore

$$\mu(y) \wedge s \leq \mu(z) \vee r, \text{ for all } x, z \in R \text{ with } z \in x \cdot y,$$

which implies that

$$\mu(y) \wedge s \leq \bigwedge_{z \in x \cdot y} (\mu(z) \vee r), \text{ for all } x \in R.$$

Hence, condition (4) of Definition 4.2.6 holds. ■

4.3 Interval-Valued Fuzzy H_v-Ideals

In 1975, Zadeh [203] introduced the concept of interval-valued fuzzy subsets, where the values of the membership functions are intervals of numbers and not just numbers. In [14], Biswas defined interval-valued fuzzy subgroups of a group. In [60], Davvaz defined interval-valued fuzzy ideals of a hyperring. Now, in this section we consider the notion of interval-valued fuzzy H_v-ideals. A necessary and sufficient condition for an interval-valued fuzzy subset to be an interval-valued fuzzy H_v-ideal is stated and some basic results are discussed.

Definition 4.3.1. Let X be a set. An *interval-valued fuzzy subset* (i.e., *i-v fuzzy subset*) F defined on X is given by

$$F = \{(x, [\mu_F^L(x), \mu_F^U(x)]) \mid x \in X\},$$

where μ_F^L and μ_F^U are two fuzzy subsets of X such that $\mu_F^L(x) \leq \mu_F^U(x)$, for all $x \in X$.

Suppose that $\hat{\mu}_F(x) = [\mu_F^L(x), \mu_F^U(x)]$ and $D[0,1]$ denotes the family of all closed subintervals of $[0,1]$. If $\mu_F^L(x) = \mu_F^U(x) = c$, where $0 \leq c \leq 1$, then we have $\hat{\mu}_F(x) = [c,c]$ which we also assume, for the sake of convenience, to belong to $D[0,1]$. Thus, $\hat{\mu}_F(x) \in D[0,1]$, for all $x \in X$. Therefore, the i-v fuzzy subset F is given by

$$F = \{(x, \hat{\mu}_F(x)\ ) | \ x \in X\}, \text{ where } \hat{\mu}_F : X \to D[0,1].$$

Definition 4.3.2. Let $D_1 = [a_1, b_1], D_2 = [a_2, b_2]$ and $D_i = [a_i, b_i]$ be elements of $D[0,1]$. Then, we define

- $r\max(D_1, D_2) = [\max(a_1, a_2), \max(b_1, b_2)]$,
- $r\min(D_1, D_2) = [\min(a_1, a_2), \min(b_1, b_2)]$,
- $r\sup\limits_i\{D_i\} = [\sup\limits_i\{a_i\}, \sup\limits_i\{b_i\}]$,
- $r\inf\limits_i\{D_i\} = [\inf\limits_i\{a_i\}, \inf\limits_i\{b_i\}]$.

We call $D_2 \leq D_1$ if and only if $a_2 \leq a_1$ and $b_2 \leq b_1$.

Definition 4.3.3. Let f be a mapping from a set X into a set Y. Let A be an i-v fuzzy subset of X. Then, the *image* of A, i.e., $f[A]$ is the i-v fuzzy subset of Y with the membership function defined by

$$\hat{\mu}_{f[A]}(y) = \begin{cases} r \sup\limits_{z \in f^{-1}(y)} \{\hat{\mu}_A(z)\} \text{ if } f^{-1}(y) \neq \emptyset \\ \\ [0,0] \qquad\qquad \text{otherwise} \end{cases}$$

for all $y \in Y$.

Definition 4.3.4. Let f be a mapping from a set X into a set Y. Let B be an i-v fuzzy subset of Y. Then, the *inverse image* of B, i.e, $f^{-1}[B]$ is the i-v fuzzy subset of X with the membership function given by

$$\hat{\mu}_{f^{-1}[B]} = \hat{\mu}_B(f(x)), \text{ for all } x \in X.$$

Definition 4.3.5. An i-v fuzzy subset F of a ring $(R, +, \cdot)$ is called a *left* (respectively, *right*) *i-v fuzzy ideal* if

(1) $\hat{\mu}_F(x + y) \geq r\min\{\hat{\mu}_F(x), \hat{\mu}_F(y)\}$, for all $x, y \in R$.
(2) $\hat{\mu}_F(-x) \geq \hat{\mu}_F(x)$, for all $x \in R$.
(3) $\hat{\mu}_F(x \cdot y) \geq \hat{\mu}_F(y)$ (respectively, $\hat{\mu}_F(x \cdot y) \geq \hat{\mu}_F(x)$, for all $x, y \in R$.

Now, we define interval-valued fuzzy H_v-ideals in an H_v-ring and then we prove some results in this connection.

Definition 4.3.6. An i-v fuzzy subset F of an H_v-ring R is called a *left* (respectively *right*) *i-v fuzzy H_v-ideal* if the following statements hold.

(1) $r\min\{\hat{\mu}_F(x), \hat{\mu}_F(y)\} \leq r \inf\limits_{\alpha \in x+y} \{\hat{\mu}_F(\alpha)\}$, for all $x, y \in R$.
(2) For all $x, a \in R$ there exists $y \in R$ such that $x \in a + y$ and

$$r\min\{\hat{\mu}_F(x), \hat{\mu}_F(a)\} \leq \hat{\mu}_F(y).$$

(3) For all $x, a \in R$ there exists $z \in R$ such that $x \in z + a$ and

$$r\min\{\hat{\mu}_F(x), \hat{\mu}_F(a)\} \leq \hat{\mu}_F(z).$$

(4) $\hat{\mu}_F(y) \leq r \inf\limits_{\alpha \in x \cdot y} \{\hat{\mu}_F(\alpha)\}$ (respectively, $\hat{\mu}_F(x) \leq r \inf\limits_{\alpha \in x \cdot y} \{\hat{\mu}_F(\alpha)\}$), for all $x, y \in R$.

Here we present all the proofs for the left H_v-ideals. For right H_v-ideals similar results hold as well.

Definition 4.3.7. Let X be a set and F be an i-v fuzzy subset of X. Then, we define $F_{[t,s]} = \{x \in X \mid \hat{\mu}_F(x) \geq [t, s]\}$.

Theorem 4.3.8. *Let R be an H_v-ring and F be an i-v fuzzy subset of R. Then, F is an i-v fuzzy H_v-ideal if and only if for every t, s, where $0 \leq t \leq s \leq 1, F_{[t,s]} \neq \emptyset$ is an H_v-ideal of R.*

Proof. Suppose that F is an i-v fuzzy H_v-ideal of R. For every $x, y \in F_{[t,s]}$ we have $\hat{\mu}_F(x) \geq [t, s]$ and $\hat{\mu}_F(y) \geq [t, s]$, hence

$$r\min\{\hat{\mu}_F(x), \hat{\mu}_F(y)\} \geq [t, s]$$

and so $r \inf\limits_{\alpha \in x+y} \{\hat{\mu}_F(\alpha)\} \geq [t, s]$. Therefore, for every $\alpha \in x + y$ we have $\alpha \in F_{[t,s]}$, so $x+y \subseteq F_{[t,s]}$. Hence, for every $a \in F_{[t,s]}$ we have $a+F_{[t,s]} \subseteq F_{[t,s]}$.

Now, let $x \in F_{[t,s]}$. Then, there exists $y \in H$ such that $x \in a+y$ and

$$r\min\{\hat{\mu}_F(x), \hat{\mu}_F(a)\} \leq \hat{\mu}_F(y).$$

Form $x \in F_{[t,s]}, a \in F_{[t,s]}$ we get $r\min\{\hat{\mu}_F(x), \hat{\mu}_F(a)\} \geq [t,s]$ and so $\hat{\mu}_F(y) \geq [t,s]$ or $y \in F_{[t,s]}$ and this proves $F_{[t,s]} \subseteq a + F_{[t,s]}$. Similarly, we get $F_{[t,s]} = F_{[t,s]} + a$.

Now, we prove that $R \cdot F_{[t,s]} \subseteq F_{[t,s]}$. For every $x \in F_{[t,s]}$ and $r \in R$ we show that $r \cdot x \subseteq F_{[t,s]}$. Since F is a left i-v fuzzy H_v-ideal we have $[t,s] \leq \hat{\mu}_F(x) \leq r \inf\limits_{\alpha \in r\cdot x} \{\hat{\mu}_F(\alpha)\}$. Therefore, for every $\alpha \in r \cdot x$ we get $\hat{\mu}_F(\alpha) \geq [t,s]$ which implies that $\alpha \in F_{[t,s]}$, so $r \cdot x \subseteq F_{[t,s]}$.

Conversely, assume that for every t, s where $0 \leq t \leq s \leq 1$, $F_{[t,s]} \neq \emptyset$ is an H_v-ideal of R. For every $x, y \in R$ if we put $[t_0, s_0] = r\min\{\hat{\mu}_F(x), \hat{\mu}_F(y)\}$, then $x \in F_{[t_0,s_0]}, y \in F_{[t_0,s_0]}$ and so $x+y \subseteq F_{[t_0,s_0]}$. Thus, for every $\alpha \in x+y$ we have $\alpha \in F_{[t_0,s_0]}$ implying $r \inf\limits_{\alpha \in x+y} \{\hat{\mu}_F(\alpha)\} \geq r\min\{\hat{\mu}_F(x), \hat{\mu}_F(y)\}$ and in this way the condition (1) of Definition 4.3.6 is verified. To verify the second condition, for every $a, x \in R$ we put $[t_1, s_1] = r\min\{\hat{\mu}_F(a), \hat{\mu}_F(x)\}$. Then, $x \in F_{[t_1,s_1]}$ and $a \in F_{[t_1,s_1]}$, so there exists $y \in F_{[t_1,s_1]}$ such that $x \in a+y$. On the other hand, since $y \in F_{[t_1,s_1]}$ then $\hat{\mu}_F(y) \geq [t_1, s_1]$ and hence $r\min\{\hat{\mu}_F(a), \hat{\mu}_F(x)\} \leq \hat{\mu}_F(y)$. In the similar way, the third condition of Definition 4.3.6 is valid. Now we prove the forth condition of Definition 4.3.6. For every $x, y \in R$, we put $\hat{\mu}_F(y) = [t_2, s_2]$. Then, $y \in F_{[t_2,s_2]}$. Since $F_{[t_2,s_2]}$ is an H_v-ideal of R, $x \cdot y \subseteq F_{[t_2,s_2]}$. Therefore, for every $\alpha \in x \cdot y$ we have $\alpha \in F_{[t_2,s_2]}$ which implies that $\hat{\mu}_F(\alpha) \geq [t_2, s_2]$. Hence, $r \inf\limits_{\alpha \in x\cdot y} \{\hat{\mu}_F(\alpha)\} \geq [t_2, s_2]$ and so $\hat{\mu}_F(y) \leq [t_2, s_2]$. Thus, $r \inf\limits_{\alpha \in x\cdot y} \{\hat{\mu}_F(\alpha)\} \geq \hat{\mu}_F(y)$. ■

Proposition 4.3.9. *$[\mu_F^L, \mu_F^U]$ is an i-v fuzzy H_v-ideal of an H_v-ring R if and only if μ_F^L and μ_F^U are fuzzy H_v-ideals of R.*

Proof. The proof is straightforward. ■

Proposition 4.3.10. *Let R_1 and R_2 be two H_v-rings and f be a homomorphism from R_1 into R_2. Let F be an i-v fuzzy H_v-ideal of R_2. Then, the inverse image $f^{-1}[F]$ of F is an i-v fuzzy H_v-ideal of R_1.*

Proof. The proof is straightforward. ■

Proposition 4.3.11. *Let R_1 and R_2 be two H_v-rings and f be an epimorphism from R_1 onto R_2 and F be an i-v fuzzy H_v-ideal of R_1. Then,*

$$(f[F])_{[t,s]} = f(F_{[t,s]}).$$

Proof. For every $y \in (f[F])_{[t,s]}$ we have $\hat{\mu}_{f[F]}(y) \geq [t,s]$ which implies that $r\sup\limits_{z \in f^{-1}(y)} \{\hat{\mu}_F(z)\} \geq [t,s]$. Therefore, there exists $z_0 \in f^{-1}(y)$ such that $\hat{\mu}_F(z_0) \geq [t,s]$ which implies that $z_0 \in F_{[t,s]}$ and so $f(z_0) \in f(F_{[t,s]})$ implying $y \in f(F_{[t,s]})$.

Now, for every $y \in f(F_{[t,s]})$ there exists $x \in F_{[t,s]}$ such that $f(x) = y$. Since $x \in F_{[t,s]}$ we have $\hat{\mu}_F(x) \geq [t,s]$ and so $r\sup_{x \in f^{-1}(y)} \{\hat{\mu}_F(x)\} \geq [t,s]$ which implies that $\hat{\mu}_{f[F]}(y) \geq [t,s]$. Therefore, $y \in (f[F])_{[t,s]}$. ■

Corollary 4.3.12. *Let R_1 and R_2 be two H_v-rings and f be an epimorphism from R_1 onto R_2. If F is an i-v fuzzy H_v-ideal of R_1, then $f[F]$ is an i-v fuzzy H_v-ideal of R_2.*

Proof. Suppose that F is an i-v fuzzy H_v-ideal of R_1. By Theorem 4.3.8, for every $t, s, 0 \leq t \leq s \leq 1$, $F_{[t,s]}(F_{[t,s]} \neq \emptyset)$ is an H_v-ideal of R_1. Thus, $f(F_{[t,s]})$ is an H_v-ideal of R_2. By Proposition 4.3.11, we obtain $(f[F])_{[t,s]}$ is an H_v-ideal of R_2. Therefore, $f[F]$ is an i-v fuzzy H_v-ideal of R_2. ■

4.4 Fuzzy Hyperrings

Using the fuzzy semihypergroup notion, introduced in [166], Leoreanu-Fotea and Davvaz defined and studied the fuzzy hyperring notion and connections with hyperrings [136]. In this section, instead of the set of all non-empty subsets of H, denoted by $\mathcal{P}^*(H)$, we consider the set of all non-zero fuzzy subsets of H, denoted by $\mathcal{F}^*(H)$. Then, we consider and study the subfuzzy-structures of a such notion and homomorphisms between fuzzy hyperrings and fundamental relations on fuzzy hyperrings. The main reference for this section is [136, 166]. First, we recall the fuzzy semihypergroup notion, introduced and studied in [166].

Let S be a non-empty set. $\mathcal{F}^*(S)$ denotes the set of all non-zero fuzzy subsets of S. A *fuzzy hyperoperation* on S is a map $\circ : S \times S \to \mathcal{F}^*(S)$, which associates a nonzero fuzzy subset $a \circ b$ with any pair (a, b) of elements of S.

The couple $(S, \circ)$ is called a *fuzzy hypergroupoid.*

A fuzzy hypergroupoid $(S, \circ)$ is called a *fuzzy semihypergroup* if for all $a, b, c \in S$, we have $a \circ (b \circ c) = (a \circ b) \circ c$, where for any $\mu \in F^*(S)$, we have $(a \circ \mu)(r) = \bigvee_{t \in S} ((a \circ t)(r) \wedge \mu(t))$ and $(\mu \circ a)(r) = \bigvee_{t \in S} (\mu(t) \wedge (t \circ a)(r))$ for all $r \in S$.

Let μ and λ be two nonzero fuzzy subsets of a fuzzy hypergroupoid $(S, \circ)$. We define $(\mu \circ \lambda)(t) = \bigvee_{p,q \in S} (\mu(p) \wedge (p \circ q)(t) \wedge \lambda(q))$, for all $t \in S$.

Let us define now the notion of a fuzzy hyperring.

Definition 4.4.1. Let R be a non-empty set and $+$, $\cdot$ be two fuzzy hyperoperations on R. The triple $(R, \oplus, \odot)$ is called a *fuzzy hyperring* if the following axioms hold.

(1) $a \oplus (b \oplus c) = (a \oplus b) \oplus c$, for all $a, b, c \in R$.
(2) $x \oplus R = R \oplus x = \chi_R$, for all $x \in R$.
(3) $a \oplus b = b + a$, for all $a, b \in R$.
(4) $a \odot (b \odot c) = (a \odot b) \odot c$, for all $a, b, c \in R$.

(5) $\odot$ is distributive over the addition $\oplus$, i.e., for all x, y, z of R we have $x \odot (y \oplus z) = x \odot y \oplus x \odot z$ and $(x \oplus y) \odot z = x \odot z \oplus y \odot z$.

The following three theorems provide us some examples of fuzzy hyperrings:

Theorem 4.4.2. *Let $(R, +, \odot)$ be a ring and μ be a fuzzy semigroup of $(R, \odot)$. Define the following fuzzy hyperoperations on R:*

$$\text{for all } a, b \in R,\ a \oplus b = \chi_{a+b} \text{ and}$$

$$(a \odot b)(t) = \begin{cases} \mu(a) \wedge \mu(b) & \text{if } t = ab \\ 0 & \text{otherwise.} \end{cases}$$

The system $(R, \oplus, \odot)$ is a fuzzy hyperring if and only if μ is constant.

Proof. It is easy to see that $(R, \oplus)$ is a commutative fuzzy hypergroup and $(R, \odot)$ is a fuzzy semihypergroup.

On the other hand, for all $a, b, c \in R$, we have

$$((a \oplus b) \odot c)(t) = (\chi_{a+b} \odot c)(t) = \bigvee_{r \in R} (\chi_{a+b}(r) \wedge (r \odot c)(t))$$
$$= ((a+b) \odot c)(t) = \begin{cases} \mu(a+b) \wedge \mu(c) \text{ if } t = (a+b)c \\ 0 \qquad \text{otherwise} \end{cases}$$

and

$$((a \odot c) \oplus (b \odot c))(t) = \bigvee_{p,q \in R} ((a \odot c)(p) \wedge (p \oplus q)(t) \wedge (b \odot c)(q))$$
$$= \mu(a) \wedge \mu(c) \wedge \mu(b) \wedge \mu(c) \wedge \chi_{ac+bc}(t)$$
$$= \begin{cases} \mu(a) \wedge \mu(b) \wedge \mu(c) \text{ if } t = ac + bc \\ 0 \qquad \text{otherwise.} \end{cases}$$

If μ is constant, then clearly the distributivity holds. Conversely, from the distributivity, it follows that for all $a, b \in R$, we have $(a \oplus b) \odot (a + b) = (a \odot (a+b)) \oplus (b \odot (a+b))$, whence $\mu(a+b) = \mu(a) \wedge \mu(b) \wedge \mu(a+b)$, which means that $\mu(a+b) \leq \mu(a) \wedge \mu(b) \leq \mu(a)$. For $x = a + b$ and $y = -b$, we obtain

$$\mu(x+y) = \mu(a) \leq \mu(x) = \mu(a+b).$$

Hence, $\mu(a) = \mu(a+b)$ for all $a, b \in R$, which means that μ is constant. ∎

Theorem 4.4.3. *Let $(S, +, \cdot)$ be a semigroup. If we define the following hyperoperations on S:*

$$\text{for all } a, b \text{ of } S,\ a \oplus b = \chi_{\{a,b\}} \text{ and } a \odot b = \chi_{ab},$$

then $(S, \oplus, \odot)$ is a fuzzy hyperring.

Proof. It is easy to see that $(S, \oplus)$ is a commutative fuzzy hypergroup and $(S, \odot)$ is a fuzzy semihypergroup. We check now the distributivity. Let $a, b, c \in S$. We have

$$\begin{aligned}((a \oplus b) \odot c)(t) &= \bigvee_{r \in S} (\chi_{\{a,b\}}(r) \wedge (r \odot c)(t)) \\ &= (a \odot c)(t) \vee (b \odot c)(t) = \chi_{ac}(t) \vee \chi_{bc}(t) \\ &= \begin{cases} 1 & \text{if } t \in \{ac, bc\} \\ 0 & \text{otherwise} \end{cases}\end{aligned}$$

and

$$\begin{aligned}((a \odot c) \oplus (b \odot c))(t) &= \bigvee_{p,q \in S} ((a \odot c)(p) \wedge (b \odot c)(q) \wedge (p \oplus q)(t)) \\ &= \bigvee_{p,q \in S} [((a \odot c)(t) \wedge (b \odot c)(q)) \vee ((a \odot c)(p) \wedge (b \odot c)(t))] \\ &= \bigvee_{p,q \in S} [(\chi_{ac}(t) \wedge \chi_{bc}(q)) \vee (\chi_{ac}(p) \wedge \chi_{bc}(t))] = \\ &= \begin{cases} 1 & \text{if } t \in \{ac, bc\} \\ 0 & \text{otherwise.} \end{cases}\end{aligned}$$

Hence, $(S, \oplus, \odot)$ is a fuzzy hyperring. ∎

Theorem 4.4.4. *Let $(S, +, \cdot)$ be a semigroup and μ be a fuzzy semigroup on S. If we define the following hyperoperations on S:*

$$\text{for all } a, b \in S,\ a \oplus b = \chi_{\{a,b\}} \text{ and}$$
$$(a \odot b)(t) = \begin{cases} \mu(a) \wedge \mu(b) & \text{if } t = ab \\ 0 & \text{otherwise} \end{cases}$$

then $(S, \oplus, \odot)$ is a fuzzy hyperring if and only if μ is constant.

Proof. Again, $(S, \oplus)$ is a commutative fuzzy hypergroup and $(S, \odot)$ is a fuzzy semihypergroup. On the other hand, for all $t \in S$, we have:

$$\begin{aligned}((a \oplus b) \odot c)(t) &= \bigvee_{r \in S} (\chi_{\{a,b\}}(r) \wedge (r \odot c)(t)) = (a \odot c)(t) \vee (b \odot c)(t) \\ &= \begin{cases} \mu(a) \wedge \mu(c) & \text{if } t = ac \\ \mu(b) \wedge \mu(c) & \text{if } t = bc \\ 0 & \text{otherwise} \end{cases}\end{aligned}$$

and

$$\begin{aligned}((a \odot c) \oplus (b \odot c))(t) &= \bigvee_{p,q \in S} ((a \odot c)(p) \wedge (p \oplus q)(t) \wedge (b \odot c)(q)) \\ &= \bigvee_{p,q \in S} ((a \odot c)(t) \wedge (b \odot c)(q)) \vee ((a \odot c)(p) \wedge (b \odot c)(t)) \\ &= \begin{cases} \mu(a) \wedge \mu(b) \wedge \mu(c) & \text{if } t = ac \\ \mu(a) \wedge \mu(b) \wedge \mu(c) & \text{if } t = bc \\ 0 & \text{otherwise.} \end{cases}\end{aligned}$$

If the distributivity holds, then for all a, b, c of R we have $\mu(a) \wedge \mu(c) = \mu(a) \wedge \mu(b) \wedge \mu(c) = \mu(b) \wedge \mu(c)$.

For $a = c$, we obtain $\mu(a) = \mu(a) \wedge \mu(b)$, whence $\mu(a) \leq \mu(b)$. From here, it follows that μ is constant.

Conversely, it is immediate. ∎

Now, we consider that $(G, \circ)$ is a fuzzy hypergroup and we define the following hyperoperation on G:

$$\text{for all } a, b \text{ of } G,\ a * b = \{x \in G \mid (a \circ b)(x) > 0\}.$$

Then,

Theorem 4.4.5. *If $(G, \circ)$ is a fuzzy hypergroup, then $(G, *)$ is a hypergroup.*

Proof. It is straightforward. ∎

Let us consider now $(R, \oplus, \odot)$ be a fuzzy hyperring. Define the following hyperoperations on R: for all a, b of R,

$$a + b = \{x \in R \mid (a \oplus b)(x) > 0\} \text{ and } a \cdot b = \{x \in R \mid (a \odot b)(x) > 0\}.$$

The next theorem represents a first connection between fuzzy hyperrings and hyperrings. More exactly, we associate a hyperring with any fuzzy hyperring.

Theorem 4.4.6. *If $(R, \oplus, \odot)$ is a fuzzy hyperring, then $(R, +, \cdot)$ is a hyperring.*

Proof. According to the above theorem, we have that $(R, +)$ is a commutative hypergroup, while $(R, \cdot)$ is a semihypergroup. Let us check now the distributivity. First of all, we check that for all a, b, c of R we have $(a+b)\cdot c \subset a\cdot c+b\cdot c$. If $x \in (a + b) \cdot c$, then there exists $y \in a + b$, such that $x \in y \cdot c$. This means that $(a \oplus b)(y) > 0$ and $(y \odot c)(x) > 0$. Hence,

$$((a \oplus b) \odot c)(x) = \bigvee_{p \in R} ((a \oplus b)(p) \wedge (p \odot c)(x)) \geq (a \oplus b)(y) \wedge (y \odot c)(x) > 0.$$

Since $(a \oplus b) \odot c = (a \odot c) \oplus (b \odot c)$, it follows that $((a \odot c) \oplus (b \odot c))(x) > 0$. This means that

$$\bigvee_{p,q \in R} (a \odot c)(p) \wedge (b \odot c)(q) \wedge (p \oplus q)(x) > 0,$$

hence there are $p, q \in R$, such that $(a \odot c)(p) > 0$, $(b \odot c)(q) > 0$ and $(p \oplus q)(x) > 0$. We have $p \in a \cdot c$, $q \in b \cdot c$ and $x \in p + q$, whence $x \in a \cdot c + b \cdot c$. Similarly, we obtain the converse inclusion and the equality $a \cdot (b + c) = a \cdot b + a \cdot c$. Therefore, $(R, +, \cdot)$ is a hyperring. ■

On the other hand, if $(G, *)$ is a hypergroup and we define the following fuzzy hyperoperation on G for all a, b of G, $a \circ b = \chi_{a*b}$, then the next result follows.

Theorem 4.4.7. *If $(G, *)$ is a hypergroup, then $(G, \circ)$ is a fuzzy hypergroup.*

Proof. It is straightforward. ■

Now, we consider $(R, +, \cdot)$ a hyperring and as above, we define

for all $a, b \in R$, $a \oplus b = \chi_{a+b}$ and $a \odot b = \chi_{a \cdot b}$.

The following theorem shows that we can associate a fuzzy hyperring with any hyperring. This represents a second connection between fuzzy hyperrings and hyperrings.

Theorem 4.4.8. *If $(R, +, \cdot)$ is a hyperring, then $(R, \oplus, \odot)$ is a fuzzy hyperring.*

Proof. According to the above theorem, $(R, \oplus)$ is a commutative fuzzy hypergroup and $(R, \odot)$ is a fuzzy semihypergroup. We have to check the distributivity. First, we show that for all a, b, c of R, we have

$$(a \oplus b) \odot c = a \odot c \oplus b \odot c.$$

For all $t \in R$, we have

$$\begin{aligned} ((a \oplus b) \odot c)(t) &= \bigvee_{r \in R} ((a \oplus b)(r) \wedge (r \odot c)(t)) \\ &= \bigvee_{r \in R} (\chi_{a+b}(r) \wedge \chi_{r \cdot c}(t)) \\ &= \begin{cases} 1 \text{ if } t \in (a+b) \cdot c \\ 0 \text{ otherwise.} \end{cases} \end{aligned}$$

On the other hand,

$$\begin{aligned} ((a \odot c) \oplus (b \odot c)(t)) &= \bigvee_{p,q \in R} (a \odot c)(p) \wedge (b \odot c)(q) \wedge (p \oplus q)(t) \\ &= \bigvee_{p,q \in R} (\chi_{a \cdot c}(p) \wedge \chi_{b \cdot c}(q) \wedge \chi_{p+q}(t)) \\ &= \begin{cases} 1 \text{ if } t \in a \cdot c + b \cdot c \\ 0 \text{ otherwise.} \end{cases} \end{aligned}$$

Since $(R, +, \cdot)$ is a hyperring, it follows that

$$(a \oplus b) \odot c = a \odot c + b \odot c.$$

Similarly, we have that $a \odot (b \oplus c) = (a \odot b) \oplus (a \odot c)$. Therefore, $(R, \oplus, \odot)$ is a fuzzy hyperring. ■

Denote by $\mathcal{FHR}$ the class of all fuzzy hyperrings and by $\mathcal{HR}$ the class of all hyperrings. We define the following two maps:

$$\varphi : \mathcal{HR} \to \mathcal{FHR},\ \varphi((R, +, \cdot)) = (R, \oplus, \odot)$$

where for all a, b of R we have $a \oplus b = \chi_{a+b}$ and $a \odot b = \chi_{a \cdot b}$ and

$$\psi : \mathcal{FHR} \to \mathcal{HR},\ \psi((R, \oplus, \odot)) = (R, +, \cdot)$$

where for all a, b of R, we have $a + b = \{x \mid (a \oplus b)(x) > 0\}$ and $a \cdot b = \{x \mid (a \odot b)(x) > 0\}$.

REMARK 3 We have $\psi\varphi = 1_{\mathcal{HR}}$. Define the following equivalence relation on $\mathcal{FHR}$:

$$(R_1, \oplus_1, \odot_1) \sim (R_2, \oplus_2, \odot_2) \text{ if and only if } \psi((R_1, \oplus_1, \odot_1)) = \psi((R_2, \oplus_2, \odot_2)).$$

Hence,

$$\begin{array}{ll} (R_1, \oplus_1, \odot_1) \sim (R_2, \oplus_2, \odot_2) & \text{if and only if} \\ (a \oplus_1 b)(x) > 0 \Leftrightarrow (a \oplus_2 b)(x) > 0 & \text{and} \\ (a \odot_1 b)(x) > 0 \Leftrightarrow (a \odot_2 b)(x) > 0. & \end{array}$$

REMARK 4 There exists a bijection between $\mathcal{FHR}/_\sim$ and $\mathcal{HR}$. According to the above notations, a bijection is

$$\overline{\psi} : \mathcal{FHR}/_\sim \to \mathcal{HR},\ \overline{\psi}((\overline{R, \oplus, \odot})) = \psi(R, \oplus, \odot),$$

where $(\overline{R, \oplus, \odot})$ is the equivalence class of $(R, \oplus, \odot)$, with respect to the equivalence relation "$\sim$".

It is natural to consider homomorphisms between fuzzy hyperrings and see if there are connections between fuzzy hyperring homomorphisms and hyperring homomorphisms. In order to do this, we need the following notion.

Definition 4.4.9. If μ_1, μ_2 are fuzzy sets on R, then we say that μ_1 is *smaller* than μ_2 and we denote $\mu_1 \leq \mu_2$ if and only if for all $a \in R$, we have $\mu_1(a) \leq \mu_2(a)$.

Let $f : R_1 \to R_2$ be a map. If μ is a fuzzy set on R_1, then we define $f(\mu) : R_2 \to [0, 1]$, as follows

$$(f(\mu))(t) = \bigvee_{r \in f^{-1}(t)} \mu(r), \text{ if } f^{-1}(t) \neq \emptyset,$$

otherwise we consider $(f(\mu))(t) = 0$.

REMARK 5 If $f : R_1 \to R_2$ is a map and $a \in R_1$, then $f(\chi_a) = \chi_{f(a)}$. Indeed, for all $t \in R_1$, we have

$$(f(\chi_a))(t) = \bigvee_{r\in f^{-1}(t)} \chi_a(r) = \begin{cases} 1 & \text{if } f(a) = t \\ 0 & \text{otherwise} \end{cases} = \chi_{f(a)}(t).$$

We can introduce now the fuzzy hyperring homomorphism notion, as follows.

Definition 4.4.10. Let $(R_1, \oplus_1, \odot_1)$ and $(R_2, \oplus_2, \odot_2)$ be two fuzzy hyperrings and $f : R_1 \to R_2$ be a map. We say that f is a *homomorphism* of fuzzy hyperrings if for all a, b of R_1, we have

$$f(a \oplus_1 b) \leq f(a) \oplus_2 f(b) \text{ and } f(a \odot_1 b) \leq f(a) \odot_2 f(b).$$

The following two theorems present some connections between fuzzy hyperring homomorphisms and hyperring homomorphisms, that we have looking for.

Theorem 4.4.11. *Let $(R_1, \oplus_1, \odot_1)$ and $(R_2, \oplus_2, \odot_2)$ be two fuzzy hyperrings and $(R_1, +_1, \cdot_1) = \psi(R_1)$, $(R_2, +_2, \cdot_2) = \psi(R_2)$ be the associated hyperrings. If $f : R_1 \to R_2$ is a homomorphism of fuzzy hyperrings, then f is a homomorphism of the associated hyperrings, too.*

Proof. For all $a, b \in R$, we have

$$f(a \oplus_1 b) \leq f(a) \oplus_2 f(b) \text{ and } f(a \odot_1 b) \leq f(a) \odot_2 f(b).$$

Let $x \in a +_1 b$, which means that $(a \oplus_1 b)(x) > 0$ and let $t = \varphi(x)$. We have

$$(f(a \oplus_1 b))(t) = \bigvee_{r\in f^{-1}(t)} (a \oplus_1 b)(r) \geq (a \oplus_1 b)(x) > 0,$$

whence $(f(a) \oplus_2 f(b))(t) > 0$. Hence $t \in f(a) +_2 f(b)$. We obtain $f(a +_1 b) \subseteq f(a) +_2 f(b)$ and similarly, $f(a \cdot_1 b) \subseteq f(a) \cdot_2 f(b)$. ■

Theorem 4.4.12. *Let $(R_1, +_1, \cdot_1)$ and $(R_2, +_2, \cdot_2)$ be two hyperrings and $(R_1, \oplus_1, \odot_1) = \varphi(R_1)$, $(R_2, \oplus_2, \odot_2) = \varphi(R_2)$ be the associated fuzzy hyperrings. The map $f : R_1 \to R_2$ is a homomorphism of hyperrings if and only if it is a homomorphism of fuzzy hyperrings.*

Proof. Suppose that f is a homomorphism of hyperrings. Let $a, b \in R_1$. For all $t \in \operatorname{Im} f$ we have

$$\begin{aligned} (f(a \oplus_1 b))(t) &= \bigvee_{r\in f^{-1}(t)} (a \oplus_1 b)(r) = \bigvee_{r\in f^{-1}(t)} \chi_{a+_1 b}(r) \\ &= \begin{cases} 1 & \text{if } (a +_1 b) \cap f^{-1}(t) \neq \emptyset \\ 0 & \text{otherwise} \end{cases} = \begin{cases} 1 \text{ if } t \in f(a +_1 b) \\ 0 \text{ otherwise} \end{cases} \\ &= \chi_{f(a+_1 b)}(t) \leq \chi_{f(a)+_2 f(b)}(t) = (f(a) \oplus_2 f(b))(t). \end{aligned}$$

If $t \notin \operatorname{Im} f$, then $(f(a \oplus_1 b))(t) = 0 \leq (f(a) \oplus_2 f(b))(t)$. Hence, $f(a \oplus_2 b) \leq f(a) \oplus_2 f(b)$. Similarly, we obtain $f(a \odot_1 b) \leq f(a) \odot_2 f(b)$. Therefore, f is a homomorphism of fuzzy hyperrings.

Conversely, suppose that f is a homomorphism of fuzzy hyperrings and $a, b \in R_1$. Then, for all $t \in R_1$, we have

$$(f(a \oplus_1 b))(t) \leq (f(a) \oplus_2 (b))(t),$$

whence we obtain

$$\chi_{f(a+_1 b)}(t) \leq \chi_{f(a)+_2 f(b)}(t),$$

which means that

$$f(a +_1 b) \subseteq f(a) +_2 f(b).$$

Similarly, from $f(a \odot_1 b) \leq f(a) \odot_2 f(b)$ we obtain $f(a \cdot_1 b) \subseteq f(a) \cdot_2 f(b)$. Hence, f is a homomorphism of hyperrings. ■

REMARK 6 We can endow $\mathcal{HR}$ and $\mathcal{FHR}$ with category structures. The objects of $\mathcal{HR}$ are all the hyperrings and the morphisms are the homomorphisms of hyperrings, while the objects of $\mathcal{FHR}$ are the fuzzy hyperrings and the morphisms are homomorphisms of fuzzy hyperrings.

Let us consider now the algebraic subhyperstructures of fuzzy hyperrings. We give the following definition.

Definition 4.4.13. Let $(R, \oplus, \odot)$ be a fuzzy hyperring. A non-empty subset S is called

(a) a *subfuzzy hyperring* if for all $s_1, s_2 \in S$, the following conditions hold:
  1. if $s_1, s_2 \in S$ and $(s_1 \oplus s_2)(x) > 0$, then $x \in S$;
  2. for all $s \in S$, we have $s \oplus S = \chi_S$;
  3. if $s_1, s_2 \in S$ and $(s_1 \odot s_2))(x) > 0$, then $x \in S$.

(b) a *subfuzzy hyperideal* if the following conditions hold:
  1. if $s_1, s_2 \in S$ and $(s_1 \oplus s_2)(x) > 0$, then $x \in S$;
  2. for all $s \in S$, we have $s \oplus S = \chi_S$;
  3. if $s \in S$, $r \in R$ and $(s \odot r)(x) > 0$, or $(r \odot s)(x) > 0$, then $x \in S$.

The next two theorems point out on the connections between subfuzzy hyperrings (subfuzzy hyperideals) and subhyperrings (hyperideals respectively).

Theorem 4.4.14. (1) *If $(S, \oplus, \odot)$ is a subfuzzy hyperring of $(R, \oplus, \odot)$, then $(S, +, \cdot)$ is a subhyperring of $(R, +, \cdot)$.*
(2) *If $(S, \oplus, \odot)$ is a subfuzzy hyperideal of $(R, \oplus, \odot)$, then $(S, +, \cdot)$ is a hyperideal of $(R, +, \cdot)$.*

Now, let $(R, +, \cdot)$ be a hyperring and $(R, \oplus, \odot) = \varphi(R)$ be the associated fuzzy hyperring. For all $a, b \in R$, we have $a \oplus b = \chi_{a+b}$ and $a \odot b = \chi_{a \cdot b}$. We obtain:

Theorem 4.4.15. (1) *$(S, +, \cdot)$ is a subhyperring of $(R, +, \cdot)$ if and only if $(S, \oplus, \odot)$ is a subfuzzy hyperring of $(R, \oplus, \odot) = \varphi(R)$;*
(2) *$(S, +, \cdot)$ is a hyperideal of $(R, +, \cdot)$ if and only if $(S, \oplus, \odot)$ is a subfuzzy hyperideal of $(R, \oplus, \odot) = \varphi(R)$.*

REMARK 7 According to the above results, we can identify any hyperring $(R, +, \cdot)$ with the associated fuzzy hyperring $\varphi(R)$. In this manner, $\mathcal{HR}$ can be seen as a full subcategory of $\mathcal{FHR}$.

Let ρ be an equivalence relation on a fuzzy semihypergroup $(S, \circ)$ and let μ, u be two fuzzy subsets on S. We say that $\mu\rho u$ if the following two conditions hold.

(1) If $\mu(a) > 0$, then there exists $b \in S$, such that $u(b) > 0$ and $a\rho b$.
(2) If $u(x) > 0$, then there exists $y \in S$, such that $\mu(y) > 0$ and $x\rho y$.

In [166] are introduced the fuzzy regular relations on fuzzy semihypergroups, as follows.

An equivalence relation ρ on a fuzzy semihypergroup $(S, \circ)$ is called a *fuzzy regular relation* (or a *fuzzy hypercongruence*) on $(S, \circ)$ if, for all $a, b, c \in S$, the following implication holds:

$$a\rho b \Rightarrow (a \circ c)\rho(b \circ c) \text{ and } (c \circ a)\rho(c \circ b).$$

Clearly, the condition is equivalent to

$$a\rho a',\ b\rho b' \text{ implies } (a \circ b)\rho(a' \circ b'), \text{ for all } a, b, a', b' \text{ of } S.$$

Now, we can introduce the fuzzy regular relations on fuzzy hyperrings, in the following manner:

Definition 4.4.16. An equivalence relation ρ on a fuzzy hyperring $(R, \oplus, \odot)$ is called a *fuzzy regular relation* on $(R, \oplus, \odot)$ if it is a fuzzy regular relation both on $(R, \oplus)$ and on $(R, \odot)$.

Let $(R, \oplus, \odot)$ be a fuzzy hyperring and let $(R, +, \cdot) = \psi(R, \oplus, \odot)$ be the associated hyperring, where, for all a, b of R, we have

$$a + b = \{x \in R \mid (a \oplus b)(x) > 0\} \text{ and } a \cdot b = \{x \in R \mid (a \cdot b)(x) > 0\}.$$

The next theorem shows that also fuzzy regular relations on fuzzy hyperrings can be connected to regular relations on corresponding hyperrings.

Theorem 4.4.17. *An equivalence relation ρ is a fuzzy regular relation on $(R, \oplus, \odot)$ if and only if ρ is a regular relation on $(R, +, \cdot)$.*

Proof. Set $a\rho b$ and $a'\rho b'$, where $a, a', b', b' \in R$. We have $(a \oplus a')\rho(b \oplus b')$ and $(a \odot a')\rho(b \odot b')$ if and only if the following conditions hold.

- If $(a \oplus a')(x) > 0$, then there exists $y \in R$, such that $(b \oplus b')(y) > 0$ and $x\rho y$.

- If $(b \oplus b')(u) > 0$, then there exists $v \in R$, such that $(a \oplus a')(v) > 0$ and $u\rho v$.
- If $(a \odot a')(x) > 0$, then there exists $y \in R$, such that $(b \odot b')(y) > 0$ and $x\rho y$.
- If $(b \odot b')(u) > 0$, then there exists $v \in R$, such that $(a \odot a')(v) > 0$ and $u\rho v$.

These are equivalent to:

- If $x \in a + a'$, then there exists $y \in b + b'$, such that $x\rho y$.
- If $u \in b + b'$, then there exists $v \in a + a'$, such that $u\rho v$.
- If $x \in a \cdot a'$, then there exists $y \in b \cdot b'$, such that $x\rho y$.
- If $u \in b \cdot b'$, then there exists $v \in a \cdot a'$, such that $u\rho v$,

which mean that $(a + a')\bar{\rho}(b + b')$ and $(a \cdot a')\bar{\rho}(b \cdot b')$. Hence, ρ is fuzzy regular on $(R, \oplus, \odot)$ if and only if ρ is regular on $(R, +, \cdot)$. ■

Similarly, as for semihypergroups, we can introduce the fuzzy strongly regular relations on fuzzy hyperrings and try to connect them with the corresponding strongly regular relations on hyperrings.

Definition 4.4.18. An equivalence relation ρ on a fuzzy semihypergroup $(S, \circ)$ is called a *fuzzy strongly regular relation* on $(S, \circ)$ if, for all a, a', b, b' of S, such that $a\rho b$ and $a'\rho b'$, the following condition holds:

$$\forall x \in S, \text{ such that } (a \circ c)(x) > 0 \text{ and } \forall y \in S, \text{ such that } (b \circ d)(y) > 0,$$
$$\text{we have } x\rho y.$$

An equivalence relation ρ on a fuzzy hyperring $(R, \oplus, \odot)$ is called a *fuzzy strongly regular relation* on $(R, \oplus, \odot)$ if it is a fuzzy strongly regular relation both on $(R, \oplus)$ and on $(R, \odot)$.

Let $(R, \oplus, \odot)$ be a fuzzy hyperring and let $(R, +, \cdot) = \psi(R, \oplus, \odot)$ be the associated hyperring. We obtain that

Theorem 4.4.19. *An equivalence relation ρ is a fuzzy strongly regular relation on $(R, \oplus, \odot)$ if and only if ρ is a strongly regular relation on $(R, +, \cdot)$.*

Proof. Set $a\rho b$ and $a'\rho b'$, where $a, a', b, b' \in R$. The relation ρ is strongly regular on $(R, \oplus, \odot)$ if and only if the following conditions are satisfied:

- $\forall x \in R$, such that $(a \oplus a')(x) > 0$ and $\forall y \in R$, such that $(b \oplus b')(y) > 0$, we have $x\rho y$.
- $\forall x \in R$, such that $(a \odot a')(x) > 0$ and $\forall y \in R$, such that $(b \odot b')(y) > 0$, we have $x\rho y$.

These conditions are equivalent to the following ones:

- $\forall x \in R$, such that $x \in a + a'$ and $\forall y \in R$, such that $y \in b + b'$, we have $x\rho y$.

- $\forall x \in R$, such that $x \in a \cdot a'$ and $\forall y \in R$, such that $y \in b \cdot b'$, we have $x\rho y$.

Which mean that $(a+a')\overline{\overline{\rho}}(b+b')$ and $(aa')\overline{\overline{\rho}}(bb')$. Hence, ρ is fuzzy strongly regular on $(R, \oplus, \odot)$ if and only if ρ is strongly regular on $(R, +, \cdot)$. ■

REMARK 8 The study of fuzzy regular and fuzzy strongly regular relations on a fuzzy hyperring is reduced to the study of regular and strongly regular relations on a hyperring.

REMARK 9 If ρ is a fuzzy strongly relation on a fuzzy semihypergroup $(S, \circ)$, then it is fuzzy regular on $(S, \circ)$.

Consequently, if ρ is a fuzzy strongly relation on a fuzzy hyperring $(R, \oplus, \odot)$, then it is fuzzy regular on $(R, \oplus, \odot)$.

Now, let ρ be an equivalence relation, which is fuzzy regular on the fuzzy hyperring $(R, \oplus, \odot)$.

We consider the following hyperoperations on the quotient set R/ρ :

$$\bar{x} \boxplus \bar{y} = \{\bar{z} \mid z \in x + y\} = \{\bar{z} \mid (x \oplus y)(z) > 0\}$$
$$\bar{x} \boxdot \bar{y} = \{\bar{z} \mid z \in x \cdot y\} = \{\bar{z} \mid (x \odot y)(z) > 0\}.$$

Theorem 4.4.20. *Let $(R, \oplus, \odot)$ be a fuzzy hyperring and $(R, +, \cdot) = \psi(R, \oplus, \odot)$ be the associated hyperring. Then, we have the following characterization.*

(1) *The relation ρ is a fuzzy regular relation on $(R, \oplus, \odot)$ if and only if $(R/\rho, \boxplus, \boxdot)$ is a hyperring.*
(2) *The relation ρ is a fuzzy strongly regular relation on $(R, \oplus, \odot)$ if and only if $(R/\rho, \boxplus, \boxdot)$ is a ring.*

Proof. It is straightforward. ■

Recall that starting with a hypergroupoid $(R, +, \cdot)$, the fundamental relation γ^* is the smallest equivalence relation such that $(R/\rho, \boxplus, \boxdot)$ is a ring, which means that γ^* is the smallest strongly regular relation on $(R, +, \cdot)$.

If we denote by $\mathcal{U}$ the set of all finite hypersums of finite hyperproducts of elements of R then we have

$$x\gamma y \Leftrightarrow \exists\, u \in \mathcal{U} : \{x, y\} \subset u.$$

Suppose that $u \in \sum_i^* \prod_j^* a_{ji}$, where $\sum^*$ denotes a finite hypersum, while $\prod^*$ denotes a finite hyperproduct.

Definition 4.4.21. An equivalence relation γ^* is called the *fundamental relation* on a fuzzy hyperring $(R, \oplus, \odot)$ if γ^* is the fundamental relation on the associated hyperring $(R, +, \cdot) = \psi(R, \oplus, \odot)$.

Hence, γ^* is the fundamental relation on a fuzzy hyperring $(R, \oplus, \odot)$ if and only if γ^* is the smallest fuzzy strongly equivalence relation on $(R, \oplus, \odot)$.

Now, recall again the hyperoperation associated with a fuzzy semihypergroup (see [166]).

Let $(S, \circ)$ be a fuzzy semihypergroup and let $*$ be the associated hyperoperation on S, defined as follows:

$$a * b = \{x \in S \mid (a \circ b)(x) > 0\}.$$

Then we have $p \in a * b * c$ if and only if $(a \circ b \circ c)(p) > 0$. Indeed, we have $p \in (a * b) * c \Leftrightarrow \exists q \in a * b : p \in q * c \Leftrightarrow (a \circ b)(q) > 0$ and $(q \circ c)(p) > 0 \Leftrightarrow \bigvee\limits_{r \in S} ((a \circ b)(r \wedge (r \circ c)(p)) > 0 \Leftrightarrow ((a \circ b) \circ c)(p) > 0$.

Denote by $\sum_{\oplus}^*$ any finite fuzzy hypersum and by $\prod_{\odot}^*$ any finite fuzzy hyperproduct of the fuzzy hyperring $(R, \oplus, \odot)$.

As above, we obtain $\left(\sum\limits_i{}_{\oplus}^* \prod\limits_j{}_{\odot}^* a_{ji}\right)(p) > 0$ if and only if $p \in \sum\limits_i{}^* \prod\limits_j{}^* a_{ji}$.

Hence, $\{x, y\} \subset \sum\limits_i{}^* \prod\limits_j{}^* a_{ji}$ if and only if $\left(\sum\limits_i{}_{\oplus}^* \prod\limits_j{}_{\odot}^* a_{ji}\right)(x) > 0$ and $\left(\sum\limits_i{}_{\oplus}^* \prod\limits_j{}_{\odot}^* a_{ji}\right)(y) > 0$.

Denote by $\mathcal{UF}$ the set of all finite fuzzy hypersums of finite fuzzy hyperproducts of elements of R.

We obtain $x\gamma y \Rightarrow \mu_f \in \mathcal{UF}$ such that $\mu_f(x) > 0$ and $\mu_f(y) > 0$.

Chapter 5
Connections between Hypergroups and Fuzzy Sets

5.1 Join Spaces and Fuzzy Sets

Prenowitz and Jantosciak [155]- [157] have given an algebraic interpretation of ordered linear geometry (known as the descriptive geometry), spherical and projective geometry, based on the new concept of "join" hyperoperation, which assigns to two distinct points a and b an appropriate element that connects them. In the linear geometry, the connective element is the segment with the endpoints a and b; in the spherical geometry, this is the minor arc of a great circle passing through a and b, in the projective geometry the hypercomposition between a and b represents the line containing both points. More exactly, a join space is a commutative hypergroup $(H, \circ)$ satisfying the "incidence" or "transposition" axiom: for any four elements $a, b, c, d \in H$, such that $a/b \cap c/d \neq \emptyset$, it follows that $a \circ d \cap b \circ c \neq \emptyset$, where the left division (or extension of a from b) is defined as $a/b = \{x \in H \mid a \in x \circ b\}$.

The join space is not just an abstract algebraic model for classical geometries, but it was also used to characterize lattices (see Varlet [180]), median algebras (see Bandelt and Hedlikova [10]), graphs (see Nieminen [151]), ecc. Throughout this section, we present several connections between join spaces and fuzzy sets.

5.1.1 Constructions Using Fuzzy Sets

In this subsection we present two different constructions of hypergroups based on fuzzy sets. The first one was introduced by Corsini [28] in 1993 and the second one by Ameri-Zahedi [2] in 1997. Both of them are remarkable in the theory of join spaces and fuzzy sets, even if the second one was studied less, but we think it could open new lines of research connected with group theory. For the completeness of our presentation, we briefly recall this construction in the second paragraph §2.

Since, in this chapter, we focus more on the further applications of these basic constructions, obtained in the last ten years, than on their properties, that are well known also from the previous book written by Corsini-Leoreanu [47], the proofs are omitted.

B. Davvaz and I. Cristea, *Fuzzy Algebraic Hyperstructures*,
Studies in Fuzziness and Soft Computing 321, DOI: 10.1007/978-3-319-14762-8_5

§1. The first connection between fuzzy sets and hyperstructures was established by Corsini [28], when he defined a hyperoperation by means of fuzzy subsets.

More exactly, let $\mu : X \to [0,1]$ be a fuzzy subset of a non-empty set X. Define on X the hyperoperation $\circ$, setting, for any $x, y \in X$,

$$x \circ y = \{z \in X \mid \mu(x) \wedge \mu(y) \leq \mu(z) \leq \mu(x) \vee \mu(y)\} \tag{5.1}$$

Theorem 5.1.1. *The associated hypergroupoid $(X, \circ)$ is a join space.*

Let us examinate now the meaning and the importance of this association. From one point of view, an algebraic one, it offers a very simple and direct method to obtain a join space from a non-empty set endowed with a fuzzy subset. Then, as we will see in Section 5.3, it stays at the basis of the construction of a sequence of join spaces associated with an initial non-empty set.

From another point of view, as Kehagias-Konstantinidou noticed in [125], it has practical implications in machine learning and computing learning theory, where the set of discourse X is an Euclidean space $\mathbb{R}^n$ and thus the hyperproduct

$$x \circ y = \{z \mid x \wedge y \leq z \leq x \vee y\} = [x, y]$$

can be viewed as the n-dimensional hyperbox with the lowest vertex in $x \wedge y$ and the highest one in $x \vee y$.

§2. Considering a fuzzy subset μ of a group $(G, \cdot)$, Ameri-Zahedi [2] defined a new hyperoperation, finding necessary and sufficient conditions (regarding the fuzzy subset μ) such that the obtained hyperstructure is a hypergroup or a join space.

Let μ be a fuzzy subset of a group $(G, \cdot)$ with the identity e. For any $x \in G$, define

- the *left fuzzy coset* of μ as the fuzzy subset $x\mu : G \to [0,1]$, where for any $g \in G$, $(x\mu)(g) = \mu(x^{-1}g)$.
- the *right fuzzy coset* of μ as the fuzzy subset $\mu x : G \to [0,1]$, where for any $g \in G$, $(\mu x)(g) = \mu(gx^{-1})$.

We say that

- μ is *symmetric* if $\mu(x) = \mu(x^{-1})$, for any $x \in G$.
- μ is *invariant* if $\mu(xy) = \mu(yx)$, for any $x, y \in G$.
- μ is *subnormal* if it is both symmetric and invariant.

It is clear that, for any $x, y, z \in G$, we have

$$x\mu = y\mu \Leftrightarrow zx\mu = zy\mu$$

and a similarly property holds also for the right fuzzy cosets.

We introduce now some notations. For any $(a,b) \in G^2$, define

$$\begin{aligned} {}^{a}\mu &= \{x \in G \mid x\mu = a\mu\}. \\ \mu^{a} &= \{x \in G \mid \mu x = \mu a\}. \\ \mu^{a}\mu^{b} &= \{xy \in G \mid x \in \mu^{a}, y \in \mu^{b}\}. \\ a\mu^{e} &= \{ax \in G \mid x \in \mu^{e}\}. \end{aligned}$$

It is clear that, if μ is invariant, then ${}^{a}\mu = \mu^{a}$, for any $a \in G$.

The next result covers some basic properties, useful in the sequel.

Proposition 5.1.2. *Let μ be a subnormal fuzzy subset of the group $(G,\cdot)$ with the identity e. The following statements hold.*

(1) $x\mu = y\mu \Leftrightarrow xz\mu = yz\mu \Leftrightarrow xy^{-1} \in G$, *for any* $x,y,z \in G$.
(2) μ^{e} *is a normal subgroup of* G.
(3) $\mu^{a} = a\mu^{e}$, *for any* $a \in G$.
(4) $\mu^{a}\mu^{b} = \mu^{ab}$, *for any* $a,b \in G$.

Now, we can endow G with the hyperoperation induced by μ, defining for any $a,b \in G$,

$$a \circ_{\mu} b = \mu^{a}\mu^{b} \tag{5.2}$$

Conditions under which $(G,\circ_{\mu})$ is a hypergroup or a join space are expressed below.

Theorem 5.1.3. *Let μ be a fuzzy subset of a group $(G,\cdot)$ with the identity e.*

(1) $(G,\circ_{\mu})$ *is a quasihypergroup.*
(2) *If μ is subnormal, then*
   (a) $(G,\circ_{\mu})$ *is a polygroup.*
   (b) $(G,\circ_{\mu})$ *is a join space if and only if $[G,G] \subseteq \mu^{e}$, where $[G,G]$ is the commutator subgroup of G.*

5.1.2 Generalizations to L-Fuzzy Sets

§1. Extending Corsini's construction to the case of L-fuzzy sets, Serafimidis et al. [169] and Kehagias et al. [125] have founded conditions such that the associated hyperstructure is a join space, when $(L,\leq)$ is a *complete lattice*, results that we present in the following.

Consider the L-fuzzy set $\mu : X \to L$, where $(L,\leq)$ is here a complete lattice, and define on X the binary relation $\sim$ as follows: $x \sim y$ if and only if $\mu(x) = \mu(y)$. It is clear that $\sim$ is an equivalence relation on X. The equivalence class of an element $x \in X$ is denoted by $\overline{x}$ and the quotient of X modulo $\sim$ by $X/\!\sim\, = \{\overline{x}\}_{x\in X}$.

Define on X the hyperoperation $\circ$ like in (5.1), and on the quotient $X/\!\sim$ the hyperoperation $\otimes$ by

$$\overline{x} \otimes \overline{y} = \{\overline{z} \mid z \in x \circ y\}, \tag{5.3}$$

with the extension hyperproduct (called also the left division) defined by $\overline{x}||\overline{y} = \{\overline{z} \mid \overline{x} \in \overline{y} \otimes \overline{z}\}$.

We notice that the two hyperoperations are equivalent: $z \in x \circ y \Leftrightarrow \overline{z} \in \overline{x} \otimes \overline{y}$. Moreover, the equivalence $\sim$ is a congruence on X, as we see here below.

Proposition 5.1.4. *For all $x, y, z \in X$ the implication $\overline{x} = \overline{y} \Rightarrow \overline{x} \otimes \overline{z} = \overline{y} \otimes \overline{z}$ holds.*

Proof. Let x, y, z be in X such that $\overline{x} = \overline{y}$, that is $\mu(x) = \mu(y)$. For any $\overline{u} \in \overline{x} \otimes \overline{z}$, it follows that $\mu(x) \wedge \mu(z) \leq \mu(u) \leq \mu(x) \vee \mu(z)$ and since $\mu(x) = \mu(y)$, we get that $\overline{u} \in \overline{y} \otimes \overline{z}$. So $\overline{x} \otimes \overline{z} \subseteq \overline{y} \otimes \overline{z}$. Similarly, the converse inclusion can be proved. ∎

Define now an L-fuzzy subset of the quotient $X/\sim$ by $\overline{\mu} : X/\sim \to L$, $\overline{\mu}(\overline{x}) = \mu(x)$. It is easy to see that $\overline{\mu}$ is well defined, one-to-one and onto $\mu(X)$.

Now, we introduce one more hyperoperation on the subset $\mu(X) \subseteq L$, setting for any $a, b \in \mu(X)$,

$$a * b = [a \wedge b, a \vee b] \cap \mu(X), \tag{5.4}$$

with the extension hyperproduct defined by $a \wr b = \{c \in \mu(X) \mid a \in b * c\}$.

The next result gives a connection between the three hyperoperations $\circ, \otimes$, and $*$ defined above.

Proposition 5.1.5. *For any $x, y, z \in X$, the following equivalences hold:*

$$z \in x \circ y \Leftrightarrow \overline{z} \in \overline{x} \otimes \overline{y} \Leftrightarrow \mu(z) \in \mu(x) * \mu(y) \Leftrightarrow \overline{\mu}(\overline{z}) \in \overline{\mu}(\overline{x}) * \overline{\mu}(\overline{y}).$$

Proof. It is straightforward. ∎

Next we introduce on the quotient $X/\sim$ an order, defining for any $x, y \in X$,

$$\overline{x} \sqsubseteq \overline{y} \Leftrightarrow \mu(x) \leq \mu(y).$$

The reflexivity and transitivity of the relation $\sqsubseteq$ follow from the fact that the relation $\leq$ is an order on L and, in particular, on $\mu(X) \subseteq L$. For the antisymmetry, we use the equivalence $\overline{x} \sqsubseteq \overline{y} \Leftrightarrow \overline{\mu}(\overline{x}) \leq \overline{\mu}(\overline{x})$ and the fact that the L-fuzzy subset $\overline{\mu} : X/\sim \to \mu(X)$ is a bijection.

We remark that, defining the relation $\preceq$ on X by $x \preceq y \Leftrightarrow \mu(x) \leq \mu(y)$, we obtain that $\preceq$ is only a preorder on X and the classes generated by it are exactly the elements of $X/\sim$.

The following result establishes an isomorphism between the domain and the range of the L-fuzzy subset $\overline{\mu}$.

Proposition 5.1.6. *The L-fuzzy subset $\overline{\mu} : (X/\sim, \sqsubseteq, \otimes) \to (\mu(X), \leq, *)$ is an order isomorphism, that is the following statements hold:*

(1) *$\overline{\mu}$ is one-to-one.*

(2) $\overline{x} \sqsubseteq \overline{y} \Leftrightarrow \mu(x) \leq \mu(y)$.
(3) $\overline{\mu}(\overline{x} \otimes \overline{y}) = \mu(x) * \mu(y)$.

Proof. The first two items are clear from the above discussion. It remains to prove the last statement. Accordingly with Proposition 5.1.5, for all $\overline{z} \in \overline{x} \otimes \overline{y}$, we have $\mu(z) \in \mu(x) * \mu(y)$, equivalently with $\overline{\mu}(\overline{z}) \in \overline{\mu}(\overline{x}) * \overline{\mu}(\overline{y})$, and thus the direct inclusion is proved.

Conversely, take $a \in \mu(x) * \mu(y)$. By relation (5.4), it follows that

$$a \in [\mu(x) \wedge \mu(y), \mu(x) \vee \mu(y)] \cap \mu(X),$$

meaning that there exists $z \in X$ such that $a = \mu(z) \in [\mu(x) \wedge \mu(y), \mu(x) \vee \mu(y)]$. Thus, $z \in x \circ y$, and again by Proposition 5.1.5, we have $\overline{z} \in \overline{x} \otimes \overline{y}$ and then $a = \overline{\mu}(\overline{z}) \in \overline{\mu}(\overline{x} \otimes \overline{y})$, therefore the second inclusion holds, too. ■

The next characterization follows immediately.

Proposition 5.1.7. *The lattice $(X/{\sim}, \sqsubseteq)$ is modular (distributive) if and only if $(\mu(X), \leq)$ is a modular (distributive) lattice.*

We are now ready to present the main results of this paragraph, containing conditions for the hyperoperations $\circ$, $\otimes$, and $*$ to be join hyperoperations (meaning that the corresponding hypergroupoids are join spaces), in the case when $(L, \leq)$ is a complete lattice.

Proposition 5.1.8. *The quotient hyperstructure $(X/{\sim}, \otimes)$ is a hypergroup (or a join space) if and only if the hyperstructure $(\mu(X), *)$, constructed on the image of X through the fuzzy subset μ, is a hypergroup (or a join space).*

Proof. Based on the definitions of the hyperoperations $\otimes$ and $*$, we know that $(X/{\sim}, \otimes)$ and $(\mu(X), *)$ are always commutative quasihypergroups.

First we prove that the associativity of $\otimes$ is equivalent with the associativity of $*$. For this, we need to show that, for any $x, y, z \in X$,

$$\overline{\mu}(\overline{x} \otimes \overline{y}) * \overline{\mu}(\overline{z}) = \overline{\mu}((\overline{x} \otimes \overline{y}) \otimes \overline{z}).$$

Indeed, for any $a \in \overline{\mu}(\overline{x} \otimes \overline{y}) * \overline{\mu}(\overline{z})$, there exists $u \in X$ such that $\overline{\mu}(\overline{u}) \in \overline{\mu}(\overline{x} \otimes \overline{y}) = \mu(x) * \mu(y)$ (by Proposition 5.1.6) and $a \in \overline{\mu}(\overline{u}) * \overline{\mu}(\overline{z})$. Since $a \in \mu(X)$, there exists $w \in X$ with $a = \mu(w)$ and then the previous two relations become $\mu(x) \wedge \mu(y) \leq \overline{\mu}(\overline{u}) = \mu(u) \leq \mu(x) \vee \mu(y)$ and $\mu(u) \wedge \mu(z) \leq a = \mu(w) \leq \mu(u) \vee \mu(z)$, meaning that, there exist $u, w \in X$ such that $\overline{u} \in \overline{x} \otimes \overline{y}$ and $\overline{w} \in \overline{u} \otimes \overline{z}$, equivalently with $\overline{w} \in (\overline{x} \otimes \overline{y}) \otimes \overline{z}$. Therefore, $a = \overline{\mu}(\overline{w}) \in \overline{\mu}((\overline{x} \otimes \overline{y}) \otimes \overline{z})$. Hence, we have that

$$\overline{\mu}(\overline{x} \otimes \overline{y}) * \overline{\mu}(\overline{z}) = \overline{\mu}((\overline{x} \otimes \overline{y}) \otimes \overline{z}).$$

Thus, by this relation and Proposition 5.1.6, it follows that

$$(\overline{\mu}(\overline{x}) * \overline{\mu}(\overline{y})) * \overline{\mu}(\overline{z}) = \overline{\mu}(\overline{x} \otimes \overline{y}) * \overline{\mu}(\overline{z}) = \overline{\mu}((\overline{x} \otimes \overline{y}) \otimes \overline{z}).$$

Since $\overline{\mu}$ is one-to-one, the last relation is equivalent with

$$\overline{\mu}^{-1}((\overline{\mu}(\overline{x}) * \overline{\mu}(\overline{y})) * \overline{\mu}(\overline{z})) = (\overline{x} \otimes \overline{y}) \otimes \overline{z}.$$

Similarly, one proves that $\overline{\mu}(\overline{x}) * (\overline{\mu}(\overline{y}) * \overline{\mu}(\overline{z})) = \overline{\mu}(\overline{x} \otimes (\overline{y} \otimes \overline{z}))$ is equivalent with

$$\overline{\mu}^{-1}(\overline{\mu}(\overline{x}) * (\overline{\mu}(\overline{y}) * \overline{\mu}(\overline{z}))) = \overline{x} \otimes (\overline{y} \otimes \overline{z}).$$

Now, it is clear the equivalence between the associativity of $\otimes$ and that of $*$.

Till now we have proven that $(X/\sim, \otimes)$ is a commutative hypergroup if and only if $(\mu(X), *)$ is a commutative hypergroup.

Similarly one proves the equivalence of the join space condition, that is the implication $\overline{x}||\overline{y} \sim \overline{z}||\overline{t} \Rightarrow \overline{x} \otimes \overline{t} \sim \overline{y} \otimes \overline{z}$, for all $\overline{x}, \overline{y}, \overline{z}, \overline{t} \in X/\sim$, is equivalent with the implication $a \wr b \sim c \wr d \Rightarrow a * d \sim b * c$, for any $a, b, c, d \in \mu(X)$. And the proof is complete now. ■

In the same manner, one obtains the following characterization.

Proposition 5.1.9. *The quotient hyperstructure $(X/\sim, \otimes)$ is a hypergroup (or a join space) if and only if the initial hyperstructure $(X, \circ)$ is a hypergroup (or a join space).*

We conclude this paragraph with an interpretation of the previous results in connection with the Corsini's result, when he took $L = [0, 1]$, the real unit interval. In this case, he proved that the initial hyperstructure $(X, \circ)$ is a join space, and therefore the quotient hyperstructure $(X/\sim, \otimes)$ and the image one $(\mu(X), *)$ are join spaces, too. But what happens in the more general case, when $(L, \leq)$ is just a complete lattice? An answer is given below. First we present an auxiliary result [1], useful in the proof of Proposition 5.1.11.

Lemma 5.1.10. *In any distributive lattice $(L, \leq)$ endowed with the join hyperoperation $*$ defined in (5.4), we have:*

(1) *For all $a, x, y \in L$, with $x \leq y$,*

$$a * [x, y] = [a \wedge x, a \vee y] \tag{5.5}$$

(2) *The relations $a \wedge d \leq b \vee c$ and $b \wedge c \leq a \vee d$ are equivalent with $a * d \sim b * c$, for any $a, b, c, d \in L$.*

Proof. (1) For any $u \in a * [x, y]$, there exists $z \in [x, y]$ such that $u \in a * z$, that is $a \wedge z \leq u \leq a \vee z$, with $x \leq z \leq y$. Then, $a \wedge x \leq a \wedge z \leq u$ and $u \leq a \vee z \leq a \vee y$, meaning that $u \in [a \wedge x, a \vee y]$.

[1] This result has been proved for the first time in 2001 by Ath. Kehagias and M. Konstantinidou in the unpublished manuscript "Lattice-ordered join space: an Applications-Oriented Example" [125]

Conversely, take $u \in [a \wedge x, a \vee y]$ and define $z = (u \vee x) \wedge y = (u \wedge y) \vee (x \wedge y) = (u \wedge y) \vee x$, by distributivity. Since $x \leq (u \wedge y) \vee x = z = (u \vee x) \wedge y \leq y$, it follows that $z \in [x, y]$.

We will prove that $u \in a * z$, that is $a \wedge z \leq u \leq a \vee z$. Now, using the distributivity, $a \wedge z = a \wedge ((u \vee x) \wedge y) = (a \wedge y) \wedge (u \vee x) = (a \wedge y \wedge u) \vee (a \wedge y \wedge x) = (a \wedge y \wedge u) \vee (a \wedge x) \leq u \vee u = u$. Similarly, $u \leq a \vee z$ and thus $u \in a * z$.

(2) Assume that $a \wedge d \leq b \vee c$ and $b \wedge c \leq a \vee d$. It follows that $a \wedge d \leq (a \wedge d) \vee (b \wedge c) \leq a \vee d$ and $b \wedge c \leq (a \wedge d) \vee (b \wedge c) \leq b \vee c$.
Similarly we get $a \wedge d \leq (a \vee d) \wedge (b \vee c) \leq a \vee d$ and $b \wedge c \leq (a \vee d) \wedge (b \vee c) \leq b \vee c$.

Moreover, from the hypothesis, we have $(a \wedge d) \vee (b \wedge c) \leq (a \vee d) \wedge (b \vee c)$.

Defining $x = (a \wedge d) \vee (b \wedge c)$ and $y = (a \vee d) \wedge (b \vee c)$, we obtain that $x \leq y$ and $[x, y] \subseteq a * d \cap b * c$, so $a * d \sim b * c$.

On the other hand, if $a * d \sim b * c$, then there exists $u \in a * d \cap b * c$ such that $a \wedge d \leq u \leq a \vee d$ and $b \wedge c \leq u \leq b \vee c$, that can be written as $a \wedge d \leq u \leq b \vee c$ and $b \wedge c \leq u \leq a \vee d$. Thus, the converse implication holds. ■

Proposition 5.1.11. *If $\mu(X)$ is a distributive sublattice of $(L, \leq)$, then all three hyperstructures $(X, \circ)$, $(X/\sim, \otimes)$ and $(\mu(X), *)$ are join spaces.*

Proof. Based on Proposition 5.1.8 and Proposition 5.1.9, it is enough to prove that $(\mu(X), *)$ is a join space. The commutative property is immediate.

Let us see now the reproducibility, i.e., $a * \mu(X) = \mu(X)$, for any $a \in \mu(X)$. It is clear that $a * \mu(X) \subseteq \mu(X)$. Conversely, $\mu(X) = \bigcup_{x \in X} \mu(x) \subseteq \bigcup_{x \in X} a * \mu(x) \subseteq a * \mu(X)$, so the equality holds.

The associativity is based on Lemma 5.1.10 (1). Therefore, accordingly with (5.4) and (5.5), we have $a * (b * c) = a * [b \wedge c, b \vee c] = [a \wedge b \wedge c, a \vee b \vee c]$, and similarly, $(a * b) * c = [a \wedge b, a \vee b] * c = [a \wedge b \wedge c, a \vee b \vee c]$, meaning that $*$ is associative. Thus, $(\mu(X), *)$ is a commutative hypergroup.

It remains to prove the join space condition: for any $a, b, c, d \in \mu(X)$, the following implication holds $a \wr b \sim c \wr d \Rightarrow a * d \sim b * c$. For proving this, we use Lemma 5.1.10 (2). So, if $a \wr b \sim c \wr d$, then there exists $u \in a \wr b \cap c \wr d$ such that $a \in [u \wedge b, u \vee b]$ and $c \in [u \wedge d, u \vee d]$. It follows that $a \wedge d \leq (u \vee b) \wedge d = (u \wedge d) \vee (b \wedge d) \leq c \vee (b \wedge d) = (c \vee b) \wedge (c \vee d) \leq c \vee b$ and $c \wedge b \leq (u \vee d) \wedge b = (u \wedge b) \vee (d \wedge b) \leq a \vee (d \wedge b) = (a \vee d) \wedge (a \vee b) \leq a \vee d$. Using Lemma 5.1.10, the relations $a \wedge d \leq c \vee b$ and $c \wedge b \leq a \vee d$ imply that $a * d \sim b * c$ and the proof is now complete. ■

We stress again that the associativity and the join space property are both based on the fact that $\mu(X)$ is a distributive lattice.

§2. The conversely problem has been investigated by Tofan and Volf [178], in the case when L is a *lattice possessing the unit* 1.

Suppose that $\mu(X)$ is a sublattice with the greatest element 1. First we need the following auxiliary property.

Proposition 5.1.12. *In the above hypothesis, there exists $\omega \in X$, such that:*

(1) *For any $a, b \in X$, the condition $a \circ \omega = b \circ \omega$ implies $a \circ a = b \circ b$.*
(2) *For any $a, b \in X$, there exist $m, M \in X$ such that $M \circ \omega = \bigcap\{x \circ \omega \mid x \in a \circ b\}$ and $m \circ \omega = \bigcap\{x \circ \omega \mid \{a, b\} \subset x \circ \omega\}$.*

Proof. Let ω be arbitrary in $\mu^{-1}(1)$. Then, $a \circ \omega = \{x \in X \mid \mu(a) \wedge \mu(\omega) \leq \mu(x) \leq \mu(a) \vee \mu(\omega)\} = \{x \in X \mid \mu(a) \leq \mu(x)\}$, since $\mu(\omega) = 1$.

(1) Since $b \in b \circ \omega = a \circ \omega$, it follows that $\mu(a) \leq \mu(b)$. Symmetrically, we have $\mu(b) \leq \mu(a)$, and thus $a \circ a = b \circ b = \{x \in X \mid \mu(a) = \mu(b) = \mu(x)\}$.

(2) Let m be arbitrary in $\mu^{-1}(\mu(a) \wedge \mu(b))$ and M arbitrary in $\mu^{-1}(\mu(a) \vee \mu(b))$. For any $x \in X$, we have $x \in a \circ \omega \cap b \circ \omega$ iff $\mu(a) \leq \mu(x)$ and $\mu(b) \leq \mu(x)$, that is $\mu(a) \vee \mu(b) \leq \mu(x)$. But $\mu(a) \vee \mu(b) = \mu(M)$, thus $a \circ \omega \cap b \circ \omega = M \circ \omega \supset \{x \circ \omega \mid x \in a \circ b\}$. Since $a, b \in a \circ b$, in order to prove the first equality, it is enough to show that $x \in a \circ b$ implies $x \circ \omega \supset M \circ \omega$. Indeed, $x \in a \circ b$ means that $\mu(x) \leq \mu(a) \vee \mu(b) = \mu(M)$, so $x \circ \omega \supset M \circ \omega$.

For the second equality, we remark that $a, b \in m \circ \omega = \{x \in X \mid \mu(m) = \mu(a) \wedge \mu(b) \leq \mu(x)\}$, so $m \circ \omega \supset \bigcap\{x \circ \omega \mid x \circ \omega \supset \{a, b\}\}$. Conversely, if $x \circ \omega \supset \{a, b\}$, then $\mu(x) \leq \mu(a)$ and $\mu(x) \leq \mu(b)$, that is $\mu(x) \leq \mu(a) \wedge \mu(b) = \mu(m)$, equivalent to $x \in m \circ \omega$. This implies that $x \circ \omega \supset m \circ \omega$. Now, the proof is complete. ■

It is quite evident that the following properties hold in $(X, \circ)$:

(1′) $a, b \in a \circ b$;
(2′) $a \circ b = b \circ a$;
(3′) $a \circ (a \circ b) = (a \circ a) \circ b = (a \circ a) \circ (b \circ b) = (a \circ b) \circ b = a \circ b$.

We start now to consider the converse problem: *given a hypergroupoid $(H, \circ)$ satisfying the properties (1′), (2′), (3′) and (1), (2) mentioned above, can one find a lattice L and a map $\mu : H \to L$ such that $\circ$ is the hyperoperation induced by μ as in relation (5.1)?*

In order to answer to this question, suppose that H satisfies the properties above. Define on H the following equivalence

$$a \sim b \text{ iff } a \circ a = b \circ b.$$

We denote by $\hat{a} = \{x \in H \mid x \sim a\}$ the equivalence class of a and let $L = H/\sim$ the quotient set of H modulo the equivalence $\sim$.

Besides, define on L a relation ρ by taking

$$\hat{a}\rho\hat{b} \Leftrightarrow b \circ b \subset a \circ \omega \tag{5.6}$$

First we prove that ρ is well defined. Indeed, if $\hat{x} = \hat{a}$ and $\hat{y} = \hat{b}$, with $\hat{a}\rho\hat{b}$, then $y \circ y = b \circ b \subset a \circ \omega = (a \circ a) \circ \omega = (x \circ x) \circ \omega = x \circ \omega$, which proves that the relation ρ does not depend on the representatives a and b.

Secondly, note that condition (5.6) is equivalent with the following one

$$\hat{a}\rho\hat{b} \Leftrightarrow b \circ \omega \subset a \circ \omega \tag{5.7}$$

Indeed, if relation (5.6) is satisfied, then $b \circ \omega = (b \circ b) \circ \omega \subset (a \circ \omega) \circ \omega = a \circ \omega$. Conversely, if relation (5.7) holds, then $b \in b \circ \omega$ implies that $b \circ b \subset b \circ \omega \subset a \circ \omega$.

Moreover, the relation ρ is an ordering on L.

- It is reflexive: since $a \in a \circ \omega$, it follows that $a \circ a \subset a \circ (a \circ \omega) = a \circ \omega$, so $\hat{a}\rho\hat{a}$.
- It is antisymmetric: let $\hat{a}\rho\hat{b}$ and $\hat{b}\rho\hat{a}$; then $b \circ \omega \subset a \circ \omega$ and $a \circ \omega \subset b \circ \omega$, so $a \circ \omega = b \circ \omega$. Accordingly with condition (1) in Proposition 5.1.12, it follows that $a \circ a = b \circ b$, that is $\hat{a} = \hat{b}$.
- It is transitive: let $\hat{a}\rho\hat{b}$ and $\hat{b}\rho\hat{c}$; then $b \circ \omega \subset a \circ \omega$ and $c \circ \omega \subset b \circ \omega$, thus $c \circ \omega \subset a \circ \omega$, that is $\hat{a}\rho\hat{c}$.

Besides, the ordered set (L, ρ) is a lattice. Indeed, for any $\hat{a}, \hat{b}$ in L, there exist $\hat{m}, \hat{M} \in L$ satisfying condition (2) of Proposition 5.1.12. We show that $\hat{m} = \hat{a} \wedge \hat{b}$ and $\hat{M} = \hat{a} \vee \hat{b}$. If $\hat{x}\rho\hat{a}$ and $\hat{x}\rho\hat{b}$, then $\{a, b\} \subset x \circ \omega$. Condition (2) ensures that $x \circ \omega \supset m \circ \omega$, so $\hat{x}\rho\hat{m}$. Obviously, $\hat{m}\rho\hat{a}$ and $\hat{m}\rho\hat{b}$, so $\hat{m} = \hat{a} \wedge \hat{b}$. On the other side, $\hat{a}\rho\hat{M}$ and $\hat{b}\rho\hat{M}$. If $\hat{a}\rho\hat{c}$ and $\hat{b}\rho\hat{c}$, then $c \circ \omega \subset a \circ \omega \cap b \circ \omega = M \circ \omega$, so $\hat{M}\rho\hat{c}$, that is $\hat{M} = \hat{a} \vee \hat{b}$. Moreover, $\hat{\omega}$ is the greatest element of the lattice L.

Define now the application $\mu : H \to L$ as the canonical projection: $\mu(a) = \hat{a}$, for any $a \in H$. The following result answers to the above question.

Proposition 5.1.13. *In the previous conditions, for any $a, b \in H$, it holds:*

$$a \circ b = \{x \in H \mid \mu(a) \wedge \mu(b) \leq \mu(x) \leq \mu(a) \vee \mu(b)\}.$$

Proof. Let m and M be given by relation (2) of Proposition 5.1.12, for the elements a and b. Since $\mu(a) \wedge \mu(b) = \mu(m)$ and $\mu(a) \vee \mu(b) = \mu(M)$ (because $\hat{a} \wedge \hat{b} = \hat{m}, \hat{a} \vee \hat{b} = \hat{M}$), we have to prove that $y \in a \circ b \Leftrightarrow M \circ \omega \subset y \circ \omega \subset m \circ \omega$.

Since $a \in m \circ \omega$ and $b \in m \circ \omega$, it follows that $a \circ b \subset m \circ \omega$, so $a \circ b \circ \omega \subset (m \circ \omega) \circ \omega = m \circ \omega$. This means that $y \circ \omega \subset m \circ \omega$, for any $y \in a \circ b$.

The condition (2) regarding M says that $M \circ \omega \subset y \circ \omega$, for any $y \in a \circ b$.

Conversely, if $M \circ \omega \subset y \circ \omega \subset m \circ \omega$, then $y \circ \omega \subset \bigcap\{x \circ \omega \mid x \circ \omega \supset \{a, b\}\} = \bigcap\{x \circ \omega \mid x \circ \omega \supset a \circ b\}$, so $y \in y \circ \omega \subset a \circ b$.

Now, the proof is complete and thus $\circ$ is the hyperoperation induced by μ as in relation (5.1). ■

5.2 *L*-Fuzzy Join Spaces

Starting from a generalized de Morgan lattice $(X, \leq, \vee, \wedge, ')$, Kehagias [123] defined a collection of hyperoperations $*_p$, defined in a such way to obtain a family of join spaces $(X, *_p)$, that can be considered as L-fuzzy join spaces in the sense given by Corsini in [28] (substituting the real interval $[0, 1]$ with a lattice L). In the following we present and discuss about the results obtained by Kehagias in the paper [123].

Definition 5.2.1. A *generalized de Morgan lattice* is a distributive lattice $(X, \leq, \vee, \wedge, ')$ with the minimum element 0 and the maximum element 1, where the symbol $'$ denotes the negation operation, and that satisfies the following properties:

(1) For all $x \in X, Y \subset X$, we have $x \wedge (\bigvee_{y \in Y} y) = \bigvee_{y \in Y} (x \wedge y)$, and $x \vee (\bigwedge_{y \in Y} y) = \bigwedge_{y \in Y} (x \vee y)$. (Complete distributivity)
(2) For all $x \in X$, $(x')' = x$. (Negation is involutory)
(3) For all $x, y \in X$, $x \leq y$ implies that $y' \leq x'$. (Negation is order reversing)
(4) For all $Y \subset X$, we have $(\bigvee_{y \in Y} y)' = \bigwedge_{y \in Y} y'$, and $(\bigwedge_{y \in Y} y)' = \bigvee_{y \in Y} y'$. (Complete de Morgan laws)

The set of crisp intervals of X is denoted by $I(X)$ and defined by $I(X) = \{[a,b]\}_{a,b \in X} \subset \mathcal{P}(X)$; the empty set $\emptyset$ is also consider a member of $I(X)$, being symbolized by $\emptyset = [a,b]$, with any $a, b \in X$, with $a \not\leq b$. $I(X)$ with the inclusion relation $\subset$ on $\mathcal{P}(X)$ is a lattice $(I(X), \subset, \bigcap, \bigcup)$, where, for any two intervals $A = [a_1, a_2]$ and $B = [b_1, b_2]$, the intersection $\bigcap$ and the union $\bigcup$ are defined as $[a_1, a_2] \cap [b_1, b_2] = [a_1 \vee b_1, a_2 \wedge b_2]$, and $[a_1, a_2] \cup [b_1, b_2] = [a_1 \wedge b_1, a_2 \vee b_2]$. Moreover, if $(X, \leq, \vee, \wedge, ')$ is a complete lattice, then the lattice $(I(X), \subset, \bigcap, \bigcup)$ is complete, too.

The following properties of $\vee$ and $\wedge$ acting on intervals of a distributive lattice are fundamental.

Proposition 5.2.2. *For all $a, b, x, y \in X$ such that $x \leq y, a \leq b$, we have*

(1) $a \vee [x, y] = [a \vee x, a \vee y]$.
(2) $a \wedge [x, y] = [a \wedge x, a \wedge y]$.
(3) $[a, b] \vee [x, y] = [a \vee x, b \vee y]$.
(4) $[a, b] \wedge [x, y] = [a \wedge x, b \wedge y]$.

For every $p \in X$, define the hyperoperation $*_p : X \times X \to I(X)$ like:

$$a *_p b = [a \wedge b \wedge p, a \vee b \vee p'],$$

which is called the *p-join hyperoperation* on X.

Proposition 5.2.3. *For all $a, b, c, d, p \in X$ such that $a \leq b, c \leq d$, we have*

(1) $[a, b] *_p c = [a \wedge c \wedge p, b \vee c \vee p']$.
(2) $[a, b] *_p [c, d] = [a \wedge c \wedge p, b \vee d \vee p']$.

Proof. Choose any $a, b, c, d, p \in X$ such that $a \leq b, c \leq d$.

(1) Set $u \in [a, b] *_p c$; then there exists $x \in [a, b]$ such that $u \in x *_p c$, that is $a \leq x \leq b$ and $x \wedge c \wedge p \leq u \leq x \vee c \vee p'$. It follows that $a \wedge c \wedge p \leq x \wedge c \wedge p \leq u \leq x \vee c \vee p' \leq b \vee c \vee p'$, which means that $u \in [a \wedge c \wedge p, b \vee c \vee p']$; thus $[a, b] *_p c \subset [a \wedge c \wedge p, b \vee c \vee p']$.

Conversely, let v be arbitrary in $[a\wedge c\wedge p, b\vee c\vee p']$ and set $z_v = (v\vee a)\wedge b = (v \wedge b) \vee a$. We have $a \leq (v \wedge b) \vee a = z_v = (v \vee a) \wedge b \leq b$, so $z_v \in [a, b]$. But, $z_v \wedge c \wedge p = (v \vee a) \wedge b \wedge c \wedge p = (v \wedge b \wedge c \wedge p) \vee (a \wedge b \wedge c \wedge p)$, where $v \wedge b \wedge c \wedge p \leq v$ and $a \wedge b \wedge c \wedge p = a \wedge c \wedge p \leq v$. Hence, $z_v \wedge c \wedge p \leq v$. Similarly, one obtains that $v \leq z_v \vee c\vee p'$ and therefore $v \in z_v *_p c \subset [a, b] *_p c$ and then we obtain that $[a \wedge c \wedge p, b \vee d \vee p'] \subset [a, b] *_p c$.

We conclude that $[a, b] *_p c = [a \wedge c \wedge p, b \vee c \vee p']$.

(2) The second relation can be proved in a similar way. ■

Next, some basic properties of the p-join hyperoperation $*_p$ are presented.

Proposition 5.2.4. *The following properties hold, for any* $a, b, c, p \in X$.

(1) $a \in a *_p a$; $a, b \in a *_p b$.
(2) $a *_p b = b *_p a$.
(3) $(a *_p b) *_p c = a *_p (b *_p c)$.
(4) $a *_p X = X$. *Moreover,* $(X, *_p)$ *is a commutative hypergroup.*
(5) $(a *_p b, *_p)$ *is a commutative subhypergroup of* $(X, *_p)$.

Proof. Let a, b, c, p be arbitrary elements in X. By the definition of the hyperoperation $*_p$, the properties (1) and (2) are obvious.

(3) Let us prove now the associativity, using Proposition 5.2.3. Note that $(a *_p b) *_p c = [a \wedge b \wedge p, a \vee b \vee p'] *_p c = [a \wedge b \wedge c \wedge p, a \vee b \vee c \vee p'] = a *_p [b \wedge c \wedge p, b \vee c \vee p'] = a *_p (b *_p c)$.

(4) In addition, $X = \bigcup_{x\in X} \{x\} \subset \bigcup_{x\in X} a *_p x = a *_p X$ and since the converse inclusion is obviously true, we get $a *_p X = X$. We can conclude now that $(X, *_p)$ is a commutative hypergroup.

(5) First, let us consider $x, y \in a *_p b$. Then, having $a\wedge b\wedge p \leq x \leq a\vee b\vee p'$ and $a\wedge b\wedge p \leq y \leq a\vee b\vee p'$, it follows that $a\wedge b\wedge p \leq x\wedge y\wedge p \leq x\vee y\vee p' \leq a \vee b \vee p'$, that is $x *_p y \subset a *_p b$.

Secondly, let x in $a *_p b$, i.e., $a \wedge b \wedge p \leq x \leq a \vee b \vee p'$, which implies that $x *_p a *_p b = [x \wedge a \wedge b \wedge p, x \vee a \vee b \vee p'] = [a \wedge b \wedge p, a \vee b \vee p'] = a *_p b$. So, it is clear that $(a *_p b, *_p)$ is a subhypergroup of $(X, *_p)$. ■

Proposition 5.2.5. *For all* $a, b, c, p, q \in X$, *we have a kind of associativity:*

$$a *_p (b *_q c) = (a *_p b) *_q c = a *_{p\wedge q} b *_{p\wedge q} c.$$

Proof. Since $p' \vee q' = (p \wedge q)'$, we have $a *_p (b *_q c) = a *_p [b \wedge c \wedge q, b \vee c \vee q'] = [a \wedge b \wedge c \wedge p \wedge q, a \vee b \vee c \vee p' \vee q'] = a *_{p\wedge q} b *_{p\wedge q} c$.

Similarly, $(a *_p b) *_q c = [a \wedge b \wedge p, a \vee b \vee p'] *_q c = [a \wedge b \wedge c \wedge p \wedge q, a \vee b \vee c \vee p' \vee q'] = a *_{p\wedge q} b *_{p\wedge q} c$. ■

Let us now investigate the p-extension hyperoperation derived from the p-join hyperoperation, i.e.,

$$a/_p b = \{x \in X \mid a \in x *_p b\} = \{x \in X \mid x \wedge b \wedge p \leq a \leq x \vee b \vee p'\}.$$

It is obvious that, for any $a, b \in X$, we have $a \in a/_p b$ and now we prove that the p-extension hyperoperation has the join property.

Proposition 5.2.6. *For any $p \in X$, the hypergroup $(X, *_p)$ is a join space.*

Proof. By Proposition 5.2.4, it is enough to prove that, for any $a, b, c, d, p \in X$, we have the following implication:

$$a/_p b \cap c/_p d \neq \emptyset \Rightarrow a *_p d \cap b *_p c \neq \emptyset.$$

Let $x \in a/_p b \cap c/_p d \neq \emptyset$. Then, $x \wedge b \wedge p \leq a \leq x \vee b \vee p'$ and $x \wedge d \wedge p \leq c \leq x \vee d \vee p'$. Since $a \leq x \vee b \vee p'$, we write $a \wedge d \leq (x \vee b \vee p') \wedge d = (x \wedge d) \vee ((b \vee p') \wedge d) \leq (x \wedge d) \vee (b \vee p')$ and now $a \wedge d \wedge p \leq ((x \wedge d) \vee (b \vee p')) \wedge p$. Since $x \wedge d \wedge p \leq c$, it follows that $x \wedge d \wedge p \leq c \wedge p \leq (c \vee p') \wedge p$. Then,

$$\begin{aligned} &a \wedge d \wedge p \leq (x \wedge d \wedge p) \vee ((b \vee p') \wedge p) \leq ((c \vee p') \wedge p) \vee ((b \vee p') \wedge p) = \\ &= ((c \vee p') \vee (b \vee p')) \wedge p \leq b \vee c \vee p'. \end{aligned} \tag{5.8}$$

Similarly, $c \leq x \vee d \vee p'$ implies that $c \wedge b \leq (x \vee d \vee p') \wedge b = (x \wedge b) \vee ((d \vee p') \wedge b) \leq (x \wedge b) \vee (d \vee p')$. Then, $c \wedge b \wedge p \leq ((x \wedge b) \vee (d \vee p')) \wedge p$. But $x \wedge b \wedge p \leq a$ implies that $x \wedge b \wedge p \leq a \wedge p \leq (a \vee p') \wedge p$ and thus

$$\begin{aligned} &c \wedge b \wedge p \leq (x \wedge b \wedge p) \vee ((d \vee p') \wedge p) \leq ((a \vee p') \wedge p) \vee ((d \vee p') \wedge p) = \\ &= ((a \vee p') \vee (d \vee p')) \wedge p \leq a \vee d \vee p'. \end{aligned} \tag{5.9}$$

Combining relations (5.8) and (5.9) we get $(a \wedge d \wedge p) \vee (c \wedge b \wedge p) \leq (a \vee d \vee p') \wedge (b \vee c \vee p')$ which means that the interval $[(a \wedge d \wedge p) \vee (c \wedge b \wedge p), (a \vee d \vee p') \wedge (b \vee c \vee p')] = a *_p d \cap b *_p c$ is non-empty, i.e., $*_p$ satisfies the join property. ■

Proposition 5.2.7. *If $(X, \leq, \vee, \wedge, ')$ is a Boolean lattice, then $a/_p b = a *_p b'$.*

Proof. Set any $a, b, p \in X$ and $x \in a/_p b$. Then, $x \wedge b \wedge p \leq a \leq x \vee b \vee p'$. Now, $a \leq x \vee b \vee p'$ implies that $a \wedge b' \wedge p \leq (x \vee b \vee p') \wedge b' \wedge p = (x \vee b \vee p') \wedge (b \vee p')' = (x \wedge (b \vee p')') \vee ((b \vee p') \wedge (b \vee p')') = (x \wedge (b \vee p')') \vee 0 = x \wedge b' \wedge p$. Thus, $a \wedge b' \wedge p \leq x \wedge b' \wedge p \leq x$.

Similarly, one finds that $x \leq a \vee b' \vee p'$ and thereby $x \in a *_p b'$, that is $a/_p b \subseteq a *_p b'$.

Conversely, take $x \in a *_p b'$. Since $a \wedge b' \wedge p \leq x$, it follows that $(a \wedge b' \wedge p) \vee (b \vee p') \leq x \vee b \vee p'$. Then, $(a \vee b \vee p') \wedge ((b' \wedge p) \vee (b \vee p')) = (a \vee b \vee p') \wedge ((b' \wedge p) \vee (b' \wedge p)') \leq x \vee b \vee p'$ which implies that $(a \vee b \vee p') \wedge 1 = a \vee b \vee p' \leq x \vee b \vee p'$, thus $a \leq x \vee b \vee p'$.

Similarly, since $x \leq a \vee b' \vee p'$, it follows that $x \wedge b \wedge p \leq (a \vee b' \vee p') \wedge (b \wedge p) = (a \wedge b \wedge p) \vee ((b' \vee p') \wedge (b \wedge p)) = (a \wedge b \wedge p) \vee ((b \wedge p)' \wedge (b \wedge p)) = (a \wedge b \wedge p) \vee 0 = a \wedge b \wedge p \leq a$. Therefore, $x \wedge b \wedge p \leq a \leq x \vee b \vee p'$, i.e., $a \in x *_p b$, so $x \in a/_p b$. Now, the inverse inclusion $a *_p b' \subset a/_p b$ is proved and it concludes the proof. ■

The next step is to prove that the hyperoperations $*_p$ and $/_p$ defined above can be considered as p-cuts of an L-fuzzy set, since they satisfy the three conditions in Proposition 2.1.5.

Proposition 5.2.8. *For all $a, b \in X$, we have*

(1) *For all $p, q \in X$, $p \leq q$ implies that $a *_q b \subseteq a *_p b$.*
(2) *For all $P \subset X$, $\bigcap_{p \in P} a *_p b = a *_{\bigvee P} b$.*
(3) $a *_0 b = X$.

Proof. Choose any $a, b \in X$.

(1) Let $p, q \in X$ be such that $p \leq q$. Then, $p \wedge a \wedge b \leq q \wedge a \wedge b$. By the order inversion property of negation, it follows that $q' \vee a \vee b \leq p' \vee a \vee b$. We can conclude that $[q \wedge a \wedge b, q' \vee a \vee b] \subseteq [p \wedge a \wedge b, p' \vee a \vee b]$, which is exactly the required result.

(2) For any subset $P \subset X$, we can write, according to the generalized de Morgan properties, that

$$\begin{aligned}\bigcap_{p \in P} a *_p b &= \bigcap_{p \in P} [a \wedge b \wedge p, a \vee b \vee p'] = \\ &= [\bigvee_{p \in P} (a \wedge b \wedge p), \bigwedge_{p \in P} (a \vee b \vee p')] = \\ &= [a \wedge b \wedge (\bigvee_{p \in P} p), a \vee b \vee (\bigwedge_{p \in P} p')] = \\ &= [a \wedge b \wedge (\bigvee P), a \vee b \vee (\bigvee P)'] = \\ &= a *_{\bigvee P} b.\end{aligned}$$

(3) $a *_0 b = [a \wedge b \wedge 0, a \vee b \vee 1] = [0, 1] = X$. ■

Proposition 5.2.9. *For all $a, b \in X$, the following statements hold:*

(1) *For all $p, q \in X$, $p \leq q$ implies that $a/_q b \subseteq a/_p b$.*
(2) *For all $P \subset X$, $\bigcap_{p \in P} a/_p b = a/_{\bigvee P} b$.*
(3) $a/_0 b = X$.

Proof. Choose any $a, b \in X$.

(1) Let $p, q \in X$ be such that $p \leq q$. For any $x \in a/_q b$ we have $b \wedge x \wedge q \leq a \leq b \vee x \vee q'$. But $p \leq q$ implies that $b \wedge x \wedge p \leq b \wedge x \wedge q$ and $b \vee x \vee q' \leq b \vee x \vee p'$. Thus, $b \wedge x \wedge p \leq b \wedge x \wedge q \leq a \leq b \vee x \vee q' \leq b \vee x \vee p'$, which means that $x \in a/_p b$. Hence, $a/_q b \subseteq a/_p b$.

(2) Choose any subset $P \subset X$. Then, we obtain the following implications, for any $p \in P$,

$$p \leq \bigvee P \Rightarrow a/_{\bigvee P} b \subseteq a/_p b \Rightarrow a/_{\bigvee P} b \subseteq \bigcap_{p \in P} a/_p b.$$

On the other hand, for any $x \in \bigcap_{p \in P} a/_p b$, we have that $x \in a/_p b$, for any $p \in P$, thus $a \in x *_p b$, that is, for any $p \in P$, $x \wedge b \wedge p \leq a \leq x \vee b \vee p'$. Then, $\bigvee_{p \in P} (x \wedge b \wedge p) \leq a \leq \bigwedge_{p \in P} (x \vee b \vee p')$, which means that $x \wedge b \wedge (\vee P) \leq$

$a \leq x \vee b \vee (\bigvee P)'$, equivalent with $a \in x *_{\bigvee P} b$, i.e., $x \in a/_{\bigvee P} b$. Hence, again by the crucial use of the generalized de Morgan properties, we have $\bigcap_{p \in P} a/_p b \subseteq a/_{\bigvee P} b$. Now, we can conclude that $\bigcap_{p \in P} a/_p b = a/_{\bigvee P} b$.

(3) It is obvious that $a/_0 b = \{x \mid a \in x *_0 b\} = \{x \mid a \in [x \wedge b \wedge 0, x \vee b \vee 1]\} = X$. ■

Based on these properties, it is natural now to construct two L-fuzzy hyperoperations, similarly to those introduced in Corsini-Tofan [45].

Definition 5.2.10. For any $a, b \in X$ define the following L-fuzzy sets, for any $x \in X$:

$$(a * b)(x) = \bigvee \{p \mid x \in a *_p b\}$$

and consequently

$$(a/b)(x) = \bigvee \{p \mid x \in a/_p b\}.$$

Since the families $\{a *_p b\}_{p \in X}$ and $\{a/_p b\}_{p \in X}$ have the p-cut properties, we can affirm that, for any $a, b \in X$ and $p \in X$, $(a *_p b)_p = a *_p b$ and $(a/b)_p = a/_p b$, meaning that the p-cuts of the L-fuzzy sets $*$ and $/$ coincide with the crisp hyperoperations $a *_p b$ and $a/_p b$.

The next goal is to prove that $(X, *)$ is an L-fuzzy hypergroup. First we introduce some auxiliary definitions, regarding the composition between an element and a fuzzy subset, or between two fuzzy subsets.

Definition 5.2.11. Given an L-fuzzy hyperoperation $\circ : X \times X \to F(X)$, for all $a \in X$ and $\mu \in F(X)$, define the L-fuzzy set $a \circ \mu \in F(X)$ by

$$(a \circ \mu)(x) = \bigvee_{b \in X, \mu(b) > 0} (a \circ b)(x)$$

and the crisp set $a \hat{\circ} \mu$ by

$$a \hat{\circ} \mu = \bigcup_{b \in X, \mu(b) > 0} \{x \mid (a \circ b)(x) = 1\}.$$

Moreover, for all $\mu, \lambda \in F(X)$, define the L-fuzzy set $\mu \circ \lambda \in F(X)$ by

$$(\mu \circ \lambda)(x) = \bigvee_{a, b \in X, \mu(a) > 0, \lambda(b) > 0} (a \circ b)(x)$$

and the crisp set

$$\mu \hat{\circ} \lambda = \bigcup_{a, b \in X, \mu(a) > 0, \lambda(b) > 0} \{x \mid (a \circ b)(x) = 1\}.$$

Proposition 5.2.12. *For all $a \in X$, $B \in \mathcal{P}(X)$ and $p \in X$, we have $a *_p B \subseteq (a * B)_p$.*

Proof. If $p = 0$, it is clear that $a *_p B = X = (a * B)_p$.

If $p > 0$, choose an arbitrary $x \in a *_p B$. It means that, there exists $b \in B$ such that $x \in a *_p b = (a * b)_p$. Hence, $(a * b)(x) \geq p > 0$. But $(a * B)(x) = \bigvee_{z \in B} (a * z)(x) \geq (a * b)(x) \geq p$, thus $x \in (a * B)_p$. Hence, $a *_p B \subseteq (a * B)_p$. ■

Now, we investigate some basic properties of the fuzzy hyperoperations $*$ and $/$.

Proposition 5.2.13. *For all $a, b, c, p \in X$ we have*

(1) $(a * b)(a) = (a * b)(b) = (a * a)(a) = 1$.
(2) $a * b = b * a$.
(3) *There exists $x \in X$ such that $((a * b) * c)(x) \wedge (a * (b * c))(x) \geq p$.*
(4) $a \hat{*} X = X$, *i.e., for all $x \in X$, there exists $y \in X$ such that $(a * y)(x) = 1$.*
(5) $a * X = X$, *i.e., for all $x \in X$, $(a * X)(x) = \bigvee_{y \in X} (a * y)(x) = 1$.*

Proof. Choose three arbitrary elements $a, b, c \in X$.

(1) Since $a \in [a \wedge b, a \vee b] = a *_1 b$, it follows that $1 \in \{p \mid a \in a *_p b\}$ and so $(a * b)(a) = \bigvee\{p \mid a \in a *_p b\} = 1$. Similarly, $(a * b)(b) = 1$. In particular, taking $a = b$, we get $(a * a)(a) = 1$.

(2) Since $a * b$ and $b * a$ have the same cuts, they are identical.

(3) If $p = 0$, then $(a * (b * c))_0 = X = ((a * b) * c)_0$, and so there exists $x \in X$ such that $(a * (b * c))(x) \wedge ((a * b) * c)(x) \geq 0$.

If $p > 0$, take an arbitrary $x \in a *_p b *_p c = a *_p (b *_p c)$; then there exists $z \in b *_p c = (b * c)_p$ such that $x \in a *_p z = (a * z)_p$. It follows that $(b * c)(z) \geq p > 0$ and $(a * z)(x) \geq p > 0$. Since $z \in \{u \mid (b * c)(u) > 0\}$, we can write that $p \leq (a * z)(x) \leq \bigvee_{u \in X, (b*c)(u) > 0} (a * u)(x) = (a * (b * c))(x)$ and then $x \in (a * (b * c))_p$, thus we proved that $a *_p b *_p c \subseteq (a * (b * c))_p$.

Similarly, we show that $a *_p b *_p c = (a *_p b) *_p c \subseteq ((a * b) * c)_p$, obtaining that $(a * (b * c))(x) \wedge ((a * b) * c)(x) \geq p > 0$.

(4) For any $x \in X$, there exists $y = x \in X$ such that $(a * y)(x) = (a * x)(x) = 1$.

(5) It is clear that $(a * X)(x) = \bigvee_{y \in X} (a * y)(x) > (a * x)(x) = 1$, so $a * X = X$. ■

In the following, for all $A, B \in F(X)$ and $p \in X$, we write $A \sim_p B$ if and only if there exists $x \in X$ such that $A(x) \wedge B(x) \geq p$.

Proposition 5.2.14. *For all $a, b, c, d, p \in X$ we have*

(1) $a * b \sim_p c * d$ *if and only if* $a *_p b \sim c *_p d$.
(2) $a/b \sim_p c/d$ *if and only if* $a/_p b \sim c/_p d$.
(3) $a/b \sim_p c/d$ *implies that* $a * d \sim_p b * c$.

Proof. (1) We prove the first item, the second one having a similar proof.

For any $a, b, c, d, p \in X$, we have $a * b \sim_p c * d$ if and only if there exists $x \in X$ such that $(a * b)(x) \geq p$ and $(c * d)(x) \geq p$. This is equivalent with $x \in (a * b)_p = a *_p b$ and $x \in (c * d)_p = c *_p d$. So, $a *_p b \cap c *_p d \neq \emptyset$. ■

(3) By Proposition 5.2.6, $(X, *_p)$ is a join space, and therefore the relation $a/b \sim_p c/d$, equivalently with $a/_p b \sim c/_p d$, implies that $a *_p d \sim b *_p c$, and then $a * d \sim_p b * c$. ■

We have already discussed, in Section 3.6 F-hypergroups, about the four types of reproducibility R_1, R_2, R_3, and R_4, defined by Corsini and Tofan [45], for the crisp hyperoperations obtained from a fuzzy hyperoperation. In the sequel, we will focus our attention only on the strongest one, the R_3-reproducibility, since it implies all the other three types.

Definition 5.2.15. Given an L-fuzzy hyperoperation $\circ : X \times X \to F(X)$, the hyperstructure $(X, \circ)$ is called *L-fuzzy-3 hypergroup* if it satisfies the following conditions:

(1) For all $a, b, c \in X$ we have $(a \circ b) \circ c = a \circ (b \circ c)$.
(2) For all $a \in X$, the R_3-reproducibility holds: $a \otimes X = X = X \otimes a$.

Definition 5.2.16. Given an L-fuzzy hyperoperation $\circ : X \times X \to F(X)$, the hyperstructure $(X, \circ)$ is called *L-fuzzy-3 $-\ p$ hypergroup* if it satisfies the following conditions:

(1) For all $a, b, c, p \in X$ we have $(a \circ_p b) \circ_p c = a \circ_p (b \circ_p c)$.
(2) For all $a \in X$, the R_3-reproducibility holds: $a \otimes X = X = X \otimes a$.

Definition 5.2.17. Given an L-fuzzy hyperoperation $\circ : X \times X \to F(X)$, the hyperstructure $(X, \circ)$ is called *L-fuzzy-3 H_v-group* if it satisfies the following conditions:

(1) For all $a, b, c, p \in X$ we have $(a \circ b) \circ c \sim_p a \circ (b \circ c)$.
(2) For all $a \in X$, the R_3-reproducibility holds: $a \otimes X = X = X \otimes a$.

Proposition 5.2.18. *The hyperstructure $(X, *)$ is an L-fuzzy-3 H_v-group and also an L-fuzzy-3 $-\ p$ hypergroup.*

Proof. The result follows immediately from Proposition 5.2.13 and the above definitions. ■

Now, we may introduce a particular type of L-fuzzy join space.

Definition 5.2.19. Given an L-fuzzy hyperoperation $\circ : X \times X \to F(X)$, the hyperstructure $(X, \circ)$ is called *L-fuzzy-3 $-\ p$ join space* if it is a commutative L-fuzzy-3 $-\ p$ hypergroup satisfying the following implication: for all $a, b, c, d, p \in X$, $a \parallel b \sim_p c \parallel d \Rightarrow a \circ d \sim_p b \circ c$, where we denote $x \parallel y = \{z \in X \mid x \in z \circ y\}$.

The hyperstructure $(X, \circ)$ is called *L-fuzzy-3 H_v-join space*, if it is a commutative L-fuzzy-3 H_v-group satisfying the previous join condition.

The following result follows now immediately.

Proposition 5.2.20. *The hyperstructure $(X, *)$ is an L-fuzzy-3 H_v-join space and also an L-fuzzy-3 $-$ p join space.*

We conclude the study of the hyperstructure $(X, *)$ with another property related to the theory of L-fuzzy sublattices. Notice that, for any two fixed elements $a, b \in X$, the families $\{(a * b)_p\}_{p\in X}$ and $\{(a/b)_p\}_{p\in X}$ are families of (crisp) closed intervals. On the other hand, $a * b$ and a/b are L-fuzzy intervals and thus, they are L-fuzzy convex sublattices. We based here on the following property: $I : X \to X$ is an L-fuzzy interval of $(X, \leq)$ if and only if, for any $p \in X$, the cut I_p is a closed interval of $(X, \leq)$.

More details about the lattice of the p-cuts of the hyperproduct $a * b$ can be founded in [123], while for basic properties of L-fuzzy intervals, the reader is refereed to [124].

5.3 Fuzzy Grade of a Hypergroupoid

Based on an iterative construction, consisting in associating to each hypergroupoid H a fuzzy subset that determines a new join space, one defines a numerical function which computes the so called *fuzzy grade* of H. This grade represents the number of the non-isomorphic join spaces associated with the hypergroupoid H in the sequence of join spaces and fuzzy sets iterated by the related construction. This idea was initiated by Corsini-Cristea in [38, 39], studied afterthat by Cristea [51, 55], Cristea-Ştefănescu [173] in the general case, and later for particular hypergroups: hypergroups with partial identities, by Corsini-Cristea [38, 39], complete hypergroups, by Cristea [50], Angheluţă-Cristea [4], or non-complete 1-hypergroups, by Corsini-Cristea [40]. Extensions to hypergraphs have been considered by Corsini et al. [42, 43], Feng et al. [106].

Here, we first present the general construction, then we determine the fuzzy grades of some finite special hypergroups, concluding with the study of the extension to the case of hypergraphs.

5.3.1 The Main Construction

For any hypergroup $(H, \circ)$, P. Corsini defined in [36] a fuzzy subset $\widetilde{\mu}$ of H in the following way: for any $u \in H$, consider:

$$\widetilde{\mu}(u) = \frac{\displaystyle\sum_{(x,y)\in Q(u)} \frac{1}{|x \circ y|}}{q(u)}, \tag{5.10}$$

where $Q(u) = \{(a, b) \in H^2 \mid u \in a \circ b\}$, $\quad q(u) = |Q(u)|$. If $Q(u) = \emptyset$, set

$\widetilde{\mu}(u) = 0$. In other words, $\widetilde{\mu}(u)$ is the average value of the reciprocals of the sizes of the hyperproducts $x \circ y$ containing u.

On the other hand, with any hypergroupoid H endowed with a fuzzy set α, one can associate a join space $(H, \circ_\alpha)$ like in (5.1): for any $(x, y) \in H^2$,

$$x \circ_\alpha y = \{z \in H \mid \alpha(x) \wedge \alpha(y) \leq \alpha(z) \leq \alpha(x) \vee \alpha(y)\}.$$

Let $({}^1H, \circ_1) = (H, \circ_\alpha)$ be the join space obtained before for the fuzzy subset $\widetilde{\mu}$. By using the same procedure as in (5.10), from ${}^1H$ we can obtain a membership function $\widetilde{\mu}_1$ and the associated join space ${}^2H$ and so on. A sequence of fuzzy sets and join spaces $(({}^iH, \circ_i), \widetilde{\mu}_i)_{i\geq 1}$ is determined in this way. We define $\widetilde{\mu}_0 = \widetilde{\mu}$, ${}^0H = H$.

Note that the support set for each join space in the above sequence is H, every hyperproduct $\circ_i$ is defined on the set H, which is denoted by ${}^iH$ at the step i, to indicate the current number of iteration. From $({}^iH, \circ_i)$, we obtain as in (5.10) the membership function ${}^{i+1}\widetilde{\mu}$.

The length of the sequence of join spaces associated with H is called the *fuzzy grade* of the hypergroupoid H, more exactly we give the following definition [39].

Definition 5.3.1. A hypergroupoid H has the *fuzzy grade* m, $m \in \mathbb{N}^*$, and we write $f.g.(H) = m$ if, for any i, $0 \leq i < m$, the join spaces ${}^iH$ and ${}^{i+1}H$ associated with H are not isomorphic (where ${}^0H = H$) and for any s, $s > m$, ${}^sH$ is isomorphic with ${}^mH$. We say that the hypergroupoid H has the *strong fuzzy grade* m and we write $s.f.g.(H) = m$ if $f.g.(H) = m$ and for all $s, s > m$, ${}^sH = {}^mH$.

This construction is important for at least two reasons: it provides examples of hypergroup structures on a given set and it gives the possibility of studying fuzzy sets in an algebraic approach. On the other hand, the construction could start either from a fuzzy subset or from a hypergroup structure on a non-empty set H, as in the following two examples.

EXAMPLE 45 Consider the hypergroupoid $H = \{a, b, c, d\}$ together with the following hyperoperation:

$\circ$	a	b	c	d
a	a	$\{a,b\}$	$\{a,b,c\}$	H
b	$\{a,b\}$	b	$\{b,c\}$	$\{b,c,d\}$
c	$\{a,b,c\}$	$\{b,c\}$	c	$\{c,d\}$
d	H	$\{b,c,d\}$	$\{c,d\}$	d

then, by formula (5.10) we calculate

$$\widetilde{\mu}(a) = \widetilde{\mu}(d) = 19/42, \quad \widetilde{\mu}(b) = \widetilde{\mu}(c) = 29/66$$

and so we construct the first join space represented by the table

$\circ_1$	a	b	c	d
a	$\{a,d\}$	H	H	$\{a,d\}$
b	H	$\{b,c\}$	$\{b,c\}$	H
c	H	$\{b,c\}$	$\{b,c\}$	H
d	$\{a,d\}$	H	H	$\{a,d\}$

Again we calculate

$$\widetilde{\mu}_1(a) = \widetilde{\mu}_1(b) = \widetilde{\mu}_1(c) = \widetilde{\mu}_1(d) = 4/12,$$

and consequently we obtain the second join space (which is clearly the total hypergroup)

$\circ_2$	a	b	c	d
a	H	H	H	H
b	H	H	H	H
c	H	H	H	H
d	H	H	H	H

Thus, any associated join space ${}^sH$, $s \geq 3$, will be identical with ${}^2H$ and thus, we conclude that $s.f.g.(H) = 2$, since we obtained only two non-isomorphic join spaces in the corresponding sequence.

EXAMPLE 46 Consider now a set $H = \{a, b, c, d\}$ endowed with the fuzzy subset μ defined as it follows:

$$\mu(a) = 0.23, \quad \mu(b) = 0.45, \quad \mu(c) = \mu(d) = 0.78;$$

then the associated join space $({}^1H, \circ_1)$ is represented by the table

$\circ_1$	a	b	c	d
a	a	$\{a, b\}$	H	H
b	$\{a, b\}$	b	$\{b, c, d\}$	$\{b, c, d\}$
c	H	$\{b, c, d\}$	$\{c, d\}$	$\{c, d\}$
d	H	$\{b, c, d\}$	$\{c, d\}$	$\{c, d\}$

We calculate again using formula (5.10)

$$\widetilde{\mu}_2(a) = 3/7, \quad \widetilde{\mu}_2(b) = 13/33, \quad \widetilde{\mu}_2(c) = \widetilde{\mu}_2(d) = 13/36,$$

following that the join space $({}^2H, \circ_2)$ is equal with the join space $({}^1H, \circ_1)$. Thereby, the sequence associated with H contains, in this case, only one join space, so $s.f.g.(H) = 1$.

For simplicity, we fix some notations. Let $(({}^iH, \circ_i), \widetilde{\mu}_i)_{i \geq 1}$ be the sequence of fuzzy sets and join spaces associated with H. Then, for any $i \geq 1$, there are $r \geq 1$, namely $r = r_i$, and a partition $\pi = \{{}^iC_j\}_{j=1}^r$ of ${}^iH$ such that, for any $j \geq 1$ and $x \in {}^iC_j$, we have ${}^iC_j = \widetilde{\mu}_{i-1}^{-1}(\widetilde{\mu}_{i-1}(x))$ (this means that $x, y \in {}^iC_j \Leftrightarrow \widetilde{\mu}_{i-1}(x) = \widetilde{\mu}_{i-1}(y)$).
On the set of classes $\{{}^iC_j\}_{j=1}^r$, define now the following ordering relation:

$$j < k \text{ if, for } x \in {}^iC_j \text{ and } y \in {}^iC_k, \text{ we have } \widetilde{\mu}_{i-1}(x) < \widetilde{\mu}_{i-1}(y).$$

In addition, for all j, s:

$$k_j = |^iC_j|,\ {}_sC = \bigcup_{1\leq j\leq s} {}^iC_j,\ {}^sC = \bigcup_{s\leq j\leq r} {}^iC_j,\ {}_sk = |{}_sC|,\ {}^sk = |{}^sC|. \quad (5.11)$$

With any ordered chain $({}^iC_1, {}^iC_2, \ldots, {}^iC_r)$ associate an ordered r-tuple $(k_1, k_2, \ldots, k_r)$, where $k_l = |{}^iC_l|$, for all $l, 1 \leq l \leq r$.

A general formula to compute the values of $\widetilde{\mu}$ is given in the following theorem [36].

Theorem 5.3.2. *For any $z \in {}^iC_s$, $s = 1, 2, \ldots, r$,*

$$\widetilde{\mu}_i(z) = \frac{2\sum\limits_{\substack{i\leq s\leq j\\ i\neq j}}^{i,j} \left(\dfrac{k_ik_j}{\sum\limits_{i\leq t\leq j} k_t}\right) + k_s}{2{}_sk{}^sk - k_s^2}. \quad (5.12)$$

If two consecutive hypergroups of the obtained sequence are isomorphic, then the sequence stops, therefore it is important to know when two consecutive join spaces in the associated sequence of a hypergroupoid are isomorphic or not. In this direction we have a first result, proved by Corsini-Leoreanu [41].

Theorem 5.3.3. *Let ${}^iH$ and ${}^{i+1}H$ be the join spaces associated with H determined by the membership functions $\widetilde{\mu}_{i-1}$ and $\widetilde{\mu}_i$, where ${}^iH = \bigcup\limits_{l=1}^{r_1} C_l$, ${}^{i+1}H = \bigcup\limits_{l=1}^{r_2} C'_l$. Moreover, $(k_1, k_2, \ldots, k_{r_1})$ is the r_1-tuple associated with ${}^iH$, and $(k'_1, k'_2, \ldots, k'_{r_2})$ the r_2-tuple associated with ${}^{i+1}H$.*
The join spaces ${}^iH$ and ${}^{i+1}H$ are isomorphic if and only if $r_1 = r_2 = r$ and $(k_1, k_2, \ldots, k_r) = (k'_1, k'_2, \ldots, k'_r)$ or $(k_1, k_2, \ldots, k_r) = (k'_r, k'_{r-1}, \ldots, k'_1)$.

Proof. We skip here the proof, since it can be found also in the book [47] or in [41]. ■

In order to use this result, we need to determine both r-tuples associated with ${}^iH$ and ${}^{i+1}H$. But Ştefănescu and Cristea [173] have given a sufficient condition such that two consecutive join spaces in the associated sequence are not isomorphic, that is the sequence doesn't stop, using only one r-tuple.

Theorem 5.3.4. *Let $(k_1, k_2, \ldots, k_r)$ be the r-tuple associated with the join space ${}^iH$. If $(k_1, \ldots, k_r) = (k_r, k_{r-1}, \ldots, k_1)$, then the join spaces ${}^iH$ and ${}^{i+1}H$ are not isomorphic.*

Proof. First, notice that the condition $(k_1, k_2, \ldots, k_r) = (k_r, k_{r-1}, \ldots, k_1)$ from hypothesis can be written in a more general form:

$$k_l = k_{r+1-l}, \quad \text{for any } l, 1 \leq l \leq \left[\frac{r}{2}\right].$$

To prove the theorem, it is enough to show that, for any $x \in C_l$ and any $y \in C_{r+1-l}$, we have $\widetilde{\mu}_i(x) = \widetilde{\mu}_i(y)$. It follows, then, that ${}^{i+1}H = \bigcup_{l=1}^{r'} C'_l$, where $r' = \left[\frac{r+1}{2}\right] \neq r$. This means, by Theorem 5.3.3, that the join space ${}^{i+1}H$ is not isomorphic with ${}^{i}H$.

Let us take two arbitrary elements $x \in C_l$ and $y \in C_{r+1-l}$. For simplicity, denote the numerator, in the general formula (5.12) of $\widetilde{\mu}_i(z)$, by $A(z)$ and the denominator by $q(z)$. First we prove that $q(x) = q(y)$, and then that $A(x) = A(y)$.

Since $k_l = k_{r+1-l}$, by Theorem 5.3.2, we obtain that $q(x) = 2 \cdot {}_lk \cdot {}^lk - k_l^2$ and $q(y) = {}_{r+1-l}k \cdot {}^{r+1-l}k - k_{r+1-l}^2$, where ${}_lk = k_1 + k_2 + \ldots + k_l = k_r + k_{r-1} + \ldots + k_{r+1-l} = {}^{r+1-l}k$ and ${}^lk = k_l + k_{l+1} + \ldots + k_r = k_{r+1-l} + k_{r-l} + \ldots + k_1 = {}_{r+1-l}k$, therefore ${}_lk = {}^{r+1-l}k$ and ${}^lk = {}_{r+1-l}k$. It is clear, now, that $q(x) = q(y)$.

Now, we prove that $A(x) = A(y)$. First, remark that, for any j, j', we have
$k_{l+j} = k_{r-(l+j)+1} = k_{r-l-j+1} = k_{(r-l+1)-j}$ and then

$$\sum_{t=j+1}^{(r-l+1)-j'} k_t = \sum_{t=l+j'}^{r-j} k_t, \tag{5.13}$$

because of the equality

$$\begin{aligned}\sum_{t=j+1}^{(r-l+1)-j'} k_t &= k_{j+1} + k_{j+2} + \ldots + k_{(r-l+1)-j'} \\ &= k_{r-j} + k_{r-j-1} + \ldots + k_{r-(r-l+1-j')+1} \\ &= k_{r-j} + k_{r-j-1} + \ldots + k_{l+j'} = \sum_{t=l+j'}^{r-j} k_t\end{aligned}$$

According to Theorem 5.3.2, one calculates:

$$\begin{aligned}A(x) &= k_l + \sum_{\substack{i \le l \le j \\ i \neq j}}^{i,j} \frac{k_i k_j}{\sum\limits_{i \le t \le j} k_t} = k_l + \sum_{i=1}^{l-1}\sum_{j=l}^{r} \frac{k_i k_j}{\sum\limits_{i \le t \le j} k_t} + \sum_{j=l+1}^{r} \frac{k_l k_j}{\sum\limits_{l \le t \le j} k_t} \\ &= k_l + \sum_{i=1}^{l-1}\sum_{j=l}^{r-l} \frac{k_i k_j}{\sum\limits_{i \le t \le j} k_t} + \sum_{i=1}^{l-1}\sum_{j=r-l+1}^{r} \frac{k_i k_j}{\sum\limits_{i \le t \le j} k_t} + \sum_{j=l+1}^{r-l} \frac{k_l k_j}{\sum\limits_{l \le t \le j} k_t} + \sum_{j=r-l+1}^{r} \frac{k_l k_j}{\sum\limits_{t=l}^{j} k_t} \\ &= k_l + \sum_{i=1}^{l}\sum_{j=r-l+1}^{r} \frac{k_i k_j}{\sum\limits_{i \le t \le j} k_t} + \sum_{i=1}^{l-1}\sum_{j=l}^{r-l} \frac{k_i k_j}{\sum\limits_{i \le t \le j} k_t} + k_l \sum_{j=l+1}^{r-l} \frac{k_j}{\sum\limits_{l \le t \le j} k_t}.\end{aligned}$$

Similarly we find

$$A(y) = k_{r-l+1} + \sum_{\substack{i\leq r-l+1\leq j\\ i\neq j}}^{i,j} \frac{k_i k_j}{\sum\limits_{i\leq t\leq j} k_t} = k_{r-l+1} + \sum_{i=1}^{r-l}\sum_{j=r-l+1}^{r} \frac{k_i k_j}{\sum\limits_{i\leq t\leq j} k_t} + \sum_{j=r-l+2}^{r} \frac{k_{r-l+1}k_j}{\sum\limits_{r-l+1\leq t\leq j} k_t}$$

$$= k_{r-l+1} + \sum_{i=1}^{l}\sum_{j=r-l+1}^{r} \frac{k_i k_j}{\sum\limits_{i\leq t\leq j} k_t} + \sum_{i=l+1}^{r-l}\sum_{j=r-l+1}^{r} \frac{k_i k_j}{\sum\limits_{i\leq t\leq j} k_t} + k_{r-l+1}\sum_{j=r-l+2}^{r} \frac{k_j}{\sum\limits_{t=r-l+1}^{j} k_t}.$$

Whence, since $k_l = k_{r-l+1}$, we conclude that $A(x) = A(y)$ if and only if

$$\sum_{i=1}^{l-1}\sum_{j=l}^{r-l} \frac{k_i k_j}{\sum\limits_{i\leq t\leq j} k_t} + k_l \sum_{j=l+1}^{r-l} \frac{k_j}{\sum\limits_{l\leq t\leq j} k_t} = \sum_{i=l+1}^{r-l}\sum_{j=r-l+1}^{r} \frac{k_i k_j}{\sum\limits_{i\leq t\leq j} k_t} + k_l \sum_{j=r-l+2}^{r} \frac{k_j}{\sum\limits_{t=r-l+1}^{j} k_t}.$$

This is equivalent with

$$k_1\left(\frac{k_l}{\sum\limits_{t=1}^{l} k_t} + \frac{k_{l+1}}{\sum\limits_{t=1}^{l+1} k_t} + \ldots + \frac{k_{r-l}}{\sum\limits_{t=1}^{r-l} k_t}\right) + k_2\left(\frac{k_l}{\sum\limits_{t=2}^{l} k_t} + \frac{k_{l+1}}{\sum\limits_{t=2}^{l+1} k_t} + \ldots + \frac{k_{r-l}}{\sum\limits_{t=2}^{r-l} k_t}\right) + \ldots +$$

$$+k_{l-1}\left(\frac{k_l}{\sum\limits_{t=l-1}^{l} k_t} + \frac{k_{l+1}}{\sum\limits_{t=l-1}^{l+1} k_t} + \ldots + \frac{k_{r-l}}{\sum\limits_{t=l-1}^{r-l} k_t}\right) + k_l\left(\frac{k_{l+1}}{\sum\limits_{t=l}^{l+1} k_t} + \frac{k_{l+2}}{\sum\limits_{t=l}^{l+2} k_t} + \ldots + \frac{k_{r-l}}{\sum\limits_{t=l}^{r-l} k_t}\right)$$

$$= k_r\left(\frac{k_{l+1}}{\sum\limits_{t=l+1}^{r} k_t} + \frac{k_{l+2}}{\sum\limits_{t=l+2}^{r} k_t} + \ldots + \frac{k_{r-l+1}}{\sum\limits_{t=r-l+1}^{r} k_t}\right) + k_{r-1}\left(\frac{k_{l+1}}{\sum\limits_{t=l+1}^{r-1} k_t} + \ldots + \frac{k_{r-l+1}}{\sum\limits_{t=r-l+1}^{r-1} k_t}\right) + \ldots +$$

$$+ k_{r-l+2}\left(\frac{k_{l+1}}{\sum\limits_{t=l+1}^{r-l+2} k_t} + \ldots + \frac{k_{r-l+1}}{\sum\limits_{t=r-l+1} r-l+2k_t}\right) + k_{r-l+1}\left(\frac{k_{l+1}}{\sum\limits_{t=l+1}^{r-l+1} k_t} + \ldots + \frac{k_{r-l}}{\sum\limits_{t=r-l}^{r-l+1} k_t}\right).$$

By relation (5.26), the two terms of the identity are equal and therefore $A(x) = A(y)$. Now, the proof is complete. ∎

The condition expressed in Theorem 5.3.4 is only a sufficient condition and not a necessary one in order that the join spaces ${}^iH$ and ${}^{i+1}H$ are not isomorphic, as we can see from the following example.

EXAMPLE 47 Let us consider the hypergroupoid $(H = \{0,1,2,3,4,5\}, \circ)$ endowed with the fuzzy set $\widetilde{\mu}$ which verifies the following relation

$$\begin{array}{ccccccc} \widetilde{\mu}(0) & < & \widetilde{\mu}(2) & < & \widetilde{\mu}(3) & < & \widetilde{\mu}(5). \\ \| & & & & \| & & \\ \widetilde{\mu}(1) & & & & \widetilde{\mu}(4) & & \end{array}$$

The hyperoperation $\circ$ is an arbitrary one. Clearly, the 4-tuple corresponding to H is $(2,1,2,1)$, which does not verify the condition in Theorem 5.3.4. Using the usual formula for calculating the membership function $\widetilde{\mu}_1$, one obtains:

$$A(0) = \frac{28}{5};\quad q(0) = 20;\quad \widetilde{\mu}_1(0) = \widetilde{\mu}_1(1) = 0,28,$$

$$A(2) = \frac{193}{30};\quad q(2) = 23;\quad \widetilde{\mu}_1(2) = 0,279,$$

$$A(3) = \frac{223}{30};\quad q(3) = 26;\quad \widetilde{\mu}_1(3) = \widetilde{\mu}_1(4) = 0,285,$$

$$A(5) = \frac{7}{2};\quad q(5) = 11;\quad \widetilde{\mu}_1(5) = 0,318.$$

Consequently we have

$$\begin{array}{ccccccc} \widetilde{\mu}_1(2) & < & \widetilde{\mu}_1(0) & < & \widetilde{\mu}_1(3) & < & \widetilde{\mu}_1(5) \\ & & \| & & \| & & \\ & & \widetilde{\mu}_1(1) & & \widetilde{\mu}_1(4) & & \end{array}$$

and therefore its associated 4-tuple is $(1, 2, 2, 1)$, verifying the condition in Theorem 5.3.4, and thus the join space ${}^2H$ is not isomorphic with ${}^1H$.

It would be useful to know which is the fuzzy grade of a hypergroupoid H on the base of the type of the n-tuple associated with it. We give some results for several types of n-tuples [4]. First, we determine the fuzzy grade of a hypergroupoid H with the property that $|H/R_{\widetilde{\mu}}| \in \{2, 3\}$, where the equivalence relation $R_{\widetilde{\mu}}$ is defined on H by

$$xR_{\widetilde{\mu}}y \Leftrightarrow \widetilde{\mu}(x) = \widetilde{\mu}(y).$$

Proposition 5.3.5. *Let $(H, \circ)$ be a hypergroupoid of cardinality n and denote by $(({}^iH, \circ_i), \widetilde{\mu}_i)_{i\geq 1}$ the sequence of fuzzy sets and of join spaces associated with H.*

(1) *If $|H/R_{\widetilde{\mu}}| = 2$, that is with H we associate the pair (k_1, k_2), and*
  (a) *if $k_1 = k_2$, then we have $s.f.g.(H) = 2$.*
  (b) *if $k_1 \neq k_2$, then we have $f.g.(H) = 1$.*

(2) *If $|H/R_{\widetilde{\mu}}| = 3$, that is with H we associate the triple (k_1, k_2, k_3), and*
  (a) *if $k_1 = k_2 = k_3$, then we have $f.g.(H) = 1$.*
  (b) *for $k_1 = k_3 \neq k_2$ we have $f.g.(H) = 2$ if $k_2 \neq 2k_3$ and $s.f.g.(H) = 3$ if $k_2 = 2k_3$.*
  (c) *if $k_1 < k_2 = k_3$, then we have $f.g.(H) = 1$.*
  (d) *for $k_1 = k_2 < k_3$ we have $f.g.(H) = 1$ if $P = 2k_3^3 - 8k_1^3 - k_1^2k_3 + 5k_1k_3^2 > 0$ and $f.g.(H) = 3$ otherwise.*
  (e) *for $k_1 \neq k_2 \neq k_3$, there is no precise order between $\widetilde{\mu}_1(x), \widetilde{\mu}_1(y)$ and $\widetilde{\mu}_1(z)$.*

Proof. (1) If the hypergroupoid H has two classes of equivalence, i.e., $H = C_1 \bigcup C_2$, where $|C_1| = k_1$ and $|C_2| = k_2$, with $k_1 + k_2 = n = |H|$, then one distinguishes the following two cases:

(a) $k_1 = k_2$; by Theorem 5.3.2, one finds that $\widetilde{\mu}_1(x) = \widetilde{\mu}_1(y)$, for any $x \in C_1$ and any $y \in C_2$. It follows that the join space ${}^2H$ is a total hypergroup and therefore $s.f.g.(H) = 2$.

(b) $k_1 < k_2$; by Theorem 5.3.2, $\widetilde{\mu}_1(x) = \dfrac{k_1 + 3k_2}{n(k_1 + 2k_2)}$, for any $x \in C_1$ and similarly, $\widetilde{\mu}_1(y) = \dfrac{k_2 + 3k_1}{n(k_2 + 2k_1)}$, for any $y \in C_2$. It is easy to see that $\widetilde{\mu}_1(x) > \widetilde{\mu}_1(y)$, which means that with ${}^1H$ one associates the pair (k_2, k_1). Thereby, by Theorem 5.3.3, it follows that ${}^2H$ is isomorphic with ${}^1H$, so $f.g.(H) = 1$.

(2) We suppose now that the hypergroupoid H has three equivalence classes: $H = C_1 \bigcup C_2 \bigcup C_3$, where $|C_1| = k_1$, $|C_2| = k_2$ and $|C_3| = k_3$, with $k_1 + k_2 + k_3 = n$. There exist the following possibilities:

(a) $k_1 = k_2 = k_3 = k$; by Theorem 5.3.2 it follows, for any $x \in C_1$, $y \in C_2$, $z \in C_3$, that $\widetilde{\mu}_1(x) = \dfrac{8}{15k}$, $\widetilde{\mu}_1(y) = \dfrac{11}{21k}$ and $\widetilde{\mu}_1(z) = \dfrac{8}{15k}$. Therefore, $\widetilde{\mu}_1(x) = \widetilde{\mu}_1(z) > \widetilde{\mu}_1(y)$ and thus, with the join space ${}^1H$ one associates the pair $(k, 2k)$. By the case (1)(b), we conclude that $f.g.(H) = 2$.

(b) $k_1 = k_3 \neq k_2$; using again Theorem 5.3.2, it results that $\widetilde{\mu}_1(x) = \widetilde{\mu}_1(z) \neq$ $\neq \widetilde{\mu}_1(y)$ as in the previous case. If $k_2 \neq 2k_3$, then $f.g.(H) = 2$; otherwise, if $k_2 = 2k_3$, it follows that the join space ${}^2H$ is a total hypergroup and thus $s.f.g.(H) = 3$.

(c) $k_1 < k_2 = k_3$; by Theorem 5.3.2 one obtains, for any $x \in C_1$, $y \in C_2$, $z \in C_3$,

$$\widetilde{\mu}_1(x) = \frac{1 + 2k_2\left(\dfrac{1}{k_1 + k_2} + \dfrac{1}{k_1 + 2k_2}\right)}{k_1 + 4k_2},$$

$$\widetilde{\mu}_1(y) = \frac{2 + 2k_1\left(\dfrac{1}{k_1 + k_2} + \dfrac{1}{k_1 + 2k_2}\right)}{4k_1 + 3k_2},$$

$$\widetilde{\mu}_1(z) = \frac{2 + 2k_1\dfrac{1}{k_1 + 2k_2}}{2k_1 + 3k_2}.$$

We verify whether $\widetilde{\mu}_1(x) > \widetilde{\mu}_1(y)$; this is equivalent with

$$\frac{1 + 2k_2\left(\dfrac{1}{k_1 + k_2} + \dfrac{1}{k_1 + 2k_2}\right)}{k_1 + 4k_2} > \frac{2 + 2k_1\left(\dfrac{1}{k_1 + k_2} + \dfrac{1}{k_1 + 2k_2}\right)}{4k_1 + 3k_2} \Leftrightarrow$$

$$8k_2^3 - 2k_1^3 - 5k_1^2k_2 + k_1k_2^2 > 0 \Leftrightarrow$$

$5k_2(k_2^2 - k_1^2) + 3k_2^3 - 2k_1^3 + k_1k_2^2 > 0$ which is obviously true for $k_1 < k_2$.

Now, we check if $\widetilde{\mu}_1(x) > \widetilde{\mu}_1(z)$; that is

$$\frac{1+2k_2\left(\dfrac{1}{k_1+k_2}+\dfrac{1}{k_1+2k_2}\right)}{k_1+4k_2} > \frac{2+2k_1\dfrac{1}{k_1+2k_2}}{2k_1+3k_2} \Leftrightarrow$$

$$8k_2^3 - 2k_1^3 - 7k_1^2k_2 + k_1k_2^2 > 0 \Leftrightarrow$$

$$(k_2-k_1)(2k_1^2+9k_1k_2+8k_2^2) > 0 \text{ which is clearly true for } k_1 < k_2.$$

Finally, we see if $\widetilde{\mu}_1(y) = \widetilde{\mu}_1(z)$, that is

$$\frac{2+2k_1\left(\dfrac{1}{k_1+k_2}+\dfrac{1}{k_1+2k_2}\right)}{4k_1+3k_2} = \frac{2+2k_1\dfrac{1}{k_1+2k_2}}{2k_1+3k_2} \Leftrightarrow$$

$$2k_1^2 + k_1k_2 - 2k_2^2 = 0.$$

Considering it as a second degree equation in $k_1 \in \mathbb{N}^*$, the solutions are $\dfrac{-k_2 \pm \sqrt{17}k_2}{2}$, which are not natural numbers, so $\widetilde{\mu}_1(y) \neq \widetilde{\mu}_1(z)$.
We conclude that $\widetilde{\mu}_1(x) > \widetilde{\mu}_1(y)$ and $\widetilde{\mu}_1(x) > \widetilde{\mu}_1(z)$, with $\widetilde{\mu}_1(y) \neq \widetilde{\mu}_1(z)$; therefore the triple associated with the join space ${}^1H$ is (k_2, k_2, k_1) and, according to Theorem 5.3.3, the join spaces ${}^2H$ and ${}^1H$ are isomorphic and thus $f.g.(H) = 1$.

(*d*) $k_1 = k_2 < k_3$; again by the same theorem one obtains, for any $x \in C_1$, $y \in C_2$, $z \in C_3$,

$$\widetilde{\mu}_1(x) = \frac{2\left(1+\dfrac{k_3}{2k_1+k_3}\right)}{3k_1+2k_3};$$

$$\widetilde{\mu}_1(y) = \frac{2\left(1+\dfrac{k_3}{k_1+k_3}+\dfrac{k_3}{2k_1+k_3}\right)}{3k_1+4k_3};$$

$$\widetilde{\mu}_1(z) = \frac{1+2k_1\left(\dfrac{1}{k_1+k_3}+\dfrac{1}{2k_1+k_3}\right)}{4k_1+k_3}.$$

After a simple computation one finds that $\widetilde{\mu}_1(x) \neq \widetilde{\mu}_1(y)$ and $\widetilde{\mu}_1(x) > \widetilde{\mu}_1(z)$.
Now, we see when $\widetilde{\mu}_1(y) > \widetilde{\mu}_1(z)$; this is equivalent with

$$P = P(k_1, k_3) = 2k_3^3 - 8k_1^3 - k_1^2k_3 + 5k_1k_3^2 > 0.$$

Since $k_3 > k_1$, set $k_3 = k_1 + l$, with $l \in \mathbb{N} \setminus \{0\}$. Then, $P = -2k_1^3 + 2l^3 + 15k_1^2l + 11k_1l^2$. It is obvious that, for $l \geq k_1$, that is equivalent to $k_3 \geq 2k_1$, one finds $P > 0$. For $l < k_1$ the result is variable: there are several values of (k_1, k_3) such that $P > 0$ and several values such that $P < 0$.

If $P > 0$, then $\widetilde{\mu}_1(z) < \widetilde{\mu}_1(y)$ and $\widetilde{\mu}_1(z) < \widetilde{\mu}_1(x)$, with $\widetilde{\mu}_1(x) \neq \widetilde{\mu}_1(y)$. As in the previous case, it results that $f.g.(H) = 1$.
If $P < 0$, then $\widetilde{\mu}_1(y) < \widetilde{\mu}_1(z) < \widetilde{\mu}_1(x)$ and the triple associated with the join space ${}^1H$ is (k_1, k_3, k_1). By the case (2)(*b*), with the join space ${}^2H$ one associates the pair $(2k_1, k_3)$, with $2k_1 \neq k_3$ (if $2k_1 = k_3$, then $P > 0$) and thus $f.g.(H) = 3$ (for example, if $k_1 = k_2 = 10$ and $k_3 = 11$, then $P < 0$ and $f.g.(H) = 3$).

(*e*) $k_1 \neq k_2 \neq k_3$; using Theorem 5.3.2, for any $x \in C_1$, $y \in C_2$, $z \in C_3$, we obtain

$$\widetilde{\mu}_1(x) = \frac{1 + \dfrac{2k_2}{k_1+k_2} + \dfrac{2k_3}{k_1+k_2+k_3}}{k_1 + 2k_2 + 2k_3};$$

$$\widetilde{\mu}_1(y) = \frac{k_2 + 2\dfrac{k_1k_2}{k_1+k_2} + 2\dfrac{k_1k_3}{k_1+k_2+k_3} + 2\dfrac{k_2k_3}{k_2+k_3}}{2k_1k_2 + 2k_1k_3 + k_2^2 + 2k_2k_3};$$

$$\widetilde{\mu}_1(z) = \frac{1 + \dfrac{2k_1}{k_1+k_2+k_3} + \dfrac{2k_2}{k_2+k_3}}{2k_1 + 2k_2 + k_3}.$$

It is simple to verify that $\widetilde{\mu}_1(x) > \widetilde{\mu}_1(y)$ and $\widetilde{\mu}_1(x) > \widetilde{\mu}_1(z)$. We can not say that for any triple (k_1, k_2, k_3) there is the same relation between $\widetilde{\mu}_1(y)$ and $\widetilde{\mu}_1(z)$. For example, if $k_1 < k_2 < k_3$,
- taking $(k_1, k_2, k_3) = (10, 11, 12)$ we have $\widetilde{\mu}_1(y) > \widetilde{\mu}_1(z)$;
- taking $(k_1, k_2, k_3) = (20, 21, 22)$ we have $\widetilde{\mu}_1(y) < \widetilde{\mu}_1(z)$.

Similarly, if $k_1 < k_3 < k_2$, then
- for $(k_1, k_2, k_3) = (1, 3, 2)$ we have $\widetilde{\mu}_1(y) < \widetilde{\mu}_1(z)$;
- for $(k_1, k_2, k_3) = (10, 20, 19)$ we have $\widetilde{\mu}_1(y) > \widetilde{\mu}_1(z)$.

So, there is no precise order between $\widetilde{\mu}_1(x), \widetilde{\mu}_1(y), \widetilde{\mu}_1(z)$. ■

We continue with a general result regarding the membership function $\widetilde{\mu}$ for a hypergroupoid with the associated r-tuple of a symmetric form. We distinguish two cases, depending on the parity of r.

Corollary 5.3.6. *Let r be an odd number and $(k_1, k_2, \ldots, k_{s-1}, k_s, k_{s-1}, \ldots, k_2, k_1)$, with $s = (r+1)/2$, be the r-tuple associated with a hypergroupoid H, that is $H = \bigcup_{i=1}^{r} C_i$, with $|C_i| = k_i$, for any $i = 1, 2, \ldots, r$. Then, for any $x \in C_i$ and $y \in C_{r-i+1}$, we have $\widetilde{\mu}_1(x) = \widetilde{\mu}_1(y)$, whenever $i = 1, 2, \ldots, s-1$.*

Proof. It follows directly from Theorem 5.3.2. ■

Corollary 5.3.7. *Let r be an even number and $(k_1, k_2, \ldots, k_{s-1}, k_s, k_s, k_{s-1}, \ldots, k_2, k_1)$, with $s = r/2$, be the r-tuple associated with a hypergroupoid H, that is $H = \bigcup_{i=1}^{r} C_i$, with $|C_i| = k_i$,* *for any* $i = 1, 2, \ldots, r$. *Then, for any* $x \in C_i$ *and* $y \in C_{r-i+1}$, *we have* $\widetilde{\mu}_1(x) = \widetilde{\mu}_1(y)$, *whenever* $i = 1, 2, \ldots, s$.

Proof. It follows directly from Theorem 5.3.2. ■

EXAMPLE 48 If we continue with Example 47, where $(1, 2, 2, 1)$ is the 4-tuple associated with the join space ${}^1H$, according with Corollary 5.3.7, it follows that $(2, 4)$ is the pair associated with the join space ${}^2H$. Now, based on Proposition 5.3.5 (1)(b), we conclude that the join space ${}^3H$ is not isomorphic with ${}^2H$, but it is isomorphic with ${}^4H$, therefore $f.g.(H) = 3$

Now, we have all the elements to present a method to construct a hypergroup H with $s.f.g.(H)$ equal with a given natural number. First, we recall the fundamental result obtained by Cristea [51] on a finite hypergroupoid verifying a certain condition.

Theorem 5.3.8. *Let* H *be the finite hypergroupoid* $H = \{x_1, x_2, \ldots, x_n\}$, *with* $n = 2^s$, $s \in \mathbb{N}^*$, *which verifies the relation* $\widetilde{\mu}(x_1) < \widetilde{\mu}(x_2) < \ldots < \widetilde{\mu}(x_n)$, *i.e. the* n*-tuple associated with* H *is* $(1, 1, \ldots, 1)$. *Then, the join space* ${}^{s+1}H$ *is a total hypergroup and by consequence* $s.f.g.(H) = s + 1$.

Proof. By Corollary 5.3.7, it follows that, with the join space $({}^1H, \widetilde{\mu}_1)$ is associated the r_1-tuple $(2, 2, \ldots, 2)$, with $r_1 = 2^{s-1}$. Then, with the join space $({}^2H, \widetilde{\mu}_2)$ is associated the r_2-tuple $(4, 4, \ldots, 4)$, with $r_2 = 2^{s-2}$ and so on; with the join space $({}^{s-1}H, \widetilde{\mu}_{s-1})$ one associates the pair $(2^{s-1}, 2^{s-1})$; thereby the join space ${}^{s+1}H$ is a total hypergroup. In conclusion, $s.f.g.(H) = s + 1$. ■

The same procedure can also be used if with the hypergroup $(H, \widetilde{\mu})$ one associates the r-tuple $(k, k, \ldots, k)$, where $r = 2^s$ and $n = |H| = 2^s k$. Similarly, we obtain $s.f.g.(H) = s + 1$. A such hypergroup always exists, as we can notice in the following result.

Theorem 5.3.9. *Given a natural number* $s \in \mathbb{N}^* \setminus \{1\}$, *there is always a hypergroup* H *such that* $s.f.g.(H) = s$. *Moreover,* H *is a join space.*

Proof. Consider the hypergroupoid $H = \{x_1, x_2, \ldots, x_n\}$, $n = 2^s$, with the hyperoperation:

$$x_i \circ x_i = x_i, \text{ for } i \in \{1, 2, \ldots, n\},$$
$$x_i \circ x_j = x_j \circ x_i = \{x_i, x_{i+1}, \ldots, x_j\}, \text{ for } 1 \leq i < j \leq n.$$

It is clear that $(H, \circ)$ is the join space ${}^1H$ associated with the hypergroupoid in Theorem 5.3.8. Thus, the join space ${}^sH$ is a total hypergroup and $s.f.g.(H) = s$. ■

In the sequel, we will discuss about the fuzzy grade of the direct product of two hypergroupoids, determining it in some particular cases. The method is the same: based on the form of the ordered tuples associated with the two hypergroupoids, one calculates the fuzzy grade of their direct product [55].

For the beginning, we recall the construction of the direct product of two hypergroupoids. Let $(H_1, \circ_1)$ and $(H_2, \circ_2)$ be two hypergroupoids. On the cartesian product $H_1 \times H_2$ define the hyperproduct

$$(a_1, a_2) \otimes (b_1, b_2) = \{(x, y) \mid x \in a_1 \circ_1 b_1, y \in a_2 \circ_2 b_2\}.$$

For any $i \in \{1, 2\}$, let $\widetilde{\mu}_i$ and $\widetilde{\mu}_\otimes$ be the membership functions associated with $(H_1, \circ_1)$, $(H_2, \circ_2)$, and $(H_1 \times H_2, \otimes)$, respectively, determined by relation (5.10). The next result gives us an useful relation between them.

Proposition 5.3.10. *If $\widetilde{\mu}_1$, $\widetilde{\mu}_2$ and $\widetilde{\mu}_\otimes$ are the membership functions associated as in* (5.10) *with H_1, H_2 and $H_1 \times H_2$ respectively, then, for any $(x, y) \in H_1 \times H_2$, the following relation holds:*

$$\widetilde{\mu}_\otimes(x, y) = \widetilde{\mu}_1(x) \cdot \widetilde{\mu}_2(y).$$

Proof. Using the notations from (5.10), we write

$$\widetilde{\mu}_1(x) = \frac{A_1(x)}{q_1(x)}, \text{ for any } x \in H_1 \text{ and } \widetilde{\mu}_2(y) = \frac{A_2(y)}{q_2(y)}, \text{ for any } y \in H_2.$$

For an arbitrary pair $(x, y) \in H_1 \times H_2$ we find:

$$\begin{aligned} q((x, y)) &= |\{((a_1, a_2), (b_1, b_2)) \mid (x, y) \in (a_1, a_2) \otimes (b_1, b_2)\}| \\ &= |\{((a_1, a_2), (b_1, b_2)) \mid x \in a_1 \circ_1 b_1, y \in a_2 \circ_2 b_2\}| \\ &= |\{(a_1, b_1) \mid x \in a_1 \circ_1 b_1\}| \cdot |\{(a_2, b_2) \mid y \in a_2 \circ_2 b_2\}| \\ &= q_1(x) \cdot q_2(y). \end{aligned}$$

Then, we obtain:

$$\begin{aligned} |(a_1, a_2) \otimes (b_1, b_2)| &= |\{(x, y) \mid x \in a_1 \circ_1 b_1, y \in a_2 \circ_2 b_2\}| \\ &= |\{x \in H_1 \mid x \in a_1 \circ_1 b_1\}| \cdot |\{y \in H_2 \mid y \in a_2 \circ_2 b_2\}| \\ &= |a_1 \circ_1 b_1| \cdot |a_2 \circ_2 b_2| \end{aligned}$$

and therefore

$$\begin{aligned} A((x, y)) &= \sum_{(x,y) \in (a_1, a_2) \otimes (b_1, b_2)} \frac{1}{|(a_1, a_2) \otimes (b_1, b_2)|} \\ &= \sum_{\substack{(x,y) \in H_1 \times H_2 \\ x \in a_1 \circ_1 b_1 \\ y \in a_2 \circ_2 b_2}} \left(\frac{1}{|a_1 \circ_1 b_1|} \cdot \frac{1}{|a_2 \circ_2 b_2|} \right) \\ &= \sum_{x \in a_1 \circ_1 b_1} \sum_{y \in a_2 \circ_2 b_2} \frac{1}{|a_1 \circ_1 b_1|} \cdot \frac{1}{|a_2 \circ_2 b_2|} \\ &= \left(\sum_{x \in a_1 \circ_1 b_1} \frac{1}{|a_1 \circ_1 b_1|} \right) \cdot \left(\sum_{y \in a_2 \circ_2 b_2} \frac{1}{|a_2 \circ_2 b_2|} \right) \\ &= A_1(x) \cdot A_2(y). \end{aligned}$$

By consequence, $\widetilde{\mu}_{\otimes}(x,y) = \widetilde{\mu}_1(x) \cdot \widetilde{\mu}_2(y)$. ∎

Consider now some particular cases of the ordered tuple associated with a hypergroupoid. We start with the case when the two hypergroupoids coincide.

Proposition 5.3.11. *Suppose that we associate with the hypergroupoid* $(H, \circ)$ *the ordered pair* (k,l). *Then, the ordered tuple associated with* $(H \times H, \otimes)$ *is the triple* $(k^2, 2kl, l^2)$.

Proof. Take $\widetilde{\mu}$ the membership function associated with H and $R_{\widetilde{\mu}}$ be the regular equivalence defined by: $xR_{\widetilde{\mu}}y \Leftrightarrow \widetilde{\mu}(x) = \widetilde{\mu}(y)$.

By hypothesis, $|H/R_{\widetilde{\mu}}| = 2$ and thus denote $H/R_{\widetilde{\mu}} = \{C_1, C_2\}$, with $C_1 = \{x_1, x_2, \ldots, x_k\}$ and $C_2 = \{x_{k+1}, x_{k+2}, \ldots, x_{k+l}\}$, respectively. We respect the convention that $\widetilde{\mu}(x_1) < \widetilde{\mu}(x_{k+1})$.

Then, we obtain

$$\begin{aligned}
&\forall\ i,j \in \{1,2,\ldots,k\},\ \ \widetilde{\mu}_{\otimes}(x_i,x_j) = \widetilde{\mu}(x_i)\cdot\widetilde{\mu}(x_j) = \widetilde{\mu}(x_1)^2\\
&\forall\ i',j' \in \{k+1,k+2,\ldots,k+l\},\ \ \widetilde{\mu}_{\otimes}(x_{i'},x_{j'}) = \widetilde{\mu}(x_{i'})\cdot\widetilde{\mu}(x_{j'}) = \widetilde{\mu}(x_{k+1})^2\\
&\forall\ i \in \{1,2,\ldots,k\}, \forall\ j' \in \{k+1,\ldots,k+l\},\ \ \widetilde{\mu}_{\otimes}(x_i,x_{j'}) = \widetilde{\mu}(x_1)\cdot\widetilde{\mu}(x_{k+1}).
\end{aligned}$$

Since $\widetilde{\mu}(x_1) < \widetilde{\mu}(x_{k+1})$ it follows that $\widetilde{\mu}(x_1)^2 < \widetilde{\mu}(x_{k+1})\cdot\widetilde{\mu}(x_1) < \widetilde{\mu}(x_{k+1})^2$, and thus $\widetilde{\mu}_{\otimes}(x_i,x_j) < \widetilde{\mu}_{\otimes}(x_i,x_{j'}) < \widetilde{\mu}_{\otimes}(x_{i'},x_{j'})$, for any $i,j \in \{1,2,\ldots,k\}$ and any $i',j' \in \{k+1,\ldots,k+l\}$; therefore with the hypergroupoid $(H\times H, \otimes)$ one associates the triple $(k^2, 2kl, l^2)$. ∎

Proposition 5.3.12. *Let* (k,l) *be the ordered pair associated with the hypergroupoid* H. *We have the following possibilities:*

(1) *If* $k = l$, *then* $s.f.g.(H \times H) = 3$.
(2) *If* $2k = l$, *then* $f.g.(H \times H) = 1$.
(3) *If* $k \neq l$ *and* $2k \neq l$, *then* $f.g.(H \times H) \in \{1,2\}$.

Proof. By Proposition 5.3.11, it follows that the triple associated with $H \times H$ is $(k^2, 2kl, l^2)$, so $|H/R_{\widetilde{\mu}}| = 2$. For determining the fuzzy grade, we can use now Proposition 5.3.5.

If $k = l$, then $s.f.g.(H) = 2$ and $s.f.g.(H \times H) = 3$.

If $k \neq l$, we have two situations (we suppose $k < l$):

- if $2k = l$, then $k^2 < 2kl = l^2$ and it is clear that $f.g.(H) = f.g.(H\times H) = 1$;
- if $2k \neq l$, then $k^2 \neq 2kl \neq l^2$ and thereby we obtain $f.g.(H) = 1$ and $f.g.(H \times H) \in \{1,2\}$. ∎

We pass now to the case when the two hypergroupoids are distinct.

Lemma 5.3.13. *Let* $(H_1, \circ_1)$ *and* $(H_2, \circ_2)$ *be two distinct hypergroupoids such that* $|H_1/R_{\widetilde{\mu}_1}| = |H_2/R_{\widetilde{\mu}_2}| = 2$. *For more precision we consider:* $H_1 = C_1 \cup C_2$, *and* $H_2 = C_1' \cup C_2'$, *with* $|C_1| = k, |C_2| = l,\ |C_1'| = m, |C_2'| = n$.
If, for any $x \in C_1, y \in C_2, x' \in C_1', y' \in C_2'$ *we have*

(1) $\widetilde{\mu}_{\otimes}(x, y') = \widetilde{\mu}_{\otimes}(y, x')$, *then* $|(H_1 \times H_2)/R_{\widetilde{\mu}_{\otimes}}| = 3$.
(2) $\widetilde{\mu}_{\otimes}(x, y') \neq \widetilde{\mu}_{\otimes}(y, x')$, *then* $|(H_1 \times H_2)/R_{\widetilde{\mu}_{\otimes}}| = 4$.

Proof. Based on the previous notations, we may write

$$H_1 \times H_2 = \{(x, x'), (x, y'), (y, x'), (y, y') \mid x \in C_1, y \in C_2, x' \in C_1', y' \in C_2'\}$$

and then $\widetilde{\mu}_{\otimes}(u, v) = \widetilde{\mu}_1(u) \cdot \widetilde{\mu}_2(v)$, for any $u \in H_1$ and any $v \in H_2$.
Using the standard convention $\widetilde{\mu}_1(x) < \widetilde{\mu}_1(y)$ and $\widetilde{\mu}_2(x') < \widetilde{\mu}_2(y')$, it is easy to obtain the relations:

$$\widetilde{\mu}_{\otimes}(x, x') < \widetilde{\mu}_{\otimes}(x, y') < \widetilde{\mu}_{\otimes}(y, y')$$
$$\widetilde{\mu}_{\otimes}(x, x') < \widetilde{\mu}_{\otimes}(y, x') < \widetilde{\mu}_{\otimes}(y, y').$$

If, for any $x \in C_1, y \in C_2, x' \in C_1', y' \in C_2'$ we have

(1) $\widetilde{\mu}_{\otimes}(x, y') = \widetilde{\mu}_{\otimes}(y, x')$, then $|(H_1 \times H_2)/R_{\widetilde{\mu}_{\otimes}}| = 3$ and the triple associated with $H_1 \times H_2$ is $(km, kn + lm, ln)$;
(2) $\widetilde{\mu}_{\otimes}(x, y') \neq \widetilde{\mu}_{\otimes}(y, x')$, then $|(H_1 \times H_2)/R_{\widetilde{\mu}_{\otimes}}| = 4$ and with $H_1 \times H_2$ could be associated the 4-tuples (km, kn, lm, ln) or (km, lm, kn, ln). ■

Proposition 5.3.14. *Let (k, k) be the pair associated with the hypergroupoid $(H_1, \circ_1)$ and (m, m) that one associated with a second hypergroupoid $(H_2, \circ_2)$. Then, $s.f.g.(H_1 \times H_2) = 3$.*

Proof. The corresponding join spaces ${}^2H_1$ and ${}^2H_2$ from the sequences associated with H_1 and H_2, respectively, are total hypergroups and thus $s.f.g.(H_1) = s.f.g.(H_2) = 2$. We keep the notations from Lemma 5.3.13.

(1) If $\widetilde{\mu}_{\otimes}(x, y') = \widetilde{\mu}_{\otimes}(y, x')$, then the triple associated with $H_1 \times H_2$ is $(kl, 2kl, kl)$ and thus the join space ${}^3(H_1 \times H_2)$ is a total hypergroup and $s.f.g.(H_1 \times H_2) = 3$ (by Proposition 5.3.5 2)b)).
(2) If $\widetilde{\mu}_{\otimes}(x, y') \neq \widetilde{\mu}_{\otimes}(y, x')$, then the 4-tuple related to $H_1 \times H_2$ has the form (kl, kl, kl, kl); then, by Corollary 5.3.7, to the join space ${}^1(H_1 \times H_2)$ corresponds the pair $(2kl, 2kl)$ and again the join space ${}^3(H_1 \times H_2)$ is a total hypergroup and $s.f.g.(H_1 \times H_2) = 3$ (by Proposition 5.3.5 (1)(a)). ■

EXAMPLE 49 Consider the hypergroupoid $H_1 = \{a, b, c, d\}$ with the following hyperoperation:

$\circ_1$	a	b	c	d
a	a	$\{a,b\}$	$\{a,b,c\}$	H
b	$\{a,b\}$	b	$\{b,c\}$	$\{b,c,d\}$
c	$\{a,b,c\}$	$\{b,c\}$	c	$\{c,d\}$
d	H	$\{b,c,d\}$	$\{c,d\}$	d

and the hypergroupoid $H_2 = \{x, y, z, u, t, v\}$ with the hyperoperation:

$\circ_2$	x	y	z	u	t	v
x	x	$\{x,y,z\}$	$\{x,y,z\}$	H	H	H
y	$\{x,y,z\}$	y	$\{x,y,z\}$	H	H	H
z	$\{x,y,z\}$	$\{x,y,z\}$	z	H	H	H
u	H	H	H	$\{u,t,v\}$	$\{u,t,v\}$	$\{u,t,v\}$
t	H	H	H	$\{u,t,v\}$	$\{u,t,v\}$	$\{u,t,v\}$
v	H	H	H	$\{u,t,v\}$	$\{u,t,v\}$	$\{u,t,v\}$

Then, we obtain

$$\widetilde{\mu}_1(a) = \widetilde{\mu}_1(d) = 19/42 > \widetilde{\mu}_1(b) = \widetilde{\mu}_1(c) = 29/66$$

and

$$\widetilde{\mu}_2(x) = \widetilde{\mu}_2(y) = \widetilde{\mu}_2(z) = 6/25 > \widetilde{\mu}_2(u) = \widetilde{\mu}_2(t) = \widetilde{\mu}_2(v) = 6/27,$$

that is the couple associated with H_1 is $(2,2)$ and the couple associated with H_2 is $(3,3)$. Using the notations of Lemma 5.3.13, we write

$$H_1 = C_1 \cup C_2, \quad \text{where} \quad C_1 = \{b,c\}, C_2 = \{a,d\},$$

$$H_2 = C_1' \cup C_2', \quad \text{where} \quad C_1' = \{u,t,v\}, C_2' = \{x,y,z\}$$

and then

$$\widetilde{\mu}_{\otimes}(a,u) = \widetilde{\mu}_1(a) \cdot \widetilde{\mu}_2(u) = 19/189 \neq 29/275 = \widetilde{\mu}_1(b) \cdot \widetilde{\mu}_2(x) = \widetilde{\mu}_{\otimes}(b,x)$$

and then, by Proposition 5.3.14 (2), it follows that the 4-tuple associated with $H_1 \times H_2$ has the form $(6,6,6,6)$; then, by Corollary 5.3.7, with the join space ${}^1(H_1 \times H_2)$ is associated the pair $(12,12)$; then with the second join space ${}^2(H_1 \times H_2)$ is associated the 1-tuple (24) and thus the join space ${}^3(H_1 \times H_2)$ is a total hypergroup and $s.f.g.(H_1 \times H_2) = 3$.

5.3.2 Intuitionistic Fuzzy Grade of a Hypergroupoid

A similar construction, like that presented in the previous section, can be realized substituting the fuzzy sets with one of their generalization, i.e., Atanassov's intuitionistic fuzzy sets, called here, for simplicity, just intuitionistic fuzzy sets. This idea was developed by Cristea-Davvaz [56], obtaining a new sequence of join spaces associated with a hypergroupoid, with the length called the *intuitionistic fuzzy grade.* New properties of these sequences have been brought to light, with respect to those determined by the fuzzy sets. For example, since an intuitionistic fuzzy set is characterized by two degrees, the membership and non-membership degree, one can associate with a non-empty set H endowed with an intuitionistic fuzzy set two sequences of join spaces, determining the so-called *lower and upper intuitionistic fuzzy grade* (using the minimum or, respectively, the maximum between the membership and the non-membership degree), that under some conditions are the same.

Another interesting property was identified for a particular class of hypergroups, the i.p.s. hypergroups (i.e. hypergroups with partial scalar identities) of order 7: for some of them, the associated sequence is cyclic or it contains a cyclic part. We will discus more about this in the next section.

Throughout this work, we consider an intuitionistic fuzzy set A in X, as an object having the form $A = \{(x, \mu_A(x), \lambda_A(x)) \mid x \in X\}$, where, for any $x \in X$, the *degree of membership* of x (namely $\mu_A(x)$) and the *degree of non-membership* of x (namely $\lambda_A(x)$) verify the relation $0 \leq \mu_A(x) + \lambda_A(x) \leq 1$. We denote it on short by $A = (\mu, \lambda)$. It is worth to mention here that this definition belongs to Atanassov [7, 8, 9], and for this reason the name "Atanassov's intuitionistic fuzzy set" is used in the literature, but here for simplicity we call it just "intuitionistic fuzzy set". More details about the terminology regarding this kind of sets are discussed in [102].

For any hypergroupoid $(H, \circ)$, Cristea-Davvaz [56] defined an intuitionistic fuzzy set $A = (\overline{\mu}, \overline{\lambda})$ in the following way: for any $u \in H$, we consider:

$$\overline{\mu}(u) = \frac{\sum\limits_{(x,y)\in Q(u)} \frac{1}{|x \circ y|}}{n^2}, \quad \overline{\lambda}(u) = \frac{\sum\limits_{(x,y)\in \bar{Q}(u)} \frac{1}{|x \circ y|}}{n^2}, \tag{5.14}$$

where $Q(u) = \{(a, b) \in H^2 \mid u \in a \circ b\}, \bar{Q}(u) = \{(a, b) \in H^2 \mid u \notin a \circ b\}$. If $Q(u) = \emptyset$, we set $\overline{\mu}(u) = 0$ and similarly, if $\overline{Q}(u) = \emptyset$ we set $\overline{\lambda}(u) = 0$. Moreover, it is clear that, for any $u \in H$, $0 \leq \overline{\mu}(u) + \overline{\lambda}(u) \leq 1$.

Now, we may associate with H two join spaces $({}_0H, \circ_{\overline{\mu}\wedge\overline{\lambda}})$ and $({}^0H, \circ_{\overline{\mu}\vee\overline{\lambda}})$, where, for any fuzzy set α on H, the hyperproduct " $\circ_\alpha$ ", introduced by Corsini [28] is defined by the same relation (5.1)

$$x \circ_\alpha y = \{u \in H \mid \alpha(x) \wedge \alpha(y) \leq \alpha(u) \leq \alpha(x) \vee \alpha(y)\}.$$

We recall that the new hypergroup $(H, \circ_\alpha)$ is a join space [28].

By using the same procedure as in (5.14), from the join space $({}_0H, \circ_{\overline{\mu}\wedge\overline{\lambda}})$ we can construct the intuitionistic fuzzy set $\overline{A}_1 = (\overline{\mu}_1, \overline{\lambda}_1)$ as in (5.14); we associate again the join space $({}_1H, \circ_{\overline{\mu}_1\wedge\overline{\lambda}_1})$, we determine, like in (5.14), its intuitionistic fuzzy set $\overline{A}_2 = (\overline{\mu}_2, \overline{\lambda}_2)$ and we construct the join space $({}_2H, \circ_{\overline{\mu}_2\wedge\overline{\lambda}_2})$ and so on. We obtain the sequence $({}_iH = ({}_iH, \circ_{\overline{\mu}_i\wedge\overline{\lambda}_i}); \overline{A}_i = (\overline{\mu}_i, \overline{\lambda}_i))_{i\geq 0}$ of join spaces, with the same support set H, and intuitionistic fuzzy sets associated with H. Similarly we may construct the second sequence $({}^iH = ({}^iH, \circ_{\overline{\mu}_i\vee\overline{\lambda}_i}); \overline{A}_i = (\overline{\mu}_i, \overline{\lambda}_i))_{i\geq 0}$ of join spaces and intuitionistic fuzzy sets.

The length of the two corresponding sequences associated with H are called the lower, and respectively, the upper intuitionistic fuzzy grade of H, more exactly:

Definition 5.3.15. (see [56]) A set H endowed with an intuitionistic fuzzy set $A = (\mu, \lambda)$ has the *lower intuitionistic fuzzy grade* m, $m \in \mathbb{N}^*$, and we write $l.i.f.g.(H) = m$ if, for any i, $0 \leq i < m-1$, the join spaces $({}_iH, \circ_{\overline{\mu}_i\wedge\overline{\lambda}_i})$

and $({}_{i+1}H, \circ_{\overline{\mu}_{i+1}\wedge\overline{\lambda}_{i+1}})$ associated with H are not isomorphic (where ${}_0H = ({}_0H, \circ_{\overline{\mu}\wedge\overline{\lambda}})$) and, for any s, $s \geq m$, ${}_sH$ is isomorphic with ${}_{m-1}H$.

Definition 5.3.16. (see [56]) A set H endowed with an intuitionistic fuzzy set $A = (\mu, \lambda)$ has the *upper intuitionistic fuzzy grade* m, $m \in \mathbb{N}^*$, and we write $u.i.f.g.(H) = m$ if, for any i, $0 \leq i < m-1$, the join spaces $({}^iH, \circ_{\overline{\mu}_i\wedge\overline{\lambda}_i})$ and $({}^{i+1}H, \circ_{\overline{\mu}_{i+1}\vee\overline{\lambda}_{i+1}})$ associated with H are not isomorphic (where ${}^0H = ({}^0H, \circ_{\overline{\mu}\vee\overline{\lambda}}))$ and for any s, $s \geq m$, ${}^sH$ is isomorphic with ${}^{m-1}H$.

For the sake of illustration we consider the following example.

EXAMPLE 50 (see [56]) On the set $H = \{a, b, c, d\}$ we consider the following intuitionistic fuzzy set:

$$\begin{array}{l} \overline{\mu}(a) = 0.25\ \overline{\mu}(b) = 0.25\ \overline{\mu}(c) = 0.30\ \overline{\mu}(d) = 0.10 \\ \overline{\lambda}(a) = 0.40\ \overline{\lambda}(b) = 0.40\ \overline{\lambda}(c) = 0.50\ \overline{\lambda}(d) = 0.90. \end{array}$$

To start with, we construct the first sequence of join spaces and we determine its length, which represents $l.i.f.g.(H)$.

For the associated join space

$\circ_{\overline{\mu}\wedge\overline{\lambda}}$	a	b	c	d
a	$\{a,b\}$	$\{a,b\}$	$\{a,b,c\}$	$\{a,b,d\}$
b	$\{a,b\}$	$\{a,b\}$	$\{a,b,c\}$	$\{a,b,d\}$
c	$\{a,b,c\}$	$\{a,b,c\}$	c	H
d	$\{a,b,d\}$	$\{a,b,d\}$	$\{a,b,d\}$	d

we calculate the intuitionistic fuzzy set associated with H as in (5.14) and we obtain the following values:

$$\begin{array}{l} \overline{\mu}_1(a) = 31/96\ \overline{\mu}_1(b) = 31/96\ \overline{\mu}_1(c) = 17/96\ \overline{\mu}_1(d) = 17/96 \\ \overline{\lambda}_1(a) = 12/96\ \overline{\lambda}_1(b) = 12/96\ \overline{\lambda}_1(c) = 26/96\ \overline{\lambda}_1(d) = 26/96 \end{array}$$

therefore $\overline{\mu}_1 \wedge \overline{\lambda}_1(a) = \overline{\mu}_1 \wedge \overline{\lambda}_1(b) < \overline{\mu}_1 \wedge \overline{\lambda}_1(c) = \overline{\mu}_1 \wedge \overline{\lambda}_1(d)$ which leads to the second join space

$\circ_{\overline{\mu}_1\wedge\overline{\lambda}_1}$	a	b	c	d
a	$\{a,b\}$	$\{a,b\}$	H	H
b	$\{a,b\}$	$\{a,b\}$	H	H
c	H	H	$\{c,d\}$	$\{c,d\}$
d	H	H	$\{c,d\}$	$\{c,d\}$

Then, for any $x \in H$, we have $\overline{\mu}_2(x) = 4/16$ and $\overline{\lambda}_2(x) = 2/16$; thus, for any $x, y \in H$, $x \circ_{\overline{\mu}_2\wedge\overline{\lambda}_2} y = H$. So the first sequence of join spaces associated with H has only 3 elements, meaning that $l.i.f.g.(H) = 3$.

Now, in order to determine $u.i.f.g.(H)$, we start with the join space

$\circ_{\overline{\mu}\vee\overline{\lambda}}$	a	b	c	d
a	$\{a,b\}$	$\{a,b\}$	$\{a,b,c\}$	H
b	$\{a,b\}$	$\{a,b\}$	$\{a,b,c\}$	H
c	$\{a,b,c\}$	$\{a,b,c\}$	c	$\{c,d\}$
d	H	H	$\{c,d\}$	d

Note that $({}_0H, \circ_{\overline{\mu}\wedge\overline{\lambda}})$ and $({}^0H, \circ_{\overline{\mu}\vee\overline{\lambda}})$ are not isomorphic. Since

$$\overline{\mu}_1(a) = 13/48 \; \overline{\mu}_1(b) = 13/48 \; \overline{\mu}_1(c) = 13/48 \; \overline{\mu}_1(d) = 9/48$$
$$\overline{\lambda}_1(a) = 9/48 \;\; \overline{\lambda}_1(b) = 9/48 \;\; \overline{\lambda}_1(c) = 9/48 \;\; \overline{\lambda}(d) = 13/48$$

it follows that, whenever $x \in H$, $\overline{\mu}_1 \vee \overline{\lambda}_1(x) = 13/48$ which means that, for any $x, y \in H$, $x \circ_{\overline{\mu}_1\vee\overline{\lambda}_1} y = H$. So the second sequence of join spaces associated with H has 2 elements, that is $u.i.f.g.(H) = 2$. .

Note that we may start the construction of the same sequences from a hypergroupoid $(H, \circ)$, instead of a set H endowed with an intuitionistic fuzzy set. In this case, we obtain only one sequence of join spaces. Indeed, since, for any $x \in H$, the sum $\overline{\mu}(x) + \overline{\lambda}(x)$ is always constant, it follows that $\overline{\mu}(x) = \overline{\mu}(y)$ if and only if $\overline{\lambda}(x) = \overline{\lambda}(y)$ and then the join spaces $({}_0H, \circ_{\overline{\mu}\wedge\overline{\lambda}})$ and $({}^0H, \circ_{\overline{\mu}\vee\overline{\lambda}})$ are always isomorphic. In order to explain this situation, we introduce a new concept.

Definition 5.3.17. We say that a hypergroupoid H has the *intuitionistic fuzzy grade m*, $m \in \mathbb{N}^*$, and we write $i.f.g.(H) = m$, if $l.i.f.g.(H) = m$.

We illustrate better this concept in the following example.

EXAMPLE 51 (see [56]) Consider the hypergroupoid $H = \{a, b, c, d\}$ with the hyperoperation defined in the table:

$\circ$	a	b	c	d
a	a	b	c	d
b	b	$\{a,c\}$	$\{b,c\}$	d
c	c	$\{b,c\}$	$\{a,b\}$	d
d	d	d	d	$\{a,b,c\}$

then

$$\overline{\mu}(a) = 14/96, \; \overline{\mu}(b) = 23/96, \; \overline{\mu}(c) = 23/96, \; \overline{\mu}(d) = 36/96$$
$$\overline{\lambda}(a) = 66/96 \;\; \overline{\lambda}(b) = 57/96 \;\; \overline{\lambda}(c) = 57/96 \;\; \overline{\lambda}(d) = 44/96$$

and so we have

$\circ_{\overline{\mu}\wedge\overline{\lambda}}$	a	b	c	d
a	a	$\{a,b,c\}$	$\{a,b,c\}$	H
b	$\{a,b,c\}$	$\{b,c\}$	$\{b,c\}$	$\{b,c,d\}$
c	$\{a,b,c\}$	$\{b,c\}$	$\{b,c\}$	$\{b,c,d\}$
d	H	$\{b,c,d\}$	$\{b,c,d\}$	d

Again we calculate

$$\begin{array}{llll} \overline{\mu}_1(a) = 34/192, & \overline{\mu}_1(b) = 62/192, & \overline{\mu}_1(c) = 62/192, & \overline{\mu}_1(d) = 34/192 \\ \overline{\lambda}_1(a) = 52/192 & \overline{\lambda}_1(b) = 24/192 & \overline{\lambda}_1(c) = 24/192 & \overline{\lambda}_1(d) = 52/192 \end{array}$$

and construct the associated join space

$\circ_{\overline{\mu}_1 \wedge \overline{\lambda}_1}$	a	b	c	d
a	$\{a,d\}$	H	H	$\{a,d\}$
b	H	$\{b,c\}$	$\{b,c\}$	H
c	H	$\{b,c\}$	$\{b,c\}$	H
d	$\{a,d\}$	H	H	$\{a,d\}$

obtaining

$$\begin{array}{llll} \overline{\mu}_2(a) = 2/8, & \overline{\mu}_2(b) = 2/8, & \overline{\mu}_2(c) = 2/8, & \overline{\mu}_2(d) = 2/8 \\ \overline{\lambda}_2(a) = 1/8 & \overline{\lambda}_2(b) = 1/8 & \overline{\lambda}_2(c) = 1/8 & \overline{\lambda}_2(d) = 1/8 \end{array}$$

At last, it results the following total hypergroup

$\circ_{\overline{\mu}_2 \wedge \overline{\lambda}_2}$	a	b	c	d
a	H	H	H	H
b	H	H	H	H
c	H	H	H	H
d	H	H	H	H

concluding that $i.f.g.(H) = 3$.

In order to establish a general formula to compute the values of the membership functions $\overline{\mu}_i$ and $\overline{\lambda}_i$, we introduce, like in the fuzzy case, several notations.

Let $({}_iH = ({}_iH, \circ_{\overline{\mu}_i \wedge \overline{\lambda}_i}); \overline{A}_i = (\overline{\mu}_i, \overline{\lambda}_i))_{i \geq 0}$ be the sequence of join spaces and intuitionistic fuzzy sets associated with a hypergroupoid H. Then, for any i, there are r, namely $r = r_i$, and a partition $\Pi = \{{}^iC_j\}_{j=1}^r$ of ${}_iH$ such that, for any $j \geq 1, x, y \in {}^iC_j \Leftrightarrow \overline{\mu}_{i-1}(x) = \overline{\mu}_{i-1}(y)$. On the set of the classes $\{{}^iC_j\}_{j=1}^r$ we define the following ordering relation: $i_j < i_k$ if for elements $x \in {}^iC_j$ and $y \in {}^iC_k$, $\overline{\mu}_{i-1}(x) < \overline{\mu}_{i-1}(y)$.
Moreover, for all j, s :

$$k_j = |{}^iC_j|, \quad {}_sC = \bigcup_{1 \leq j \leq s}^{i} C_j, \quad {}^sC = \bigcup_{s \leq j \leq r}^{i} C_j, \quad {}_sk = |{}_sC|, \quad {}^sk = |{}^sC|.$$

With any ordered chain $({}^iC_1, {}^iC_2, \ldots, {}^iC_r)$ we associate an ordered r-tuple $(k_1, k_2, \ldots, k_r)$, where $k_l = |{}^iC_l|$, for all l, $1 \leq l \leq r$.

As in Theorem 5.3.2 we obtain the following formula.

Theorem 5.3.18. *For any* $z \in^i C_s$, $i \geq 0$, $s = 1, 2, \ldots, r$,

$$\overline{\mu}_i(z) = \frac{k_s + 2 \sum\limits_{\substack{l \leq s \leq m \\ l \neq m}} \frac{k_l k_m}{\sum\limits_{l \leq t \leq m} k_t}}{n^2}.$$

In a similar way, we find the next result.

Theorem 5.3.19. *For any* $z \in^i C_s$, $i \geq 0$, $s = 1, 2, \ldots, r$,

$$\overline{\lambda}_i(z) = \frac{\sum\limits_{l \neq s} k_l + 2 \sum\limits_{s \leq l < m \leq r} \frac{k_l k_m}{\sum\limits_{l \leq t \leq m} k_t} + 2 \sum\limits_{1 \leq l < m \leq s} \frac{k_l k_m}{\sum\limits_{l \leq t \leq m} k_t}}{n^2}.$$

Proof. Since, for any $z \in^i C_s$, $s = 1, 2, \ldots, r$, the following formula

$$\overline{\mu}_i(z) + \overline{\lambda}_i(z) = \frac{\sum\limits_{l=1}^{r} k_l + 2 \sum\limits_{1 \leq l < m \leq r} \frac{k_l k_m}{\sum\limits_{l \leq t \leq m} k_t}}{n^2}$$

holds, the statement of this theorem follows immediately from Theorem 5.3.18. ■

It is worth to mention that, the conditions under which two consecutive join spaces in the related sequences are isomorphic (Theorem 5.3.3 and Theorem 5.3.4) are valid also in the intuitionistic fuzzy case. In particular it holds the following.

Theorem 5.3.20. *Let* ${}_iH$ *and* ${}_{i+1}H$ *be the join spaces associated with* H *determined by the membership functions* $\overline{\mu}_i \wedge \overline{\lambda}_i$ *and* $\overline{\mu}_{i+1} \wedge \overline{\lambda}_{i+1}$, *where* ${}_iH = \bigcup\limits_{l=1}^{r_1} C_l$, ${}_{i+1}H = \bigcup\limits_{l=1}^{r_2} C'_l$ *and* $(k_1, k_2, \ldots, k_{r_1})$ *is the* r_1*-tuple associated with* ${}_iH$, $(k'_1, k'_2, \ldots, k'_{r_2})$ *is the* r_2*-tuple associated with* ${}_{i+1}H$. *The join spaces* ${}_iH$ *and* ${}_{i+1}H$ *are isomorphic if and only if* $r_1 = r_2$ *and* $(k_1, k_2, \ldots, k_{r_1}) = (k'_1, k'_2, \ldots, k'_{r_1})$ *or* $(k_1, k_2, \ldots, k_{r_1}) = (k'_{r_1}, k'_{r_1 - 1}, \ldots, k'_1)$.

Similarly to the fuzzy case, the following corollaries are valid .

Corollary 5.3.21. *Let* r *be an odd number and* $(k_1, k_2, \ldots, k_{s-1}, k_s, k_{s-1}, \ldots, k_2, k_1)$, *with* $s = (r+1)/2$, *be the* r*-tuple associated with a hypergroupoid* H *with respect to the equivalence relation* $R_{\overline{\mu} \wedge \overline{\lambda}}$, *that is* $H = \bigcup\limits_{i=1}^{r} C_i$, *with* $|C_i| = k_i$, *for any* $i = 1, 2, \ldots, r$. *Then, for any* $x \in C_i$ *and* $y \in C_{r-i+1}$, *we have* $\overline{\mu}_1 \wedge \overline{\lambda}_1(x) = \overline{\mu}_1 \wedge \overline{\lambda}_1(y)$, *whenever* $i = 1, 2, \ldots, s-1$.

Proof. It follows directly from Theorems 5.3.18, 5.3.19. ∎

Corollary 5.3.22. *Let r be even and $(k_1, k_2, \ldots, k_{s-1}, k_s, k_s, k_{s-1}, \ldots, k_2, k_1)$, with $s = r/2$, be the r-tuple associated with a hypergroupoid H, with respect to the equivalence relation $R_{\overline{\mu}\wedge\overline{\lambda}}$, that is $H = \bigcup_{i=1}^{r} C_i$, with $|C_i| = k_i$, for any $i = 1, 2, \ldots, r$. Then, for any $x \in C_i$ and $y \in C_{r-i+1}$, we have $\overline{\mu}_1 \wedge \overline{\lambda}_1(x) = \overline{\mu}_1 \wedge \overline{\lambda}_1(y)$, whenever $i = 1, 2, \ldots, s$.*

Proof. It follows directly from Theorems 5.3.18, 5.3.19. ∎

It is interesting notice that the hypergroup considered in Theorem 5.3.9 (which is also a join space) has the intuitionistic fuzzy grade equal with the strong fuzzy grade, being an example of a hypergroup having these two grades equal with a given natural number grater than 1. For proving this, first we need some auxiliar results, presenting in the following lemmas.

Lemma 5.3.23. *On the set $H = \{x_1, x_2, \ldots, x_n\}$, where $n = 2^p$, $p \in \mathbb{N}^*$, define the hyperproduct*

$$x_i \circ x_i = x_i, 1 \le i \le n,$$
$$x_i \circ x_j = x_j \circ x_i = \{x_i, x_{i+1}, \ldots, x_j\}, 1 \le i < j \le n.$$

Then, for any $s \in \{1, 2, \ldots, n/2\}$, we obtain

$$\overline{\mu}(x_s) = \overline{\mu}(x_{n-s+1}), \quad \overline{\lambda}(x_s) = \overline{\lambda}(x_{n-s+1}).$$

Proof. The method we use here consists in defining some blocks (from the table of the hypergroup $(H, \circ)$) associated with every $x_s \in H$.

The table of the commutative hyperoperation "$\circ$" defined in the lemma is the following one:

H	x_1	x_2	x_3	$\ldots$	x_s	x_{s+1}	$\ldots$	x_{n-1}	x_n
x_1	x_1	x_1, x_2	$x_1 \to x_3$	$\ldots$	$x_1 \to x_s$	$x_1 \to x_{s+1}$	$\ldots$	$x_1 \to x_{n-1}$	H
x_2		x_2	x_2, x_3	$\ldots$	$x_2 \to x_s$	$x_2 \to x_{s+1}$	$\ldots$	$x_2 \to x_{n-1}$	$x_2 \to x_n$
x_3			x_3	$\ldots$	$x_3 \to x_s$	$x_3 \to x_{s+1}$	$\ldots$	$x_3 \to x_{n-1}$	$x_3 \to x_n$
$\vdots$				$\ddots$	$\ldots$	$\ldots$	$\ldots$	$\ldots$	$\ldots$
x_s					x_s	x_s, x_{s+1}	$\ldots$	$x_s \to x_{n-1}$	$x_s \to x_n$
x_{s+1}						x_{s+1}	$\ldots$	$x_{s+1} \to x_{n-1}$	$x_{s+1} \to x_n$
$\vdots$							$\ddots$	$\ldots$	$\ldots$
x_{n-1}								x_{n-1}	x_{n-1}, x_n
x_n									x_n

where we use the notation $x_i \to x_j = \{x_i, x_{i+1}, \ldots, x_j\}$, with $i < j$.

One finds the element x_s only in two blocks: one formed with the lines $1^{st}, 2^{nd}, \ldots, s^{th}$ and the columns $s^{th}, (s+1)^{th}, \ldots, n^{th}$ of the previous table and denoted by $A(x_s)$, and the other formed with the lines

$s^{th}, (s+1)^{th}, \ldots, n^{th}$ and the columns $1^{st}, 2^{nd}, \ldots, s^{th}$ of the same table and denoted by $A'(x_s)$.

Take $A(x_s) = (a_{lj})_{\substack{1 \le l \le s \\ 1 \le j \le n-s+1}} \in \mathcal{M}_{s,n-s+1}(\mathcal{P}^*(H))$, where $a_{lj} = x_l \circ x_{s+j-1} = \{x_l, x_{l+1}, \ldots, x_s, \ldots, x_j\}$.

For any matrix $A(x_s)$, $s \in \{1, 2, \ldots, n\}$, we construct another matrix $B(x_s) = (b_{lj})_{\substack{1 \le l \le s \\ 1 \le j \le n-s+1}} \in \mathcal{M}_{s,n-s+1}(\mathbb{N}^*)$, where $b_{lj} = |a_{lj}| = |x_l \circ x_{s+j-1}|$. The matrix $B(x_s)$ is the following one:

$$\begin{pmatrix} s & s+1 & s+2 & \ldots & 2s-2 & 2s-1 & \ldots & n-1 & n \\ s-1 & s & s+1 & \ldots & 2s-3 & 2s-2 & \ldots & n-2 & n-1 \\ s-2 & s-1 & s & \ldots & 2s-4 & 2s-3 & \ldots & n-3 & n-2 \\ \vdots & \vdots & \vdots & \ddots & \vdots & \vdots & \ldots & \vdots & \vdots \\ 2 & 3 & 4 & \ldots & s & s+1 & \ldots & n-s+1 & n-s+2 \\ 1 & 2 & 3 & \ldots & s-1 & s & \ldots & n-s & n-s+1 \end{pmatrix}$$

It is a routine verifying that, for every $s \in \{1, 2, \ldots, n/2\}$, the matrices $B(x_s)$ and $B(x_{n-s+1})$ have the same elements distributed in a similar order, meaning that if $B(x_s) = (b_{lj})_{\substack{1 \le l \le s \\ 1 \le j \le n-s+1}}$ and $B(x_{n-s+1}) = (\tilde{b}_{ik})_{\substack{1 \le i \le n-s+1, \\ 1 \le k \le s}}$ then there is the relation $\tilde{b}_{ik} = b_{s-k+1, n-s-i+2}$.

Since

$$Q(x_s) = \{(x_l, x_{s+j-1}), (x_{s+l'-1}, x'_j), 1 \le l, j' \le s, 1 \le, l', j \le n-s+1\},$$

it follows that

$$\sum_{(a,b) \in Q(x_1)} \frac{1}{|a \circ b|} = 1 + 2\left(\frac{1}{2} + \frac{1}{3} + \ldots + \frac{1}{n}\right) \tag{5.15}$$

and, whenever $s \ge 2$,

$$\sum_{(a,b) \in Q(x_s)} \frac{1}{|a \circ b|} = \left(2 \sum_{\substack{1 \le l \le s \\ 1 \le j \le n-s+1}}^{l,j} \frac{1}{b_{lj}}\right) - 1 =$$
$$= 1 + 2\left(\frac{2}{2} + \frac{3}{3} + \ldots + \frac{s}{s} + \frac{s}{s+1} + \ldots + \frac{s}{n-s+1} + \frac{s-1}{n-s+2} + \ldots + \frac{2}{n-1} + \frac{1}{n}\right). \tag{5.16}$$

In other words, in order to compute the numerator of $\overline{\mu}(x_s)$ we add twice all the inverses $1/b_{lj}$ of the elements b_{lj} of the matrix $B(x_s)$ and then substract 1 (since $1 = |x_s \circ x_s|$ is the unique common element of the matrices $B(x_s)$ and its symmetric obtained from $A'(x_s)$). The simplest method to compute these operations is to cross the matrix $B(x_s)$, starting with the lower left-hand corner, along its "principal pseudo-diagonals" that contain only the elements 1, or 2, or 3 and so on.

Since the matrices $B(x_s)$ and $B(x_{n-s+1})$, for $s \in \{1, 2, \ldots, n/2\}$, contain the same elements, it is clear that $\overline{\mu}(x_s) = \overline{\mu}(x_{n-s+1})$.

On the other hand, since, for any $x \in H$, the sum $\overline{\mu}(x) + \overline{\lambda}(x)$ is constant, it follows that $\overline{\mu}(x) = \overline{\mu}(y)$ if and only if $\overline{\lambda}(x) = \overline{\lambda}(y)$ and thereby $\overline{\lambda}(x_s) = \overline{\lambda}(x_{n-s+1})$. Now, the proof is complete. ■

Lemma 5.3.24. *Let $(H, \circ)$ be the hypergroupoid defined in Lemma 5.3.23. Then, for any $s \in \{1, 2, \ldots, n/2\}$, the following formulas hold:*

$$\overline{\mu}(x_s) < \overline{\mu}(x_{s+1}), \quad \overline{\lambda}(x_s) > \overline{\lambda}(x_{s+1}).$$

Proof. Rewriting relations (5.15) and (5.16) in the form

$$\overline{\mu}(x_1) = \frac{1}{n^2}\left(1 + 2\sum_{l=2}^{n} \frac{1}{l}\right),$$

$$\overline{\mu}(x_s) = \frac{1}{n^2}\left[1 + 2(s-1) + 2\sum_{l=1}^{n-2s+1} \frac{s}{s+l} + 2\sum_{l=1}^{s-1} \frac{l}{n-l+1}\right], 2 \leq s \leq n/2,$$

it results that

$$\overline{\mu}(x_{s+1}) = \overline{\mu}(x_s) + \frac{2}{n^2}\sum_{l=s+1}^{n-s} \frac{1}{l}, \quad \text{for any} \ \ 1 \leq s \leq n/2.$$

In a similar way, in virtue of Theorem 5.3.19, we find

$$\overline{\lambda}(x_1) = \frac{1}{n^2}\left(n - 1 + 2\sum_{l=2}^{n-1} \frac{n-l}{l}\right),$$

$$\overline{\lambda}(x_s) = \frac{1}{n^2}\left(n-1 + 2\sum_{l=2}^{s} \frac{n-2l+1}{l} + 2\sum_{l=s+1}^{n-s} \frac{n-s+1-l}{l}\right), 2 \leq s \leq n/2$$

and therefore

$$\overline{\lambda}(x_s) = \overline{\lambda}(x_{s+1}) + \frac{2}{n^2}\sum_{l=s+1}^{n-s} \frac{1}{l}, \quad \text{for any} \ \ 1 \leq s \leq n/2.$$

Then, $\overline{\mu}(x_s) < \overline{\mu}(x_{s+1})$ and $\overline{\lambda}(x_s) > \overline{\lambda}(x_{s+1})$, whenever $s \in \{1, 2, \ldots, n/2\}$. ■

Lemma 5.3.25. *Let $(H, \circ)$ be the hypergroupoid defined in Lemma 5.3.23. Then, for any $s \in \{1, 2, \ldots, n/2\}$ and $n \geq 6$, we obtain*

$$\overline{\mu}(x_s) < \overline{\lambda}(x_s).$$

Proof. It is obvious that, for any $n \geq 6$, $\overline{\mu}(x_1) < \overline{\lambda}(x_1)$ if and only if

$$1+2\sum_{l=2}^{n}\frac{1}{l} < n-1+2\sum_{l=2}^{n-1}\frac{n-l}{l} \Leftrightarrow n-2+2\left(\frac{n-3}{2}+\frac{n-4}{3}+\ldots+\frac{1}{n-2}\right)-\frac{2}{n}>0$$

which is satisfied.

Now we prove that, for $s \in \{2,3,\ldots,n/2\}$ and $n \geq 6$, we obtain $\overline{\mu}(x_s) < \overline{\lambda}(x_s)$, that is

$$s+\sum_{l=1}^{n-2s+1}\frac{s}{s+l}+\sum_{l=1}^{s-1}\frac{l}{n-l+1} < \frac{n}{2}+\sum_{l=2}^{s}\frac{n-2l+1}{l}+\sum_{l=s+1}^{n-s}\frac{n-s+1-l}{l}. \quad (5.17)$$

Since

$$\sum_{l=1}^{n-2s+1}\frac{s}{s+l} < \frac{s(n-2s+1)}{s+1}$$

$$\sum_{l=1}^{s-1}\frac{l}{n-l+1} < \frac{s(s-1)}{2(n-s+2)}$$

denoting the first member of the inequality (5.17) by M_I, we find that:

$$M_I < s+\frac{s(n-2s+1)}{s+1}+\frac{s(s-1)}{2(n-s+2)}.$$

Since

$$\sum_{l=2}^{s}\frac{n-2l+1}{l} > \frac{(s-2)(n-s)}{s-1}$$

$$\sum_{l=s+1}^{n-s}\frac{n-s+1-l}{l} > \frac{(n-2s+1)(n-2s+2)}{2(n-s)}$$

denoting the second member of the inequality (5.17) by M_{II}, we obtain that:

$$M_{II} > \frac{n}{2}+\frac{(s-2)(n-s)}{s-1}+\frac{(n-2s+1)(n-2s+2)}{2(n-s)}.$$

Then, $M_I < M_{II}$ if and only if

$$s+\frac{s(n-2s+1)}{s+1}+\frac{s(s-1)}{2(n-s+2)} < \frac{n}{2}+\frac{(s-2)(n-s)}{s-1}+\frac{(n-2s+1)(n-2s+2)}{2(n-s)}. \quad (5.18)$$

For $s=2$ the equation (5.18) becomes

$$\frac{2n^2+3}{3n} < \frac{2n-3}{2} \Leftrightarrow 2n^2-9n-6>0$$

which holds for any $n \geq 6$. For $s \geq 3$ the equation (5.18) is equivalent with

$$f(n) = 2n^3s^2 - 6n^3 - 7n^2s^3 + 3n^2s^2 + 23n^2s - 15n^2 + 8ns^4 - 10ns^3 - 28ns^2 + \\ + 42ns - 8n - 3s^5 + 9s^4 + 5s^3 - 25s^2 + 14s - 4 > 0, \quad \text{whenever} \quad s \leq n/2.$$

Calculating the first three derivatives of the function f with respect to n, we obtain that f is a monotonic increasing function, so

$$f(n) > f(2s) = s^5 + s^4 - 7s^3 - s^2 - 2s - 4 > 0, \quad \text{for any} \quad s \geq 3$$

and the required result is now proved. ■

Now, we can give the main result.

Theorem 5.3.26. *Let $H = \{x_1, x_2, \ldots, x_n\}$, where $n = 2^p$, $p \in \mathbb{N}^* \setminus \{1, 2\}$, be the hypergroupoid defined by the hyperproduct*

$$x_i \circ x_i = x_i, 1 \leq i \leq n,$$
$$x_i \circ x_j = x_j \circ x_i = \{x_i, x_{i+1}, \ldots, x_j\}, 1 \leq i < j \leq n.$$

Then, $i.f.g.(H) = p$.

Proof. We prove this statement in several steps, inductively by the index $i \geq 0$; we show that, for any $s \in \{1, 2, \ldots, n/2^{i+1}\}$, we obtain

- $\overline{\mu}_i(x_s) = \overline{\mu}_i(x_{n-s+1})$, $\overline{\lambda}_i(x_s) = \overline{\lambda}_i(x_{n-s+1})$;
- $\overline{\mu}_i(x_s) < \overline{\mu}_i(x_{s+1})$, $\overline{\lambda}_i(x_s) > \overline{\lambda}_i(x_{s+1})$;
- $\overline{\mu}_i(x_s) < \overline{\lambda}_i(x_s)$; therefore $(\overline{\mu}_i \wedge \overline{\lambda}_i)(x_s) < (\overline{\mu}_i \wedge \overline{\lambda}_i)(x_{s+1})$.

Thus, with any join space $({}_iH, \circ_{\overline{\mu}_i \wedge \overline{\lambda}_i})$ we associate the r_i-tuple $(2^{i+1}, 2^{i+1}, \ldots, 2^{i+1})$, with $r_i = n/2^{i+1}$.

Step 1. According to Lemmas 5.3.23, 5.3.24 and 5.3.25, we obtain the previous three relations for $i = 0$. Thus, with the join space $({}_0H, \circ_{\overline{\mu} \wedge \overline{\lambda}})$ we associate the r_0-tuple $(2, 2, \ldots, 2)$, where $r_0 = n/2$.

Step 2. Assume that with the join space ${}_{i-1}H$ we associate the $r = r_{i-1}$-tuple $(2^i, 2^i, \ldots, 2^i)$, with $r_{i-1} = n/2^i$ and prove that, for any $s \in \{1, 2, \ldots, n/2^{i+1}\}$,

- $\overline{\mu}_i(x_s) = \overline{\mu}_i(x_{n-s+1})$, $\overline{\lambda}_i(x_s) = \overline{\lambda}_i(x_{n-s+1})$;
- $\overline{\mu}_i(x_s) < \overline{\mu}_i(x_{s+1})$, $\overline{\lambda}_i(x_s) > \overline{\lambda}_i(x_{s+1})$;
- $\overline{\mu}_i(x_s) < \overline{\lambda}_i(x_s)$; therefore $(\overline{\mu}_i \wedge \overline{\lambda}_i)(x_s) < (\overline{\mu}_i \wedge \overline{\lambda}_i)(x_{s+1})$.

Indeed, by Lemma 5.3.23 and based on the fact that, for any $z \in H$, the sum $\overline{\mu}_i(z) + \overline{\lambda}_i(z)$ is a constant, we obtain the first relation.
Then, by Theorem 5.3.18, we know

$$\overline{\mu}_i(x_1) = \frac{2^i}{n^2}\left(1 + 2\sum_{l=2}^{r}\frac{1}{l}\right) \tag{5.19}$$

$$\overline{\mu}_i(x_s) = \frac{2^i}{n^2}\left(1 + 2\sum_{l=1}^{s-1}\sum_{j=s}^{r}\frac{1}{j-l+1} + 2\sum_{j=s+1}^{r}\frac{1}{j-s+1}\right) = \tag{5.20}$$

$$= \frac{2^i}{n^2}\left(1+2(s-1)+2\sum_{l=1}^{r-2s+1}\frac{s}{s+l}+2\sum_{l=1}^{s-1}\frac{l}{r-l+1}\right), 2 \leq s \leq r/2, \tag{5.21}$$

and thus, like in Lemma 5.3.24,

$$\overline{\mu}_i(x_{s+1}) - \overline{\mu}_i(x_s) = \frac{2^{i+1}}{n^2}\sum_{l=s+1}^{r-s}\frac{1}{l}.$$

Similarly, by Theorem 5.3.19 we obtain

$$\overline{\lambda}_i(x_1) = \frac{2^i}{n^2}\left(r - 1 + 2\sum_{l=2}^{r-1}\frac{r-l}{l}\right) \tag{5.22}$$

$$\overline{\lambda}_i(x_s) = \frac{2^i}{n^2}\left(r-1+2\sum_{l=2}^{s}\frac{r-2l+1}{l}+2\sum_{l=s+1}^{r-s}\frac{r-s+1-l}{l}\right), 2 \leq s \leq r/2, \tag{5.23}$$

consequently,

$$\overline{\lambda}_i(x_s) - \overline{\lambda}_i(x_{s+1}) = \frac{2^{i+1}}{n^2}\sum_{l=s+1}^{r-s}\frac{1}{l}.$$

So, $\overline{\mu}_i(x_s) < \overline{\mu}_i(x_{s+1})$, $\overline{\lambda}_i(x_s) > \overline{\lambda}_i(x_{s+1})$, for any $s \in \{1, 2, \ldots, r/2\}$.

It remains to prove that $\overline{\mu}_i(x_s) < \overline{\lambda}_i(x_s)$. For this, note that the formulas for $\overline{\mu}_i(x_s)$ and $\overline{\lambda}_i(x_s)$ $(1 \leq s \leq r/2)$ are similar to those for $\overline{\mu}(x_s)$ and $\overline{\lambda}(x_s)$ $(1 \leq s \leq n/2)$ and thus, according to Lemma 5.3.25, we can state that

$$\overline{\mu}_i(x_s) < \overline{\lambda}_i(x_s), \quad 1 \leq s \leq r/2, \quad \text{for any } r \geq 6, \tag{5.24}$$

and therefore

$$(\overline{\mu}_i \wedge \overline{\lambda}_i)(x_s) < (\overline{\mu}_i \wedge \overline{\lambda}_i)(x_{s+1}), \quad 1 \leq s \leq r/2, \quad r \geq 6.$$

Moreover, as in the next example, the relation $(\overline{\mu}_i \wedge \overline{\lambda}_i)(x_1) < (\overline{\mu}_i \wedge \overline{\lambda}_i)(x_2)$, holds too.

Step 3. We can now conclude the proof as follows: for any i, $1 \leq i \leq p$, with the join space ${}_iH$ we associate the r_i-tuple $(2^{i+1}, 2^{i+1}, \ldots, 2^{i+1})$, with $r_i = n/2^{i+1}$. So, with the join space ${}_{p-1}H$ we associate the 1-tuple (2^p), which leads to the fact that the join space ${}_{p-1}H$ is the total hypergroup and thus $i.f.g.(H) = p$. ∎

EXAMPLE 52 With the notations of the previous theorem, consider $r = n/2^i = 2^{p-i} = 4$; thus, with the join space ${}_{p-3}H$ we associate the quadruple of the form $(2^{p-2}, 2^{p-2}, 2^{p-2}, 2^{p-2})$. By relations (5.19) and (5.22) for $r = 4$ we obtain

$$\overline{\mu}_{p-2}(x_1) = \frac{2^{p-2}}{n^2}\left[1 + 2\left(\frac{1}{2} + \frac{1}{3} + \frac{1}{4}\right)\right] = \frac{2^{p-2}}{n^2} \cdot \frac{19}{6},$$

$$\overline{\lambda}_{p-2}(x_1) = \frac{2^{p-2}}{n^2}\left[3 + 2\left(\frac{2}{2} + \frac{1}{3}\right)\right] = \frac{2^{p-2}}{n^2} \cdot \frac{17}{3}.$$

Moreover, by relations (5.20) and (5.23) for $r = 4$ it follows that

$$\overline{\mu}_{p-2}(x_2) = \frac{2^{p-2}}{n^2}\left[1 + 2(2-1) + 2\frac{2}{3}\right] = \frac{2^{p-2}}{n^2} \cdot \frac{29}{6}$$

$$\overline{\lambda}_{p-2}(x_2) = \frac{2^{p-2}}{n^2}\left(3 + 2\frac{1}{2}\right) = \frac{2^{p-2}}{n^2} \cdot 4.$$

Now, it is obvious that

$$(\overline{\mu}_{p-2} \wedge \overline{\lambda}_{p-2})(x_1) < (\overline{\mu}_{p-2} \wedge \overline{\lambda}_{p-2})(x_2)$$

which means that with the join space ${}_{p-2}H$ we associated the pair $(2^{p-1}, 2^{p-1})$ and then to the join space ${}_{p-1}H$ we associate the 1-tuple (2^p); thus ${}_{p-1}H$ is the total hypergroup and $i.f.g.(H) = p$.

Defining on a hypergroupoid $(H, \circ)$ the equivalence realtion $R_{\bar{\mu}\wedge\bar{\lambda}}$ by

$$x \ R_{\overline{\mu}\wedge\overline{\lambda}} \ y \Leftrightarrow (\overline{\mu} \wedge \overline{\lambda})(x) = (\overline{\mu} \wedge \overline{\lambda})(y),$$

we will compute the intuitionistic fuzzy grade of H, when the cardinality of the quotient hypergroupoid $R_{\overline{\mu}\wedge\overline{\lambda}}$ is 2 or 3 [5].

Proposition 5.3.27. *Let $(H, \circ)$ be a hypergroupoid of cardinality n. If $|H/R_{\overline{\mu}\wedge\overline{\lambda}}| = 2$, that is with H one associates the pair (k_1, k_2), and*

(1) *if $k_1 = k_2$, then we have $i.f.g.(H) = 2$.*
(2) *if $k_1 \neq k_2$, then we have $i.f.g.(H) = 1$.*

Proof. In virtue of the assumption, we may write H as the union $H = C_1 \cup C_2$, with $|C_1| = k_1$, $|C_2| = k_2$, $k_1 + k_2 = n$ and $k_1 \leq k_2$, where, for any $x, x' \in C_i$, $i = 1, 2$, $(\overline{\mu} \wedge \overline{\lambda})(x) = (\overline{\mu} \wedge \overline{\lambda})(x')$. According to Theorems 5.3.18 and 5.3.19, it follows, for each $x \in C_1$ any $y \in C_2$,

$$\overline{\mu}_1(x) = (k_1 + 2k_1k_2/n)/n^2, \ \ \overline{\lambda}_1(x) = k_2/n^2;$$

$$\overline{\mu}_1(y) = (k_2 + 2k_1k_2/n)/n^2, \ \ \overline{\lambda}_1(y) = k_1/n^2.$$

(1) If $k_1 = k_2$, then $(\overline{\mu}_1 \wedge \overline{\lambda}_1)(x) = (\overline{\mu}_1 \wedge \overline{\lambda}_1)(y)$ and thereby the join space ${}_1H$ is a total hypergroup, meaning that $i.f.g.(H) = 2$.

(2) If $k_1 < k_2$, then $(\overline{\mu}_1 \wedge \overline{\lambda}_1)(x) > (\overline{\mu}_1 \wedge \overline{\lambda}_1)(y)$ and thereby the join space ${}_1H$ is isomorphic with ${}_0H$; by Theorem 5.3.20, it follows that $i.f.g.(H) = 1$. ■

Proposition 5.3.28. *Let $(H, \circ)$ be a finite hypergroupoid of cardinality n. If $|H/R_{\overline{\mu}\wedge\overline{\lambda}}| = 3$, that is with H one associates the ternary (k_1, k_2, k_3), and*

(1) *if $k_1 = k_2 = k_3$, then $i.f.g.(H) = 2$;*
(2) *if $k_1 = k_3 \neq k_2$, then $i.f.g.(H) = 2$, whenever $2k_1 \neq k_2$, and $i.f.g.(H) = 3$, whenever $2k_1 = k_2$;*
(3) *if $k_1 < k_2 = k_3$, then $i.f.g.(H) = 1$, whenever $2k_2^2 > 3k_1k_2 + 3k_1^2$ and $i.f.g.(H) = 3$, otherwise;*
(4) *if $k_1 = k_2 < k_3$, then $i.f.g.(H) = 1$, whenever $k_3 \neq 2k_1$ and $i.f.g.(H) = 2$, whenever $k_3 = 2k_1$;*
(5) *if $k_1 \neq k_2 \neq k_3$, then the value of $i.f.g.(H)$ is variable.*

Proof. Since $|H/R_{\overline{\mu}\wedge\overline{\lambda}}| = 3$, we may write H as the union $H = C_1 \cup C_2 \cup C_3$, with $|C_1| = k_1$, $|C_2| = k_2$, $|C_3| = k_3$ and $k_1 + k_2 + k_3 = n$, where, for any $x, x' \in C_i$, $i = 1, 2, 3$, $(\overline{\mu} \wedge \overline{\lambda})(x) = (\overline{\mu} \wedge \overline{\lambda})(x')$. Moreover, for any $x \in C_1, y \in C_2, z \in C_3$, we have

$$\overline{\mu}(x) \wedge \overline{\lambda}(x) < \overline{\mu}(y) \wedge \overline{\lambda}(y) < \overline{\mu}(z) \wedge \overline{\lambda}(z).$$

By Theorems 5.3.18 and 5.3.19 we obtain, for any $x \in C_1, y \in C_2, z \in C_3$, that

$$\begin{aligned}
\overline{\mu}_1(x) &= \left(k_1 + 2\frac{k_1k_2}{k_1+k_2} + 2\frac{k_1k_3}{k_1+k_2+k_3}\right)/n^2, \overline{\lambda}_1(x) = \left(k_2+k_3+2\frac{k_2k_3}{k_2+k_3}\right)/n^2\\
\overline{\mu}_1(y) &= \left(k_2 + 2\frac{k_1k_2}{k_1+k_2} + 2\frac{k_1k_3}{k_1+k_2+k_3} + 2\frac{k_2k_3}{k_2+k_3}\right)/n^2, \overline{\lambda}_1(y) = (k_1+k_3)/n^2\\
\overline{\mu}_1(z) &= \left(k_3 + 2\frac{k_1k_3}{k_1+k_2+k_3} + 2\frac{k_2k_3}{k_2+k_3}\right)/n^2, \overline{\lambda}_1(z) = \left(k_1+k_2+2\frac{k_1k_2}{k_1+k_2}\right)/n^2.
\end{aligned} \tag{5.25}$$

We analyze the following cases:

(1) If $k_1 = k_2 = k_3 = k$, then one finds $\overline{\mu}_1(x) \wedge \overline{\lambda}_1(x) = \overline{\mu}_1(z) \wedge \overline{\lambda}_1(z) = 8/(27k) \wedge 1/(3k) = 8/(27k)$ and $\overline{\mu}_1(y) \wedge \overline{\lambda}_1(y) = 11/(27k) \wedge 2/(9k) = 2/(9k)$. It follows immediately that

$$\overline{\mu}_1(x) \wedge \overline{\lambda}_1(x) = \overline{\mu}_1(z) \wedge \overline{\lambda}_1(z) > \overline{\mu}_1(y) \wedge \overline{\lambda}_1(y),$$

therefore, with the join space ${}_1H$ one associates the pair $(k, 2k)$ and in virtue of Proposition 5.3.27 (2) we obtain $i.f.g.(H) = 2$.

(2) If $k_1 = k_3 \neq k_2$, then, using equation (5.25), one obtains

$$\overline{\mu}_1(x) \wedge \overline{\lambda}_1(x) = \overline{\mu}_1(z) \wedge \overline{\lambda}_1(z) \neq \overline{\mu}_1(y) \wedge \overline{\lambda}_1(y),$$

therefore, with the join space ${}_1H$ one associates the pair $(k_2, 2k_1)$ or the pair $(2k_1, k_2)$. By Proposition 5.3.27 b) we obtain $i.f.g.(H) = 2$, whenever $2k_1 \neq k_2$, and $i.f.g.(H) = 3$, whenever $2k_1 = k_2$.

(3) If $k_1 < k_2 = k_3$, then, using the equation (5.25), one calculates

$$\overline{\mu}_1(x)=\left(k_1+2\frac{k_1k_2}{k_1+k_2}+2\frac{k_1k_2}{k_1+2k_2}\right)/(k_1+2k_2)^2,\ \ \overline{\lambda}_1(x)=3k_2/(k_1+2k_2)^2$$

$$\overline{\mu}_1(y)=\left(2k_2+2\frac{k_1k_2}{k_1+k_2}+2\frac{k_1k_2}{k_1+2k_2}\right)/(k_1+2k_2)^2,\ \ \overline{\lambda}_1(y)=(k_1+k_2)/(k_1+2k_2)^2$$

$$\overline{\mu}_1(z)=\left(2k_2+2\frac{k_1k_2}{k_1+2k_2}\right)/(k_1+2k_2)^2,\ \ \overline{\lambda}_1(z)=\left(k_1+k_2+2\frac{k_1k_2}{k_1+k_2}\right)/(k_1+2k_2)^2.$$

It is easy to see that $\overline{\mu}_1(x) \wedge \overline{\lambda}_1(x) = \overline{\mu}_1(x)$, $\overline{\mu}_1(y) \wedge \overline{\lambda}_1(y) = \overline{\lambda}_1(y)$ and $\overline{\mu}_1(z) \wedge \overline{\lambda}_1(z) = \overline{\lambda}_1(z)$, whenever $2k_2^2 > 3k_1k_2 + 3k_1^2$, and $\overline{\mu}_1(z) \wedge \overline{\lambda}_1(z) = \overline{\mu}_1(z)$, whenever $2k_2^2 < 3k_1k_2 + 3k_1^2$ (for $k_1, k_2 \in \mathbb{N}^*$, it results that $2k_2^2 \neq 3k_1k_2 + 3k_1^2$).

(a) If $2k_2^2 > 3k_1k_2 + 3k_1^2$, then it follows that

$$\overline{\mu}_1(x) \wedge \overline{\lambda}_1(x) < \overline{\mu}_1(y) \wedge \overline{\lambda}_1(y) < \overline{\mu}_1(z) \wedge \overline{\lambda}_1(z),$$

thereby the join spaces ${}_1H$ and ${}_0H$ are isomorphic. Thus, $i.f.g.(H) = 1$.

(b) If $2k_2^2 < 3k_1k_2 + 3k_1^2$, then it follows that

$$\overline{\mu}_1(y) \wedge \overline{\lambda}_1(y) < \overline{\mu}_1(x) \wedge \overline{\lambda}_1(x) < \overline{\mu}_1(z) \wedge \overline{\lambda}_1(z);$$

thus with the join space ${}_1H$ one associates the ternary (k_2, k_1, k_2). Since $k_1 < k_2$, we have $k_1 \neq k_2$ and according with the case (2), it follows that $i.f.g.(H) = 3$.

(4) If $k_1 = k_2 < k_3$, then, using the equation (5.25), one calculates

$$\overline{\mu}_1(x)=\left(2k_1+2\frac{k_1k_3}{2k_1+k_3}\right)/(2k_1+k_3)^2, \overline{\lambda}_1(x)=\left(k_1+k_3+2\frac{k_1k_3}{k_1+k_3}\right)/(2k_1+k_3)^2$$

$$\overline{\mu}_1(y)=\left(2k_1+2\frac{k_1k_3}{2k_1+k_3}+2\frac{k_1k_3}{k_1+k_3}\right)/(2k_1+k_3)^2,\ \ \overline{\lambda}_1(y)=(k_1+k_3)/(2k_1+k_3)^2$$

$$\overline{\mu}_1(z)=\left(k_3+2\frac{k_1k_3}{2k_1+k_3}+2\frac{k_1k_3}{k_1+k_3}\right)/(2k_1+k_3)^2,\ \ \overline{\lambda}_1(z)=3k_1/(2k_1+k_3)^2.$$

One verifies that $\overline{\mu}_1(x) \wedge \overline{\lambda}_1(x) = \overline{\mu}_1(x)$.

(a) If $k_1 < k_3 < 2k_1$, then it follows that $\overline{\mu}_1(y) \wedge \overline{\lambda}_1(y) = \overline{\lambda}_1(y)$ and

$$\overline{\mu}_1(y) \wedge \overline{\lambda}_1(y) < \overline{\mu}_1(x) \wedge \overline{\lambda}_1(x) < \overline{\mu}_1(z) \wedge \overline{\lambda}_1(z).$$

Thus, with the join space ${}_1H$ one associates the ternary (k_1, k_1, k_3), that is the join spaces ${}_1H$ and ${}_0H$ are isomorphic. Thereby $i.f.g.(H) = 1$.

(b) If $k_3 > 2k_1$, then it follows that

$$\overline{\mu}_1(z) \wedge \overline{\lambda}_1(z) < \overline{\mu}_1(x) \wedge \overline{\lambda}_1(x) < \overline{\mu}_1(y) \wedge \overline{\lambda}_1(y).$$

Therefore, with the join space ${}_1H$ one associates the ternary (k_3, k_1, k_1), so the join spaces ${}_1H$ and ${}_0H$ are isomorphic (by Theorem 5.3.20). Thus, $i.f.g.(H) = 1$.

(*c*) If $k_3 = 2k_1$, then it follows that

$$(\overline{\mu}_1 \wedge \overline{\lambda}_1)(x) = (\overline{\mu}_1 \wedge \overline{\lambda}_1)(y) = (\overline{\mu}_1 \wedge \overline{\lambda}_1)(z)$$

and therefore the join space ${}_1H$ is the total hypergroup, so $i.f.g.(H) = 2$.

(5) If $k_1 \neq k_2 \neq k_3$, then there is no precise order between $\overline{\mu}_1(x) \wedge \overline{\lambda}_1(x), \overline{\mu}_1(y) \wedge \overline{\lambda}_1(y), \overline{\mu}_1(z) \wedge \overline{\lambda}_1(z)$ as we can see from the following examples. If $k_1 < k_2 < k_3$, then taking $(k_1, k_2, k_3) = (10, 11, 12)$ we obtain $\overline{\mu}_1(x) \wedge \overline{\lambda}_1(x) = 0.025$, $\overline{\mu}_1(y) \wedge \overline{\lambda}_1(y) = 0.020$, $\overline{\mu}_1(z) \wedge \overline{\lambda}_1(z) = 0.028$ and thus

$$\overline{\mu}_1(y) \wedge \overline{\lambda}_1(y) < \overline{\mu}_1(x) \wedge \overline{\lambda}_1(x) < \overline{\mu}_1(z) \wedge \overline{\lambda}_1(z).$$

But if we take $(k_1, k_2, k_3) = (1, 2, 10)$, it results that $\overline{\mu}_1(x) \wedge \overline{\lambda}_1(x) = 0.09$, $\overline{\mu}_1(y) \wedge \overline{\lambda}_1(y) = 0.06$, $\overline{\mu}_1(z) \wedge \overline{\lambda}_1(z) = 0.02$ and therefore

$$\overline{\mu}_1(z) \wedge \overline{\lambda}_1(z) < \overline{\mu}_1(y) \wedge \overline{\lambda}_1(y) < \overline{\mu}_1(x) \wedge \overline{\lambda}_1(x).$$

Similar examples can be found if we consider another ordering between k_1, k_2, k_3.

The proof is now complete. ■

Next, we calculate the intuitionistic fuzzy grade of a hypergroupoid H, with $|H/R_{\overline{\mu}\wedge\overline{\lambda}}| = 3$ (it means that H may be written as the union of three subsets $H = C_1 \cup C_2 \cup C_3$, $|C_i| = k_i$, $i = 1, 2, 3$), considering some particular situations [5].

Proposition 5.3.29. *Let H be a hypergroupoid of cardinality $2k$, $k \geq 3$, with the ternary associated with H of the type $(k, k-1, 1)$. Then, $i.f.g.(H) = 1$.*

Proof. According to equation (5.25), after simple computations, we find, for any $x \in C_1, y \in C_2$, and $z \in C_3$, the following values for the membership functions $\overline{\mu}_1$ and $\overline{\lambda}_1$:

$$\overline{\mu}_1(x) = (4k^2 - k - 1)/[4k^2(2k-1)], \quad \overline{\lambda}_1(x) = (k^2 + 2k - 2)/4k^3,$$

$$\overline{\mu}_1(y) = (4k^3 - k^2 - 6k + 2)/[4k^3(2k-1)], \quad \overline{\lambda}_1(y) = (k+1)/4k^2,$$

$$\overline{\mu}_1(z) = (2k-1)/2k^3, \quad \overline{\lambda}_1(z) = (6k^2 - 6k + 1)/[4k^2(2k-1)].$$

Thereby, whenever $k \geq 3$, we obtain

$$\overline{\mu}_1(z) \wedge \overline{\lambda}_1(z) < \overline{\mu}_1(y) \wedge \overline{\lambda}_1(y) < \overline{\mu}_1(x) \wedge \overline{\lambda}_1(x),$$

that means that with the join space ${}_1H$ one associates the ternary $(1, k-1, k)$. From Theorem 5.3.20, it follows that the join spaces ${}_0H$ and ${}_1H$ are isomorphic, thus $i.f.g.(H) = 1$. ■

A generalization of Proposition 5.3.29 is giving in the following result. The proof is similar to that of previous proposition, therefore we omit it here.

Proposition 5.3.30. *Let $(pk, p(k-1), p)$, with $p, k \in \mathbb{N} \setminus \{0,1\}$, be the ternary associated with the hypergroupoid H. Then, $i.f.g.(H) = 1$.*

Let us consider another type of ternary associated with H, depending on two independent variables.

Proposition 5.3.31. *Let $(h, k, 1)$, with $2 \leq h \leq k$, be the ternary associated with the hypergroupoid H.*

(1) *If $h = 2$ and*
 (a) *$k = 2$, then $i.f.g.(H) = 3$;*
 (b) *$k = 3$, then $i.f.g.(H) = 2$;*
 (c) *$k \geq 4$, then $i.f.g.(H) = 3$ and the sequence of join spaces associated with H is cyclic.*

(2) *If $h > 2$, then $i.f.g.(H) = 1$, whenever $k \geq h$.*

Proof. By equation (5.25), it follows, for any $x \in C_1, y \in C_2, z \in C_3$, that

$$\overline{\mu}_1(x) = \frac{h^3 + 4h^2k + 3h^2 + 3hk^2 + 5hk}{(h+k)(h+k+1)^3}, \quad \overline{\lambda}_1(x) = \frac{k^2 + 4k + 1}{(k+1)(h+k+1)^2},$$

$$\overline{\mu}_1(y) = \frac{4hk^3 + h^4 + 4k^3 + 12hk^2 + 3k^2 + 3h^2k^2 + 7h^2k + 7hk + 2h^2}{(h+k)(k+1)(h+k+1)^3},$$

$$\overline{\lambda}_1(y) = \frac{h+1}{(h+k+1)^2},$$

$$\overline{\mu}_1(z) = \frac{3k^2 + 5hk + 4k + 3h + 1}{(k+1)(h+k+1)^3},$$
$$\overline{\lambda}_1(z) = \frac{k^2 + 4hk + h^2}{(h+k)(h+k+1)^2}.$$

After simple computations (that we have to omit for lack of space), we obtain

$$\overline{\mu}_1(y) \wedge \overline{\lambda}_1(y) = \overline{\lambda}_1(y) < \overline{\mu}_1(x) \wedge \overline{\lambda}_1(x)$$

and

$$\overline{\mu}_1(z) \wedge \overline{\lambda}_1(z) = \overline{\mu}_1(z) < \overline{\mu}_1(x) \wedge \overline{\lambda}_1(x).$$

Now, $\overline{\mu}_1(z) < \overline{\lambda}_1(y)$ if and only if $(h-2)k^2 + (h^2 - 2h - 2)k + h^2 - h > 0$, which is true whenever $h \geq 3$. Therefore, we distinguish the following three cases.

(1) If $3 \leq h \leq k$, then

$$\overline{\mu}_1(z) \wedge \overline{\lambda}_1(z) < \overline{\mu}_1(y) \wedge \overline{\lambda}_1(y) < \overline{\mu}_1(x) \wedge \overline{\lambda}_1(x);$$

by consequence the join spaces ${}_1H$ and ${}_0H$ are isomorphic, thus $i.f.g.(H) = 1$.

(2) If $h = k = 2$, then $(2, 2, 1)$ is the ternary associated with the join space ${}_0H$. By Theorem 5.3.20, ${}_0H$ is isomorphic with a hypergroup that has

$(1, 2, 2)$ as the associated ternary and thus, according to Proposition 5.3.28 (3), it follows that $i.f.g.(H) = 3$.

(3) If $h = 2$ and $k \geq 3$, then

$$\overline{\mu}_1(y) \wedge \overline{\lambda}_1(y) < \overline{\mu}_1(z) \wedge \overline{\lambda}_1(z) < \overline{\mu}_1(x) \wedge \overline{\lambda}_1(x)$$

and $(k, 1, 2)$ is the ternary associated with ${}_1H$. We can write now $H = C'_1 \cup C'_2 \cup C'_3$, with $x \in C'_3$, $y \in C'_1$, $z \in C'_2$. According to equation (5.25), we find

$$\overline{\mu}_2(x) = (\frac{10}{3} + \frac{4k}{k+3})/(k+3)^2, \ \overline{\lambda}_2(x) = (k+1+\frac{2k}{k+1})/(k+3)^2,$$

$$\overline{\mu}_2(y) = (k + \frac{2k}{k+1} + \frac{4k}{k+3})/(k+3)^2, \ \overline{\lambda}_2(y) = (\frac{13}{3})/(k+3)^2,$$

$$\overline{\mu}_2(z) = (\frac{7}{3} + \frac{2k}{k+1} + \frac{4k}{k+3})/(k+3)^2, \ \overline{\lambda}_2(z) = (k+2)/(k+3)^2.$$

After simple computations, we obtain the following relations:

$$\overline{\mu}_2(x) \wedge \overline{\lambda}_2(x) = \overline{\mu}_2(x) > \overline{\mu}_2(y) \wedge \overline{\lambda}_2(y) = \overline{\lambda}_2(y)$$

and $\overline{\mu}_2(x) < \overline{\mu}_2(z)$, whenever $h = 2$ and $k \geq 3$, but $\overline{\mu}_2(x) < \overline{\lambda}_2(z)$ only for $h = 2$ and $k > 3$.

If $h = 2$ and $k = 3$, then

$$\overline{\mu}_2(y) \wedge \overline{\lambda}_2(y) < \overline{\mu}_2(z) \wedge \overline{\lambda}_2(z) < \overline{\mu}_2(x) \wedge \overline{\lambda}_2(x)$$

and thus the join spaces ${}_2H$ and ${}_1H$ are isomorphic, so $i.f.g.(H) = 2$.

If $h = 2$ and $k > 3$, then

$$\overline{\mu}_2(y) \wedge \overline{\lambda}_2(y) < \overline{\mu}_2(x) \wedge \overline{\lambda}_2(x) < \overline{\mu}_2(z) \wedge \overline{\lambda}_2(z).$$

We write now H as the union $H = C''_1 \cup C''_2 \cup C''_3$, with $x \in C''_2$, $y \in C''_1$, $z \in C''_3$. According to equation (5.25), we find

$$\overline{\mu}_3(x) = (\frac{10}{3} + \frac{2k}{k+3} + \frac{4k}{k+2})/(k+3)^2, \ \overline{\lambda}_3(x) = (k+1)/(k+3)^2,$$

$$\overline{\mu}_3(y) = (k + \frac{2k}{k+3} + \frac{4k}{k+2})/(k+3)^2, \ \overline{\lambda}_3(y) = \frac{13}{3}/(k+3)^2,$$

$$\overline{\mu}_3(z) = (\frac{7}{3} + \frac{2k}{k+3})/(k+3)^2, \ \overline{\lambda}_3(z) = (k+2+\frac{4k}{k+2})/(k+3)^2$$

and we obtain

$$\overline{\mu}_3(z) \wedge \overline{\lambda}_3(z) < \overline{\mu}_3(y) \wedge \overline{\lambda}_3(y) < \overline{\mu}_3(x) \wedge \overline{\lambda}_3(x).$$

It is clear that the join spaces ${}_3H$ and ${}_0H$ are isomorphic, thereby the associated sequence of join spaces is cyclic, with

$$_{3s}H \simeq {}_0H, \quad {}_{3s+1}H \simeq {}_1H, \quad {}_{3s+2}H \simeq {}_2H, \quad \text{for any} \quad s \in \mathbb{N},$$

thus $i.f.g.(H) = 3$. ■

Now, we give an example of a hypergroupoid with a cyclic sequence of associated join spaces.

Proposition 5.3.32. *Let $(2k, k, 1)$, with $k \geq 2$, be the ternary associated with the hypergroupoid H. If $k = 2$, then $i.f.g.(H) = 3$ and the sequence of join spaces is cyclic, that is $_{3l}H \simeq {}_0H$, $_{3l+1}H \simeq {}_1H$, $_{3l+2}H \simeq {}_2H$, for any $l \in \mathbb{N}$. If $k > 2$, then $i.f.g.(H) = 3$.*

Proof. Using equation (5.25), one obtains, for any $x \in C_1, y \in C_2, z \in C_3$, the following results

$$\overline{\mu}_1(x) = \frac{30k^2 + 22k}{3(3k+1)^3}, \quad \overline{\lambda}_1(x) = \frac{k^2 + 4k + 1}{(k+1)(3k+1)^2},$$

$$\overline{\mu}_1(y) = \frac{21k^3 + 58k^2 + 25k}{3(k+1)(3k+1)^3}, \quad \overline{\lambda}_1(y) = \frac{2k+1}{(3k+1)^2},$$

$$\overline{\mu}_1(z) = \frac{13k^2 + 10k + 1}{(k+1)(3k+1)^3}, \quad \overline{\lambda}_1(z) = \frac{13k}{3(3k+1)^2}.$$

After some computations, one notices that

$$\overline{\mu}_1(x) \wedge \overline{\lambda}_1(x) = \overline{\lambda}_1(x), \quad \overline{\mu}_1(y) \wedge \overline{\lambda}_1(y) = \overline{\lambda}_1(y), \quad \overline{\mu}_1(z) \wedge \overline{\lambda}_1(z) = \overline{\mu}_1(z),$$

with

$$\overline{\mu}_1(z) < \overline{\lambda}_1(x) < \overline{\lambda}_1(y).$$

Therefore, with the join space $_1H$ one associates the triple $(1, 2k, k)$.
Now, we may write $H = C'_1 \cup C'_2 \cup C'_3$, with $x \in C'_2$, $y \in C'_3$ and $z \in C'_1$. Again by equation (5.25), it follows that

$$\overline{\mu}_2(x) = \frac{60k^3 + 98k^2 + 28k}{3(2k+1)(3k+1)^3}, \quad \overline{\lambda}_2(x) = \frac{k+1}{(3k+1)^2},$$

$$\overline{\mu}_2(y) = \frac{39k^2 + 19k}{3(3k+1)^3}, \quad \overline{\lambda}_2(y) = \frac{4k^2 + 8k + 1}{(2k+1)(3k+1)^2},$$

$$\overline{\mu}_2(z) = \frac{22k^2 + 11k + 1}{(2k+1)(3k+1)^3}, \quad \overline{\lambda}_2(z) = \frac{13k}{3(3k+1)^2}.$$

Moreover, $\overline{\mu}_2(x) \wedge \overline{\lambda}_2(x) = \overline{\lambda}_2(x) = (k+1)/(3k+1)^2$, $\overline{\mu}_2(y) \wedge \overline{\lambda}_2(y) = \overline{\lambda}_2(y) = (4k^2 + 8k + 1)/[(2k+1)(3k+1)^2]$ and $\overline{\mu}_2(z) \wedge \overline{\lambda}_2(z) = \overline{\mu}_2(z) = (22k^2 + 11k + 1)/[(2k+1)(3k+1)^3]$.

If $k \geq 3$, then $\overline{\mu}_2(z) \wedge \overline{\lambda}_2(z) < \overline{\mu}_2(x) \wedge \overline{\lambda}_2(x) < \overline{\mu}_2(y) \wedge \overline{\lambda}_2(y)$ and therefore the join spaces $_1H$ and $_2H$ are isomorphic, thus $i.f.g.(H) = 2$.

If $k = 2$, then $\overline{\mu}_2(x) \wedge \overline{\lambda}_2(x) < \overline{\mu}_2(z) \wedge \overline{\lambda}_2(z) < \overline{\mu}_2(y) \wedge \overline{\lambda}_2(y)$. It means that with ${}_2H$ one associates the ternary $(2k, 1, k) = (4, 1, 2)$. Considering $H = C_1'' \cup C_2'' \cup C_3''$, with $x \in C_1''$, $y \in C_3''$ and $z \in C_2''$, it follows by equation (5.25) that $\overline{\mu}_3(x) \wedge \overline{\lambda}_3(x) < \overline{\mu}_3(y) \wedge \overline{\lambda}_3(y) < \overline{\mu}_3(z) \wedge \overline{\lambda}_3(z)$. Thereby the join space ${}_3H$ is isomorphic to the join space ${}_0H$; then $i.f.g.(H) = 3$ and the sequence of join spaces associated with H is cyclic: ${}_{3l}H \simeq {}_0H$, ${}_{3l+1}H \simeq {}_1H$ and ${}_{3l+2}H \simeq {}_2H$, for any $l \in \mathbb{N}$.

The proof is now complete. ∎

5.3.3 Some Examples: Complete Hypergroups and i.p.s. Hypergroups

§1. **Complete Hypergroups**

The complete hypergroups have been defined by Koskas [129], using the fundamental relation β. But they can be also characterized using the structure of a group, proving again their closed connection with groups. Moreover, a commutative complete hypergroup is a join space. Migliorato [146] introduced the n-complete hypergroups, while De Salvo-Lo Faro [98] the n^*-complete hypergroups, proving that each finite hypergroup is n^*-complete and the class of n-complete hypergroups is contained in one of the n^*-complete hypergroups. Later on, after the introduction, by Freni [107], of the relation γ on a hypergroup, new types of complete hypergroups have been defined by Davvaz-Karimian [84, 86], called γ_n-complete and γ_n^*-complete hypergroups, respectively. It is proved that, for a commutative hypergroup, the notions of n-complete and γ_n-complete are equivalent. Furthermore, these classes of complete hypergroups are interested for $n = 2$, because 2-complete and γ_2-complete hypergroup means a complete hypergroup; 1-hypergroups, join spaces, canonical hypergroups, or feebly canonical hypergroups are examples of γ_2^*-complete hypergroups.

First, we briefly recall some notions and results concerning complete hypergroups and 1-hypergroups, and then we determine the fuzzy grade and the intuitionistic fuzzy grade of some particular types of them.

We start with the definition of a complete hypergroup, using the fundamental relation β, with the transitive closure β^*.

Definition 5.3.33. A hypergroup $(H, \circ)$ is called a *complete hypergroup* if, for any $x, y \in H$, we have $x \circ y = \beta^*(x \circ y)$.

As we have already mentioned before, the complete hypergroups can be also characterized by meaning of groups, as in the following result [27].

Theorem 5.3.34. *A hypergroup $(H, \circ)$ is complete if H can be written as a union of its subsets $H = \bigcup_{g \in G} A_g$, where*

(1) *$(G, \cdot)$ is a group.*
(2) *For any $(g_1, g_2) \in G^2$, with $g_1 \neq g_2$, we have $A_{g_1} \cap A_{g_2} = \emptyset$.*

(3) *If* $(a, b) \in A_{g_1} \times A_{g_2}$, *then* $a \circ b = A_{g_1 g_2}$.

The main properties of the complete hypergroups are recalled in the following result and their proofs can be found in [27].

Proposition 5.3.35. *Let H be a complete hypergroup.*

(1) *Its heart ω_H is the set of all identities of H.*
(2) *H is a regular reversible hypergroup.*
(3) *If H is commutative, then it is a join space.*

Proposition 5.3.36. *Let H be a complete hypergroup with the decomposition $H = \bigcup_{g \in G} A_g$, with $(G, \cdot)$ a group with the identity element e. Then, $\omega_H = A_e$.*

Proof. We know that, for a complete hypergroup H, its heart ω_H is the set of all identities of H.

First we prove that $A_e \subset \omega_H$. For any $x \in A_e$ and for any $a \in A_g, g \in G$, we have $a \circ x = A_{ge} = A_g \ni a$ and $x \circ a = A_{eg} = A_g \ni a$, therefore x is an identity for H and thereby $x \in \omega_H$.

Conversely, for any $x \in \omega_H$ and for any $a \in H$, we have $a \in a \circ x \cap x \circ a$. Assuming that $x \in A_{g_1}$ and $a \in A_{g_2}$, we obtain $a \in A_{g_1 \cdot g_2} \cap A_{g_2 \cdot g_1}$ and by the definition of a complete hypergroup, $g_1 \cdot g_2 = g_2 \cdot g_1$, for any $g_2 \in G$; thus $g_1 \in Z(G)$, the center of the group G. More over, $A_{g_2} \cap A_{g_1 \cdot g_2} \neq \emptyset$ and then $g_1 \cdot g_2 = g_2$, for any $g_2 \in G$, which means $g_1 = e$ and $x \in A_e$. ■

Notice that, if G is a commutative group, then H is also commutative, that is a join space. Moreover, if $|H| = |G| = n$, then $A_{g_i} = \{a_i\}$, for all $i = 1, 2, \ldots, n$, with $a_i \in H$, and thus, for any $(a, b) \in H^2$, $|a \circ b| = 1$. Therefore, H is a group. Conversely, any group can be viewed as a complete hypergroup, that is why in the following we will consider only finite complete hypergroups which are not groups, so with $m = |G| < |H| = n$.

Take $G = \{g_1, g_2, \ldots, g_m\}$. By Theorem 5.3.34, it results that the structure of any complete hypergroup H of cardinality n is determined by an m-tuple, $2 \leq m \leq n - 1$, denoted by $[k_1, k_2, \ldots, k_m]$, where, for any $i \in \{1, 2, \ldots, m\}$, $k_i = |A_{g_i}|$ and $k_2 \leq k_3 \leq \ldots \leq k_m$. If H is a complete 1-hypergroup, meaning that $|A_{g_1}| = |\omega_H| = 1$, where g_1 is the identity of the group G, then H is of the type $[1, k_2, k_3, \ldots, k_m]$.

This property suggests us to introduce the following concept.

Definition 5.3.37. [50] Let n be a positive integer, $n \geq 3$. An m-decomposition of n, $2 \leq m \leq n-1$, is an ordered system of natural numbers $(k_1, k_2, \ldots, k_m)$ such that $k_i \geq 1$, for any i, $1 \leq i \leq m$, $k_1 + k_2 + \ldots + k_m = n$, $k_2 \leq k_3 \leq \ldots \leq k_m$.

The main scope of this paragraph is to calculate the fuzzy grade and the intuitionistic fuzzy grade for some particular types of complete hypergroups, based on their decomposition, that is on their related m-tuple $[k_1, k_2, \ldots, k_m]$. The first step is to calculate the general form of the membership functions $\widetilde{\mu}$, and respectively $\overline{\mu}$ and $\overline{\lambda}$, defined in relations (5.10) and (5.14), respectively.

Theorem 5.3.38. [28], [50] [56] *Let $H = \bigcup_{g\in G} A_g$ be a complete hypergroup of cardinality n. Define on H the following equivalence $u \sim v \Leftrightarrow \exists g \in G : u, v \in A_g$. Then, for any $u \in H$,*

$$\widetilde{\mu}(u) = \frac{1}{|A_{g_u}|} \tag{5.26}$$

and

$$\overline{\mu}(u) = \frac{q(u)}{|A_{g_u}|} \cdot \frac{1}{n^2}, \ \overline{\lambda}(u) = \left(\sum_{v\notin \hat{u}} \frac{q(v)}{|A_{g_v}|}\right) \cdot \frac{1}{n^2}. \tag{5.27}$$

Proof. By the characterization theorem of a complete hypergroup, it follows that, for any $u \in H$, there exists a unique $g_u \in G$, such that $u \in A_{g_u}$ and then, $u \in x \circ y$ is equivalent with $u \in A_{g_x g_y} = A_{g_u}$. Now, the statement holds clearly from relations (5.10) and (5.14) ■

REMARK 10 If G_1 and G_2 are non-isomorphic groups of the same cardinality m and H_1, H_2 are the complete hypergroups obtained with G_1 and, respectively, with G_2, then $f.g.(H_1) = f.g.(H_2)$. Consequently, the fuzzy grade of a complete hypergroup H does not depend on the considered group G, but only on the m-decomposition of n.

On the other side, the intuitionistic fuzzy grade of a complete hypergroup does depend on the group G, as we notice in the following example.

EXAMPLE 53 [93] Let $[1, 1, 2, 2]$ be the 4-tuple representing the complete hypergroup H with 6 elements, constructed from a group G of cardinality 4. If G is isomorphic with the additive group $(\mathbb{Z}_4, +)$, then $i.f.g.(H) = 3$. If the group G is isomorphic with the Klein four-group, then $i.f.g.(H) = 1$.

In the next proposition we calculate the fuzzy grade of a complete hypergroup H of cardinality n, constructed from a group G with two or three elements.

Proposition 5.3.39. *Let $H = \bigcup_{g\in G} A_g$ be a complete hypergroup of cardinality n.*

(1) *If $|G| = 2$, then $f.g.(H) = 1$.*
(2) *If $|G| = 3$, then the fuzzy grade of H depends on the form of the triple $[k_1, k_2, k_3]$ associated with H: it can be $1, 2$, or there is no rule to determine it.*

Proof. **Case** (1): $|G| = 2$; $H = A_{g_1} \cup A_{g_2}$, with $|A_{g_i}| = k_i$, $i = 1, 2$.
The structure of H is represented by $[k_1, k_2], k_1 \neq k_2$, if n is an odd number and thus $\widetilde{\mu}(x) = 1/k_1$, $\widetilde{\mu}(y) = 1/k_2$, for any $x \in A_{g_1}$ and $y \in A_{g_2}$. By Proposition 5.3.5 it follows that $f.g.(H) = 1$.

On the other hand, if n is an even number, H is represented by $[k_1, k_2], k_1 \neq k_2$, and $[n/2, n/2]$. In both situations, by Proposition 5.3.5 it follows that $f.g.(H) = 1$; in the last one, the associated join space ${}^1H$ is a total hypergroup.

Case (2): $|G| = 3$; $H = A_{g_1} \cup A_{g_2} \cup A_{g_3}$, with $|A_{g_i}| = k_i$, $i = 1, 2, 3$. The structure of H is represented by $[k_1, k_2, k_3]$, with $n = k_1 + k_2 + k_3$, $k_2 \leq k_3$. We have to study four possibilities:

(a) $k_1 = k_2 = k_3 = k \geq 2$, then $\widetilde{\mu}(x) = \frac{1}{k}$, for any $x \in H$, so the associated join space ${}^1H$ is a total hypergroup and thus $f.g.(H) = 1$.

(b) Two of the numbers k_i are equal to the half of the third one; with the hypergroup $(H, \widetilde{\mu})$ we associate the pair (k, k). Then, the join space ${}^2H$ is a total hypergroup and thus $f.g.(H) = 2$.

(c) Two of the numbers k_i are equal and different by the half of the third one; then the pair associated with the hypergroup $(H, \widetilde{\mu})$ has the form (k, l), $k \neq l$; therefore, by Proposition 5.3.5 it follows that $f.g.(H) = 1$.

(d) $k_1 \neq k_2 \neq k_3$; in this case with the hypergroup $(H, \widetilde{\mu})$ it is associated the triple (k_i, k_j, k_l), $i \neq j \neq l \in \{1, 2, 3\}$, with $k_i \neq k_j \neq k_l$. According to Proposition 5.3.5, we have no general rule for determining the fuzzy grade of a such hypergroup. ∎

A similar result has been established in [56] for the intuitionistic fuzzy grade of a complete hypergroup constructed from a group with two elements.

Proposition 5.3.40. *Let* $H = \bigcup_{g \in G} A_g$ *be a complete hypergroup of cardinality* n*. If the group* G *is isomorphic with the additive group* $\mathbb{Z}_2$*, then* $i.f.g.(H) = 1$.

Proof. If $G \simeq \mathbb{Z}_2$, then we write $H = A_{g_1} \cup A_{g_2}$, $g_1 \neq g_2 \in G$. Therefore, for any $u \in A_{g_1}$ and $v \in A_{g_2}$, by (5.27), we get

$$\overline{\mu}(u) = \frac{q(u)}{|A_{g_1}|} \cdot \frac{1}{n^2}, \quad \overline{\mu}(v) = \frac{q(v)}{|A_{g_2}|} \cdot \frac{1}{n^2}$$

and

$$\overline{\lambda}(u) = \frac{q(v)}{|A_{g_2}|} \cdot \frac{1}{n^2}, \quad \overline{\lambda}(v) = \frac{q(u)}{|A_{g_1}|} \cdot \frac{1}{n^2}.$$

Thereby $\overline{\mu}(u) = \overline{\lambda}(v)$ and $\overline{\lambda}(u) = \overline{\mu}(v)$, implying that, for any $x \in H$, $\overline{\mu}(x) \wedge \overline{\lambda}(x)$ is constant; so the hypergroup $(H, \circ_{\overline{\mu} \wedge \overline{\lambda}})$ is the total hypergroup and $i.f.g.(H) = 1$. ∎

Cristea [50] has determined the fuzzy grade of all complete hypergroups of order less than 7, together with the sequences of join spaces and fuzzy sets associated with them, while later on, Davvaz et al. [94] determined their intuitionistic fuzzy grade. We recall here these results, omitting the proof that is technically.

Theorem 5.3.41. *Let* H *be a complete hypergroup of order* $n \leq 6$.

(1) *There are two non isomorphic complete hypergroups of order three that have* $s.f.g.(H) = i.f.g.(H) = 1$.
(2) *Among the five non isomorphic complete hypergroups of order four, three of them have* $s.f.g.(H) = i.f.g.(H) = 1$ *and two of them have* $s.f.g.(H) = i.f.g.(H) = 2$.
(3) *There are* 12 *non isomorphic complete hypergroups of order* 5. *All of them have* $s.f.g.(H) = 1$. *Nine of them have* $i.f.g.(H) = 1$ *and the other three have* $i.f.g.(H) = 3$.
(4) *There are* 21 *non isomorphic complete hypergroups of order* 6*:*
- 17 *with* $s.f.g.(H) = 1$ *and* 4 *with* $s.f.g.(H) = 2$.
- 16 *with* $i.f.g.(H) = 1$, 3 *with* $i.f.g.(H) = 2$ *and* 2 *with* $i.f.g.(H) = 3$.

In the last part of this section, we will consider some types of finite complete 1-hypergroups, investigated by Angheluţă - Cristea in [4, 5]. Then, we will generalize their structures, obtaining other complete hypergroups that are no longer 1-hypergroups.

Proposition 5.3.42. *If* H *is a complete* 1*-hypergroup of the type* $[\underbrace{1, 1, \ldots, 1}_{k\ times}, k]$, *where* $n = |H| = 2k$, *then* $s.f.g.(H)$*=2.*
More general, if the structure of the complete hypergroup H*, that is no longer an* 1*-hypergroup, is represented by* $[\underbrace{p, p, \ldots, p}_{k\ times}, kp]$, *where* $n = |H| = 2kp$, *then* $s.f.g.(H)$*=2.*

Proof. We suppose that the structure of H is determined by the $(k+1)$-tuple $[\underbrace{p, p, \ldots, p}_{k\ times}, kp]$. By Theorem 5.3.38, it results that the pair associated with the hypergroup $(H, \widetilde{\mu})$ is (kp, kp), therefore, by Proposition 5.3.5, we conclude that the associated join space ${}^2H$ is a total hypergroup and thus $s.f.g.(H) = 2$. ■

Proposition 5.3.43. *Let* H *be a complete* 1*-hypergroup of the following type:* $[\underbrace{1, 1, \ldots, 1}_{l\ times}, \underbrace{k, k, \ldots, k}_{p\ times}, l]$, *with* $1 < k < l$, $p \geq 1$, $n = |H| = kp + 2l$.

(1) *If* $kp = 2l$, *then* $s.f.g.(H) = 3$.
(2) *If* $kp \neq 2l$, *then* $f.g.(H) = 2$.

Proof. We use again Theorem 5.3.38.

(1) If $kp = 2l$, then with the hypergroupoid $(H, \widetilde{\mu})$ it is associated the triple $(l, 2l, l)$ and thus, by Proposition 5.3.5, it follows that $s.f.g.(H) = 3$, with the associated join space ${}^3H$ a total hypergroup.
(2) If $kp \neq 2l$, then the triple associated with the hypergroup $(H, \widetilde{\mu})$ is (l, kp, l). According to Proposition 5.3.5, it follows that $f.g.(H) = 2$. ■

REMARK 11 Generalizing Proposition 5.3.43 for the complete hypergroups of the type $[\underbrace{p, p, \ldots, p}_{s\ times}, \underbrace{k, k, \ldots, k}_{t\ times}, ps]$, with $2 \leq p < k < ps$, $n = |H| = 2ps + kt$, that are no longer 1-hypergroups, we obtain, for $n = 4ps$, $s.f.g.(H) = 3$, and for $kt \neq 2ps$, $f.g.(H) = 2$.

Proposition 5.3.44. *Let H be a complete 1-hypergroup of the following type: $[\underbrace{1, 1, \ldots, 1}_{l\ times}, \underbrace{k, k, \ldots, k}_{p\ times}, \underbrace{p, p, \ldots, p}_{k\ times}, l]$, with $1 < k < p < l$, $n = |H| = 2(l + pk)$.*

(1) *If $pk = l$, then $s.f.g.(H) = 3$.*
(2) *If $pk \neq l$, then $f.g.(H) = 2$.*

Proof. From Theorem 5.3.38, with the hypergroupoid $(H, \widetilde{\mu})$ it is associated the 4-tuple (l, pk, pk, l) and then, from Proposition 5.3.5, with the join space ${}^1H$ it is associated the pair $(2l, 2pk)$.

If $l = pk$ then with the join space ${}^2H$ it is associated $(4l)$; thus ${}^3H$ is a total hypergroup and $s.f.g.(H) = 3$.

If $pk \neq l$, by Proposition 5.3.5, we obtain $f.g.(H) = 2$. ∎

REMARK 12 Generalizing Proposition 5.3.44 for the complete hypergroups of the type $[\underbrace{k, k, \ldots, k}_{l\ times}, \underbrace{p, p, \ldots, p}_{s\ times}, \underbrace{s, s, \ldots, s}_{p\ times}, \underbrace{l, l, \ldots, l}_{k\ times}]$, (which are not 1-hypergroups), with $2 \leq k < p < s < l$, $n = |H| = 2(ps + kl)$, we obtain, for $kl = ps$, that $s.f.g.(H) = 3$ and, for $kl \neq ps$, that $f.g.(H) = 2$.

What can we say about their intuitionistic fuzzy grade? Till now we have just a partial answer, found by Angheluţă - Cristea in [5] and reported here below.

Proposition 5.3.45. *Let H be a complete 1-hypergroup of the type $[\underbrace{1, 1, \ldots, 1}_{k\ times}, k]$, where $n = |H| = 2k$, constructed from a group G isomorphic with the additive group $(\mathbb{Z}_{k+1}, +)$. If $k = 2$, then $i.f.g.(H) = 2$. If $k \geq 3$, then $i.f.g.(H) = 1$.*

Proof. First we suppose $k \geq 3$. For simplicity, we consider H as the union of $k + 1$ subsets, i.e., $H = \bigcup_{i=0}^{k} A_i$, with $A_0 = \{e\}$, $A_1 = \{a_1\}$, $A_2 = \{a_2\}, \ldots, A_{k-1} = \{a_{k-1}\}$, $A_k = \{a_k, a_{k+1}, \ldots, a_{2k-1}\}$. Then, the hypergroup H is represented by the following table.

H	e	a_1	a_2	$\ldots$	a_{k-1}	a_k	a_{k+1}	$\ldots$	a_{2k-1}
e	e	a_1	a_2	$\ldots$	a_{k-1}	A_k	A_k	$\ldots$	A_k
a_1		a_2	a_3	$\ldots$	A_k	e	e	$\ldots$	e
a_2			a_4	$\ldots$	e	a_1	a_1	$\ldots$	a_1
$\vdots$				$\ddots$	$\vdots$	$\vdots$	$\vdots$	$\vdots$	$\vdots$
a_{k-1}					a_{k-3}	a_{k-2}	a_{k-2}	$\ldots$	a_{k-2}
a_k						a_{k-1}	a_{k-1}	$\ldots$	a_{k-1}
a_{k+1}							a_{k-1}	$\ldots$	a_{k-1}
$\vdots$								$\ddots$	$\vdots$
a_{2k-1}									a_{k-1}

Using Theorems 5.3.18 and 5.3.19, one calculates the membership functions $\bar{\mu}$ and $\bar{\lambda}$ associated with H, obtaining: $\overline{\mu}(e) = \overline{\mu}(a_i) = (3k-1)/4k^2$, $\overline{\lambda}(e) = \overline{\lambda}(a_i) = (4k^3-6k^2+5k-1)/4k^3$, for any $i = 1, 2, \ldots, k-2$; $\bar{\mu}(a_{k-1}) = (k+1)/4k$, $\overline{\lambda}(a_{k-1}) = (3k^3-4k^2+4k-1)/4k^3$; $\overline{\mu}(a_j) = (3k-1)/4k^3$, $\overline{\lambda}(a_j) = (4k^2-3k+1)/4k^2$, for any $j = k, k+1, \ldots, 2k-1$.

After some computations, one obtains the following relation:

$$(\overline{\mu} \wedge \overline{\lambda})(a_{k-1}) \geq (\overline{\mu} \wedge \overline{\lambda})(e) = (\overline{\mu} \wedge \overline{\lambda})(a_i) \geq (\overline{\mu} \wedge \overline{\lambda})(a_j),$$

for any $i = 1, 2, \ldots, k-2$ and $j = k, k+1, \ldots, 2k-1$. It means that with the hypergroup H one associates the ternary $(k, k-1, 1)$ and, according to Proposition 5.3.29, it follows that $i.f.g.(H) = 1$.

Finally, let us consider the case when $k = 2$, that is when $(2, 1, 1)$ is the ternary associated with H. By Theorem 5.3.20, without altering the value of $i.f.g.(H)$, we may consider that $(1, 1, 2)$ is the ternary associated with H. Thus, according with Proposition 5.3.28, it follows that $i.f.g.(H) = 2$. ∎

A natural generalization of the previous proposition is considered in the following result.

Proposition 5.3.46. *Let H be a complete hypergroup of the type $[\underbrace{p, p, \ldots, p}_{k \text{ times}}, pk]$, where $k, p > 1$, $n = |H| = 2pk$, constructed from a group G isomorphic with the additive group $(\mathbb{Z}_{k+1}, +)$. Then, $i.f.g.(H)$=1, for $k \geq 3$, and $i.f.g.(H) = 2$, for $k = 2$.*

Proof Assume that $k \geq 3$. Let decompose H as the union $H = \bigcup_{i=0}^{k} A_i$, with $A_0 = \{a_{1,1}, \ldots, a_{1,p}\}$, $A_1 = \{a_{2,1}, \ldots, a_{2,p}\}, \ldots, A_{k-1} = \{a_{k,1}, \ldots, a_{k,p}\}$, $A_k = \{a_{k+1,1}, \ldots, a_{k+1,pk}\}$. The hypergroup H is represented by the following table.

H	$a_{1,1}$	...	$a_{1,p}$	$a_{2,1}$	...	$a_{2,p}$	...	$a_{k,1}$	...	$a_{k,p}$	$a_{k+1,1}$	...	$a_{k+1,pk}$
$a_{1,1}$	A_0	...	A_0	A_1	...	A_1	...	A_{k-1}	...	A_{k-1}	A_k	...	A_k
$\vdots$		$\ddots$	$\vdots$	$\vdots$	...	$\vdots$	...	$\vdots$	...	$\vdots$	$\vdots$	...	$\vdots$
$a_{1,p}$			A_0	A_1	...	A_1	...	A_{k-1}	...	A_{k-1}	A_k	...	A_k
$a_{2,1}$				A_2	...	A_2	...	A_k	...	A_k	A_0	...	A_0
$\vdots$					$\ddots$	$\vdots$	...	$\vdots$	...	$\vdots$	$\vdots$	...	$\vdots$
$a_{2,p}$						A_2	...	A_k	...	A_k	A_0	...	A_0
$\vdots$							$\ddots$	$\vdots$	...	$\vdots$	$\vdots$	...	$\vdots$
$a_{k,1}$								A_{k-3}	...	A_{k-3}	A_{k-2}	...	A_{k-2}
$\vdots$									$\ddots$	$\vdots$	$\vdots$	...	$\vdots$
$a_{k,p}$										A_{k-3}	A_{k-2}	...	A_{k-2}
$a_{k+1,1}$											A_{k-1}	...	A_{k-1}
$\vdots$												$\ddots$	$\vdots$
$a_{k+1,pk}$													A_{k-1}

From Theorems 5.3.18 and 5.3.19, one finds the membership functions $\overline{\mu}$ and $\overline{\lambda}$ associated with H: for any $i = 1, 2, \ldots, k-1$, $j = 1, 2, \ldots, p$, $l = 1, 2, \ldots, pk$, $\overline{\mu}(a_{i,j}) = (3k-1)/4pk^2$, $\overline{\lambda}(a_{i,j}) = (4k^3 - 6k^2 + 5k - 1)/4pk^3$; $\overline{\mu}(a_{k,j}) = (k+1)/4pk$, $\overline{\lambda}(a_{k,j}) = (3k^3 - 4k^2 + 4k - 1)/4pk^3$; $\overline{\mu}(a_{k+1,l}) = (3k-1)/4pk^3$, $\overline{\lambda}(a_{k+1,l}) = (4k^2 - 3k + 1)/4pk^2$. It follows that the values of the membership functions $\overline{\mu}$ and $\overline{\lambda}$ are similar to that obtained in Proposition 5.3.45, therefore it results that

$$(\overline{\mu} \wedge \overline{\lambda})(a_{k,j}) \geq (\overline{\mu} \wedge \overline{\lambda})(a_{i,j}) \geq (\bar{\mu} \wedge \bar{\lambda})(a_{k+1,l})$$

and thus, with H one associates the ternary $(pk, p(k-1), p)$.

By Proposition 5.3.30, for $k \geq 3$, it follows that $i.f.g.(H) = 1$.

If $k = 2$, then the ternary becomes $(2p, p, p)$ and from Proposition 5.3.28 it follows that $i.f.g.(H) = 2$. ■

We analyze now a complete hypergroup with a similar associated ordered tuple.

Proposition 5.3.47. *Let H be a complete hypergroup of the type $[k, \underbrace{1, 1, \ldots, 1}_{l \text{ times}}]$, where $k, l \geq 2$, $n = |H| = k + l$, constructed from the group G isomorphic with the additive group $(\mathbb{Z}_{l+1}, +)$.*

(1) *If $k \neq l$, then $i.f.g.(H) = 1$.*
(2) *If $k = l$, then $i.f.g.(H) = 2$.*

Proof. Writing the hypergroup $H = \{a_1, a_2, \ldots, a_k, a_{k+1}, \ldots, a_{k+l}\}$ as the union of $l+1$ subsets, $H = \bigcup_{i=0}^{l} A_i$, with $A_0 = \{a_1, a_2, \ldots, a_k\}$, $A_1 = \{a_{k+1}\}$, $A_2 = \{a_{k+2}\}, \ldots, A_{l-1} = \{a_{k+l-1}\}$, $A_l = \{a_{k+l}\}$, the table that represents

H is the following one:

H	a_1	a_2	$\dots$	a_k	a_{k+1}	a_{k+2}	$\dots$	a_{k+l-1}	a_{k+l}
a_1	A_0	A_0	$\dots$	A_0	A_1	A_2	$\dots$	A_{l-1}	A_l
a_2		A_0	$\dots$	A_0	A_1	A_2	$\dots$	A_{l-1}	A_l
$\vdots$			$\ddots$	$\vdots$	$\vdots$	$\vdots$	$\vdots$	$\vdots$	$\vdots$
a_k				A_0	A_1	A_2	$\dots$	A_{l-1}	A_l
a_{k+1}					A_2	A_3	$\dots$	A_l	A_0
a_{k+2}						A_4	$\dots$	A_0	A_1
$\vdots$							$\ddots$	$\vdots$	$\vdots$
a_{k+l-1}								A_{l-3}	A_{l-2}
a_{k+l}									A_{l-1}

From Theorems 5.3.18 and 5.3.19, one obtains the following membership functions: for each $i = 1, 2, \dots, k$, $\overline{\mu}(a_i) = (k+l/k)/(k+l)^2$, $\overline{\lambda}(a_i) = l(2k+l-1)/(k+l)^2$, and for each $j = k+1, k+2, \dots, k+l$, $\overline{\mu}(a_j) = (2k+l-1)/(k+l)^2$, $\overline{\lambda}(a_j) = [k+l/k+(2k+l-1)(l-1)]/(k+l)^2$. It is clear now that

$$\overline{\mu}(a_i) \wedge \overline{\lambda}(a_i) < \overline{\mu}(a_j) \wedge \overline{\lambda}(a_j),$$

for any $i = 1, 2, \dots, k$ and $j = 1, 2, \dots, l$. The pair associated with the hypergroup H is (k, l) and therefore, according to Proposition 5.3.27, it follows that $i.f.g.(H) = 1$ for $k \neq l$ and $i.f.g.(H) = 2$ for $k = l$. ■

§2. **I.p.s. Hypergroups**

The hypergroups with partial scalar identities, called by short *i.p.s. hypergroups* are finite canonical hypergroups $(H, \circ)$ satisfying a certain condition: for any $x, y \in H$, if $x \in x \circ y$, then $x \circ y = x$. The canonical hypergroups have been defined by Krasner [131] like the additive hyperstructure of hyperrings and hyperfields, that are now called, by his name, Krasner hyperrings and Krasner hyperfields. Then, Mittas [147] started their independently study, continued later by Roth [161, 162] in the theory of characters of finite groups, or by Corsini [32] and Bonansinga [15, 16], by Prenowitz and Jantosciak [157] with applications in geometry (since the canonical hypergroups are also join spaces), while McMullen and Price [144, 145] applied them in harmonic analysis. Several generalizations have been later developed, like feebly canonical hypergroups (that are particular cases of join spaces), introduced by Corsini [32] and strictly containing canonical and commutative complete hypergroups; i.p.s. hypergroups defined by Corsini [30], determining all of them of order less or equal to 8 [29, 31]; quasicanonical hypergroups analyzed by Bonansinga [15, 16]. Furthermore Corsini established a connection between i.p.s. hypergroups and strongly canonical hypergroups, proving that all i.p.s. hypergroups of order less or equal to 8 are strongly canonical, but there exist i.p.s. hypergroups of order 9 or 10 that are not strongly canonical.

Based on the hypercompositional tables, Corsini and Cristea [38, 39] calculated the fuzzy grade for all i.p.s. hypergroups of order less than 8, while their intuitionistic fuzzy grade have been determined by Davvaz, Hassan-Sadrabadi and Cristea [92, 93]. Here we summarize these results, without reporting the long technical proofs, due to lack of space. The two papers of Corsini-Cristea on the fuzzy grade of the i.p.s. hypergroups represented the starting point in the study of two numerical functions, namely *fuzzy grade* and *intuitionistic fuzzy grade*, associated with a hypergroupoid.

Theorem 5.3.48. (1) *There exists one i.p.s. hypergroup H of order 3 and $f.g.(H) = i.f.g.(H) = 1$.*

(2) *There exist 3 i.p.s. hypergroups of order 4 with $f.g.(H_1) = i.f.g.(H_1) = 1$, $f.g.(H_2) = 1, i.f.g.(H_2) = 2$, $f.g.(H_3) = i.f.g.(H_3) = 2$.*

(3) *There exist 8 i.p.s. hypergroups of order 5: one has $f.g.(H) = 2$, all the others have $f.g.(H) = 1$; four of them have $i.f.g.(H) = 1$, three have $i.f.g.(H) = 2$ and one is with $i.f.g.(H) = 3$.*

(4) *There exist 19 i.p.s. hypergroups of order 6: 14 of them have $f.g.(H) = 1$, 4 have $f.g.(H) = 2$ and one has $f.g.(H) = 3$;*
ten of them have $i.f.g.(H) = 1$, eight of them have $i.f.g.(H) = 2$ and only for one of them we find $i.f.g.(H) = 3$.

(5) *There exist 36 i.p.s. hypergroups of order 7: 27 of them with $f.g.(H) = 1$, 8 with $f.g.(H) = 2$, 1 with $f.g.(H) = 3$;*
10 of them with $i.f.g.(H) = 1$, 10 with $i.f.g.(H) = 2$, 9 with $i.f.g.(H) = 3$, 5 with $i.f.g.(H) = 4$, 2 with $i.f.g.(H) = 5$.
Moreover, for 18 of them we find that the associated sequences of join spaces and intuitionistic fuzzy sets are cyclic.

For each of these i.p.s. hypergroups, the sequences of the associated join spaces have been also constructed. It is obvious that the longer the sequence is, the more interesting the hypergroup's structure is. Taking this into account, we remark that there exist more i.p.s. hypergroups of order $4, 5, 6$, or 7 with intuitionistic fuzzy grade grater than 1, than those with the fuzzy grade grater than 1. It also worth to stress that there exist eighteen i.p.s. hypergroups of order 7 having a cyclic associated sequence of join spaces, just those obtained with the intuitionistic fuzzy sets. Some sequences have completed cycles, other have one or more join spaces out of the cyclic part. Therefore, it is natural to devide any sequence into two parts: the cyclic (or periodical) one, that contains an infinite number of join spaces, and the non-cyclic (or non-periodical) part, formed with a finite number of terms (join spaces). This should suggest us to associate with any hypergroup two new indexes that measure the length (so the number of the terms) of these two parts. Moreover, it is interested to see if the cyclicity property is a characteristic also for other classes of hypergroups, not only for the i.p.s. hypergroups. The answer is positive and in the next section we will find a new class of hypergroups having cyclic the associated sequence of join spaces.

5.3.4 Extensions to Hypergraphs

The sequences of join spaces and (intuitionistic) fuzzy sets can be constructed and investigated for any hypergroupoid, in particular also for the hypergroupoids obtained from hypergraphs. We may construct in different ways new hypergroupoids associated with hypergraphs, but the most known way is that proposed by Corsini [25]. Based on this construction, Corsini et al. [42, 43] have studied the sequence of join spaces and fuzzy sets determined by a particular hypergraph. Later on, using the same type of hypergraph, Davvaz et al. [95] have extended the problem to the case of intuitionistic fuzzy sets. In this section, after a brief introduction concerning hypergraphs, we present the mentioned researches.

A hypergraph is a generalization of a graph, where the edges are substituted by *hyperedges*: in a graph, an edge connects two vertices, while in a hypergraph, a hyperedge can connect any number of vertices.

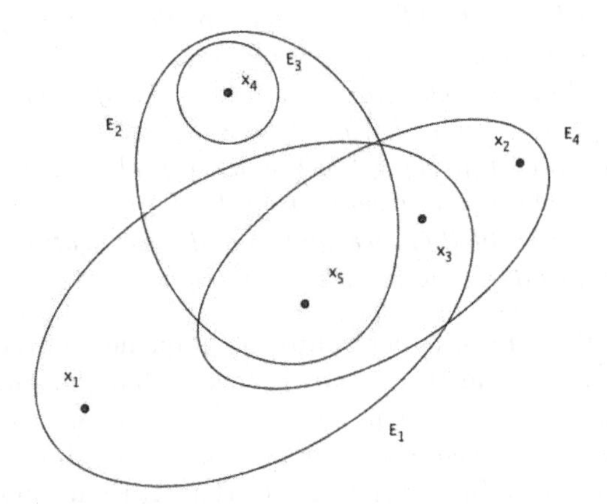

Fig. 5.1. A hypergraph with 5 vertices and 4 hyperedges

Formally, a *hypergraph* is a pair $\Gamma = (H; E)$, where H is a finite set, with the elements called *vertices*, and $E = \{E_1, E_2, \ldots, E_n\}$ is a family of non-empty subsets of H, called *hyperedges*, such that $H = \bigcup_{i=1}^{n} E_i$. Let x, y be arbitrary vertices of H. A *path* of length k between x and y is a sequence $\{E_1, \ldots, E_k\}$ of distinct hyperedges satisfying the following conditions:

- $x \in E_1$, $y \in E_k$;
- $E_i \cap E_{i+1} \neq \emptyset$, for $1 \leq i \leq k-1$.

Two vertices x and y are called *connected* if there is at least one path between them. If two vertices are connected by a path of length 1 (i.e. by a hyperedge), then they are called *adjacent*. Two hyperedges E_i and E_j are called *adjacent* if $E_i \cap E_j \neq \emptyset$. A hypergraph Γ is said to be *connected* if every pair of distinct vertices in Γ is connected. The *distance* $d_\Gamma(x, y)$ between two vertices x and

y is defined like the length of the shortest path between x and y. If there is no path between x and y, we set $d_\Gamma(x,y) = \infty$. We make the convention that, for any $x \in \Gamma$, $d_\Gamma(x,x) = 0$.

In the following, we will present a particular type of hypergraph, according with [42].

Set $E_1 = \{1,2\}$, $E_2 = \{2,3,4\}$, and $f_2 = 2$. For all $s \in \mathbb{N}, s \geq 3$, define $f_s = f_{s-1} + (s-1) = (s^2 - s)/2 + 1$, $E_s = \{f_s, f_s+1, f_s+2, \ldots, f_s+s\}$, and $H_s = \bigcup_{1 \leq i \leq s} E_i$, for $s \geq 2$, obtaining a family of hypergraphs $\{\Gamma_s = (H_s; \{E_i\}_{i=1}^s)\}_{s \geq 2}$.

With any hypergraph Γ_s we associate a hypergroupoid $(H_s, \circ)$, defining the hyperoperation $\circ$ on H_s by:

$$x \circ x = \bigcup_{x \in E_i} E_i, \quad x \circ y = x \circ x \cup y \circ y.$$

We are interested in computing the fuzzy grade and the intuitionistic fuzzy grade of the hypergroupoid $(H_s, \circ)$, for several values of $s \geq 2$.

The proofs of the two next theorems are technically, all the computations can be found in the original articles [42, 95], that is why here we report just the main steps.

Theorem 5.3.49. *Let $(H_s, \circ)$, $s \geq 2$ be the hypergroupoid constructed above.*

(1) *For $s = 2$, $s.f.g.(H_s) = 3$.*
(2) *For $s \in \{3,4,5\}$, $s.f.g.(H_s) = 1$.*

Proof. (1) Take $s = 2$. Then, the support set is $H = H_2 = E_1 \cup E_2 = \{1,2,3,4\}$, where $E_1 = \{1,2\}$ and $E_2 = \{2,3,4\}$. The associated hypergroupoid (which is clearly commutative) is represented by the following commutative table:

H	1	2	3	4
1	E_1	H	H	H
2		H	H	H
3			E_2	E_2
4				E_2

Based on formula (5.10), we obtain the first membership function

$$\widetilde{\mu}(1) = 0.270, \widetilde{\mu}(2) = 0.286, \widetilde{\mu}(3) = \widetilde{\mu}(4) = 0.272.$$

Thus, with the hypergroupoid H we associate the triple $(1,2,1)$. According with Proposition 5.3.5 item (2)(b), it follows that $s.f.g.(H) = 3$.

(2) Take $s = 3$. Now, the support set is $H = H_3 = E_1 \cup E_2 \cup E_3 = \{1,2,3,4,5,6,7\}$, where $E_1 = \{1,2\}$, $E_2 = \{2,3,4\}$ and $E_3 = \{4,5,6,7\}$. The table of the associated commutative hypergroupoid is the following one:

H	1	2	3	4	5	6	7
1	E_1	$E_1 \cup E_2$	$E_1 \cup E_2$	H	$H \setminus \{3\}$	$H \setminus \{3\}$	$H \setminus \{3\}$
2		$E_1 \cup E_2$	$E_1 \cup E_2$	H	H	H	H
3			E_2	$H \setminus \{1\}$	$H \setminus \{1\}$	$H \setminus \{1\}$	$H \setminus \{1\}$
4				$H \setminus \{1\}$	$H \setminus \{1\}$	$H \setminus \{1\}$	$H \setminus \{1\}$
5					E_3	E_3	E_3
6						E_3	E_3
7							E_3

Calculating the values of the first membership function $\widetilde{\mu}$, we get: $\widetilde{\mu}(1) = 0.195, \widetilde{\mu}(2) = 0.188, \widetilde{\mu}(3) = 0.182, \widetilde{\mu}(4) = 0.193, \widetilde{\mu}(5) = \widetilde{\mu}(6) = \widetilde{\mu}(7) = 0.179$. Therefore, with the hypergroupoid H we associate the 5-tuple $(1, 1, 1, 1, 3)$. The first join space in the associated sequence is the following one

H	1	2	3	4	5	6	7
1	1	1, 2, 4	1, 2, 3, 4	1, 4	H	H	H
2		2	2, 3	2, 4	$H \setminus \{1,4\}$	$H \setminus \{1,4\}$	$H \setminus \{1,4\}$
3			3	2, 3, 4	3, 5, 6, 7	3, 5, 6, 7	3, 5, 6, 7
4				4	$H \setminus \{1\}$	$H \setminus \{1\}$	$H \setminus \{1\}$
5					5, 6, 7	5, 6, 7	5, 6, 7
6						5, 6, 7	5, 6, 7
7							5, 6, 7

Using again formula (5.10), we obtain that the values of the second membership function $\widetilde{\mu}_1$ in the vertices of H verifies the same order relations like those for the membership function $\widetilde{\mu}$, meaning that the second join space in the associated sequence is equal with the first one, that is why $s.f.g.(H) = 1$.

Take now $s = 4$. The support set is $H = H_4 = E_1 \cup E_2 \cup E_3 \cup E_4 = \{1, 2, 3, 4, 5, 6, 7, 8, 9, 10, 11\}$, where $E_1 = \{1, 2\}$, $E_2 = \{2, 3, 4\}$, $E_3 = \{4, 5, 6, 7\}$ and $E_4 = \{7, 8, 9, 10, 11\}$. Based on formula (5.10), one calculates the values of the first membership function $\widetilde{\mu}$, obtaining the following relation

$$\begin{array}{lllllll}
\widetilde{\mu}(8) & < \widetilde{\mu}(5) & < \widetilde{\mu}(7) & < \widetilde{\mu}(4) & < \widetilde{\mu}(3) & < \widetilde{\mu}(2) & < \widetilde{\mu}(1) \\
\| & \| & & & & & \\
\widetilde{\mu}(9) & \widetilde{\mu}(6) & & & & & \\
\| & & & & & & \\
\widetilde{\mu}(10) & & & & & & \\
\| & & & & & & \\
\widetilde{\mu}(11) & & & & & &
\end{array}$$

meaning that the associated 7-tuple with H has the form $(1, 1, 1, 1, 1, 2, 4)$. Constructing the first join space in the sequence and calculating the values of the second membership function $\widetilde{\mu}_1$, one gets the same relation as before, so the same 7-tuple. Thus, the second join space is equal with the first one, therefore $s.f.g.(H) = 1$.

Take now $s = 5$. The support set is $H = H_5 = E_1 \cup E_2 \cup E_3 \cup E_4 \cup E_5 = \{1, 2, 3, 4, 5, 6, 7, 8, 9, 10, 11, 12, 13, 14, 15, 16\}$, where $E_1 = \{1, 2\}$,

$E_2 = \{2,3,4\}$, $E_3 = \{4,5,6,7\}$, $E_4 = \{7,8,9,10,11\}$ and finally $E_5 = \{11,12,13,14,15,16\}$. Calculating the values of the first membership function $\widetilde{\mu}$, one gets:

$$\begin{array}{lllllllll}
\widetilde{\mu}(12) < & \widetilde{\mu}(8) < & \widetilde{\mu}(11) < & \widetilde{\mu}(7) < & \widetilde{\mu}(5) < & \widetilde{\mu}(4) < & \widetilde{\mu}(3) < & \widetilde{\mu}(2) < & \widetilde{\mu}(1) \\
\| & \| & & & \| & & & & \\
\widetilde{\mu}(13) & \widetilde{\mu}(9) & & & \widetilde{\mu}(6) & & & & \\
\| & \| & & & & & & & \\
\widetilde{\mu}(14) & \widetilde{\mu}(10) & & & & & & & \\
\| & & & & & & & & \\
\widetilde{\mu}(15) & & & & & & & & \\
\| & & & & & & & & \\
\widetilde{\mu}(16) & & & & & & & &
\end{array}$$

and similarly for the second membership function $\widetilde{\mu}_1$. It follows again that $s.f.g.(H) = 1$.

For bigger values of s, the laborious computations should be performed with a general algorithm and for now this remains an open problem. ■

In the following, we will present the same issue in the case of intuitionistic fuzzy sets, for the same values of s, according with [95]. Since the computations are too long (and can be found in the cited article), we will skip them for the cases $s = 4$ and $s = 5$.

Theorem 5.3.50. *Let $(H_s, \circ)$, $s \geq 2$, be the hypergroupoid constructed above.*

(1) *For $s = 2$ or $s = 3$, $i.f.g.(H_s) = 3$.*
(2) *For $s = 4$, $i.f.g.(H_s) = 4$ and the sequence of associated join spaces is cyclic.*
(3) *For $s = 5$, $i.f.g.(H_s) = 12$ and the sequence of associated join spaces is cyclic.*

Proof. (1) We start with the case $s = 2$. Then, the support set is $H = H_2 = E_1 \cup E_2 = \{1,2,3,4\}$, where $E_1 = \{1,2\}$ and $E_2 = \{2,3,4\}$. The associated hypergroupoid $(H_2, \circ)$ is the same like in Theorem 5.3.49 (1). Using now formula (5.14), one obtsains the following relation between the values of the membership function $\overline{\mu} \wedge \overline{\lambda}$:

$$\begin{array}{lll}
\overline{\mu} \wedge \overline{\lambda}(0) < & \overline{\mu} \wedge \overline{\lambda}(3) < & \overline{\mu} \wedge \overline{\lambda}(1) \\
& \| & \\
& \overline{\mu} \wedge \overline{\lambda}(4) &
\end{array}$$

Thus, the associated triple has the form $(1,2,1)$ and according with Proposition 5.3.28 we conclude that $i.f.g.(H) = 3$.

Let's consider now the case $s = 3$; the support set is $H = H_3 = E_1 \cup E_2 \cup E_3 = \{1,2,3,4,5,6,7\}$, where $E_1 = \{1,2\}$, $E_2 = \{2,3,4\}$ and $E_3 = \{4,5,6,7\}$. The table of the associated commutative hypergroupoid is the same as in Theorem 5.3.49 (2). Based on formula (5.14), one gets that the membership function $\overline{\mu} \wedge \overline{\lambda}$ satisfies the relation:

$$\begin{array}{c} \overline{\mu}\wedge\overline{\lambda}(4) < \overline{\mu}\wedge\overline{\lambda}(2) < \overline{\mu}\wedge\overline{\lambda}(5) < \overline{\mu}\wedge\overline{\lambda}(3) < \overline{\mu}\wedge\overline{\lambda}(1) \\ \| \\ \overline{\mu}\wedge\overline{\lambda}(6) \\ \| \\ \overline{\mu}\wedge\overline{\lambda}(7) \end{array}$$

It results that the associated 5-tuple is $(1,1,3,1,1)$, satisfying the hypothesis of Corollary 5.3.22. It follows that the triple associated with the second join space in the sequence is $(2,3,2)$ and again according with Corollary 5.3.22, it follows that the pair $(4,3)$ is associated with the third join space. Now, based on Proposition 5.3.27 (2), the fourth join space in the sequence is isomorphic with the third one, thereby $i.f.g.(H) = 3$.

(2) In the case $s = 4$, the support set is $H = H_4 = E_1 \cup E_2 \cup E_3 \cup E_4 = \{1,2,3,4,5,6,7,8,9,10,11\}$, where $E_1 = \{1,2\}$, $E_2 = \{2,3,4\}$, $E_3 = \{4,5,6,7\}$ and $E_4 = \{7,8,9,10,11\}$. The table of the associated commutative hypergroupoid is the same as in Theorem 5.3.49 (2) for $s = 4$. Calculating the values of the first membership function $\overline{\mu}\wedge\overline{\lambda}$, we obtain that the first associated 7-tuple is $(1,1,1,4,1,1,2)$ and similarly, the following 7-tuples are $(1,2,1,1,1,1,4)$, $(1,1,1,1,4,2,1)$, $(1,1,1,2,1,1,4)$, and next $(1,1,1,4,1,1,2)$. It means that the fifth join space is isomorphic with the first one, so the associated sequence of join spaces is cyclic, having a complete cycle of length 4.

(3) For $s = 5$, the support set is $H = H_5 = E_1 \cup E_2 \cup E_3 \cup E_4 \cup E_5 = \{1,2,3,4,5,6,7,8,9,10,11,12,13,14,15,16\}$, where $E_1 = \{1,2\}$, $E_2 = \{2,3,4\}$, $E_3 = \{4,5,6,7\}$, $E_4 = \{7,8,9,10,11\}$ and $E_5 = \{11,12,13,14,15,16\}$. In this case, the associated 9-tuples are, respectivelly:
$(1,1,1,1,2,1,1,3,5)$, $(1,1,1,1,5,1,1,2,3)$, $(1,1,3,1,1,1,2,1,5)$,
$(1,1,1,5,1,3,1,1,2)$, $(1,2,1,1,1,1,1,3,5)$, $(1,2,1,5,1,1,1,1,3)$,
$(1,3,2,1,1,1,1,1,5)$, $(1,3,1,5,1,1,1,2,1)$, $(1,1,2,1,1,3,1,1,5)$,
$(1,1,1,5,2,1,1,1,3)$, $(1,1,1,3,1,1,1,2,5)$, $(1,1,1,5,1,1,2,1,3)$,
and finally, $(1,1,1,3,1,1,1,2,5)$, which is equal with the eleventh one. We conclude that, the sequence of join spaces of length 12 has a cycle of length 2 and an uncyclic part containing 10 non-isomorphic join spaces. ∎

REMARK 13 The hypergroup associated with a hypergraph as in Theorems 5.3.49 and 5.3.50 is, among the i.p.s. hypergroups, another example of hypergroup having cyclic the sequence of associated join spaces and intuitionistic fuzzy sets.

5.4 Hypergroupoids and Fuzzy Sets Endowed with Two Membership Functions

§1. The first association between hypergroups and fuzzy sets with two membership functions belongs to Corsini [37].

Consider on a non-empty set H a fuzzy set endowed with two membership functions $\mu, \lambda : H \to [0,1]$. There is no relation between the two membership functions as in the case of Atanassov's intuitionistic fuzzy set.

Define on H the following hyperproduct. For any $x, y \in H$, set

$$x \circ_{\mu\lambda} y = \left\{ u \in H \mid \begin{array}{l} \mu(x) \wedge \mu(y) \wedge \lambda(x) \wedge \lambda(y) \leq \mu(u) \wedge \lambda(u) \\ \mu(x) \vee \mu(y) \vee \lambda(x) \vee \lambda(y) \geq \mu(u) \vee \lambda(u) \end{array} \right\}.$$

It is obvious that, for any $x, y \in H$, $\{x, y\} \subset x \circ_{\mu\lambda} y$, thereby $(H, \circ_{\mu\lambda})$ is a quasi-hypergroup.

Let us investigate more properties of this hyperstructure.

Theorem 5.4.1. *The hypergroupoid $(H, \circ_{\mu\lambda})$ is a commutative hypergroup.*

Proof. First we fix some notations. For any $(x, y, z) \in H^3$, set

$$P(x,y,z) = \left\{ w \in H \mid \begin{array}{l} \mu(x) \wedge \lambda(x) \wedge \mu(y) \wedge \lambda(y) \wedge \mu(z) \wedge \lambda(z) \leq \mu(w) \wedge \lambda(w) \\ \mu(x) \vee \lambda(x) \vee \mu(y) \vee \lambda(y) \vee \mu(z) \vee \lambda(z) \geq \mu(w) \vee \lambda(w) \end{array} \right\}.$$

We start to prove that $(x \circ_{\mu\lambda} y) \circ_{\mu\lambda} z = P(x, y, z)$, for any $(x, y, z) \in H^3$. By the definition of the hyperproduct $\circ_{\mu\lambda}$, we have

$$(x \circ_{\mu\lambda} y) \circ_{\mu\lambda} z = \{w \in H \mid \exists v \in x \circ_{\mu\lambda} y : w \in v \circ_{\mu\lambda} z\} =$$

$$= \left\{ w \in H \mid \exists v \in H : \begin{array}{l} \mu(x) \wedge \mu(y) \wedge \lambda(x) \wedge \lambda(y) \leq \mu(v) \wedge \lambda(v) \\ \mu(x) \vee \mu(y) \vee \lambda(x) \vee \lambda(y) \geq \mu(v) \vee \lambda(v) \\ \mu(v) \wedge \mu(z) \wedge \lambda(v) \wedge \lambda(z) \leq \mu(w) \wedge \lambda(w) \\ \mu(v) \vee \mu(z) \vee \lambda(v) \vee \lambda(z) \geq \mu(w) \vee \lambda(w) \end{array} \right\}.$$

It follows that

$$\mu(w) \wedge \lambda(w) \geq \mu(x) \wedge \lambda(x) \wedge \mu(y) \wedge \lambda(y) \wedge \mu(z) \wedge \lambda(z)$$

and

$$\mu(w) \vee \lambda(w) \leq \mu(x) \vee \lambda(x) \vee \mu(y) \vee \lambda(y) \vee \mu(z) \vee \lambda(z),$$

whence $w \in P(x, y, z)$, that is

$$(x \circ_{\mu\lambda} y) \circ_{\mu\lambda} z \subset P(x, y, z). \tag{5.28}$$

Conversely, set $w \in P(x, y, z)$; it follows that

$$\mu(x) \wedge \lambda(x) \wedge \mu(y) \wedge \lambda(y) \wedge \mu(z) \wedge \lambda(z) \leq \mu(w) \wedge \lambda(w) \tag{5.29}$$

and

$$\mu(x) \vee \lambda(x) \vee \mu(y) \vee \lambda(y) \vee \mu(z) \vee \lambda(z) \geq \mu(w) \vee \lambda(w) \tag{5.30}$$

Denote

$$m = \mu(x) \wedge \lambda(x) \wedge \mu(y) \wedge \lambda(y) \wedge \mu(z) \wedge \lambda(z)$$
$$M = \mu(x) \vee \lambda(x) \vee \mu(y) \vee \lambda(y) \vee \mu(z) \vee \lambda(z)$$

and since $m \leq \mu(w) \wedge \lambda(w)$, $M \geq \mu(w) \vee \lambda(w)$, we obtain that

$$m = \mu(x) \wedge \lambda(x) \wedge \mu(y) \wedge \lambda(y) \wedge \mu(z) \wedge \lambda(z) \wedge \mu(w) \wedge \lambda(w)$$
$$M = \mu(x) \vee \lambda(x) \vee \mu(y) \vee \lambda(y) \vee \mu(z) \vee \lambda(z) \vee \mu(w) \vee \lambda(w).$$

Because the hyperproduct $\circ_{\mu\lambda}$ is commutative and μ and λ are interchangeable, we can limit our study of the possible different relations between the values of μ and λ in x, y, z, to the following five cases.

Case (1): $m = \mu(x)$ and $M = \mu(z)$.

Then, according to (5.29) and (5.30), we have

$$\begin{aligned}\mu(x) &= \mu(x) \wedge \lambda(x) \wedge \mu(z) \wedge \lambda(z) = \\ &= \mu(x) \wedge \lambda(x) \wedge \mu(y) \wedge \lambda(y) \wedge \mu(z) \wedge \lambda(z) \leq \mu(w) \wedge \lambda(w) \\ \mu(z) &= \mu(x) \vee \lambda(x) \vee \mu(z) \vee \lambda(z) = \\ &= \mu(x) \vee \lambda(x) \vee \mu(y) \vee \lambda(y) \vee \mu(z) \vee \lambda(z) \geq \mu(w) \vee \lambda(w),\end{aligned}$$

which means that $w \in x \circ_{\mu\lambda} z$, with $x \in x \circ_{\mu\lambda} y$, that is $w \in (x \circ_{\mu\lambda} y) \circ_{\mu\lambda} z$; thus $P(x, y, z) \subset (x \circ_{\mu\lambda} y) \circ_{\mu\lambda} z$.

Case (2): $m = \mu(x)$ and $M = \lambda(x)$.

As in the previous case, we obtain that $w \in x \circ_{\mu\lambda} z$, with $x \in x \circ_{\mu\lambda} y$, and therefore $w \in (x \circ_{\mu\lambda} y) \circ_{\mu\lambda} z$, so $P(x, y, z) \subset (x \circ_{\mu\lambda} y) \circ_{\mu\lambda} z$.

Case (3): $m = \mu(x)$ and $M = \mu(y)$.

Then

$$\mu(x) = \mu(x) \wedge \lambda(x) \wedge \mu(y) \wedge \lambda(y) \leq \mu(w) \wedge \lambda(w)$$
$$\mu(y) = \mu(x) \vee \lambda(x) \vee \mu(y) \vee \lambda(y) \geq \mu(w) \vee \lambda(w)$$

which imply that $w \in x \circ_{\mu\lambda} y$. Moreover, $w \in w \circ_{\mu\lambda} z$, so $w \in (x \circ_{\mu\lambda} y) \circ_{\mu\lambda} z$. We obtain again that $P(x, y, z) \subset (x \circ_{\mu\lambda} y) \circ_{\mu\lambda} z$.

Case (4): $m = \mu(z)$ and $M = \mu(x)$.

Since

$$m = \mu(z) = \mu(x) \wedge \lambda(x) \wedge \mu(z) \wedge \lambda(z) \leq \mu(w) \wedge \lambda(w)$$
$$M = \mu(x) = \mu(x) \vee \lambda(x) \vee \mu(z) \vee \lambda(z) \geq \mu(w) \vee \lambda(w)$$

it follows that $w \in x \circ_{\mu\lambda} z \subset (x \circ_{\mu\lambda} y) \circ_{\mu\lambda} z$. Therefore, $P(x, y, z) \subset (x \circ_{\mu\lambda} y) \circ_{\mu\lambda} z$.

Case (5): $m = \mu(z)$ and $M = \lambda(z)$.

Similarly to the previous case, $w \in x \circ_{\mu\lambda} z \subset (x \circ_{\mu\lambda} y) \circ_{\mu\lambda} z$, meaning that $P(x, y, z) \subset (x \circ_{\mu\lambda} y) \circ_{\mu\lambda} z$.

So we conclude that, for any $x, y, z \in H$,

$$P(x,y,z) \subset (x \circ_{\mu\lambda} y) \circ_{\mu\lambda} z. \tag{5.31}$$

Now, from (5.28) and (5.31) it results that $P(x,y,z) = (x \circ_{\mu\lambda} y) \circ_{\mu\lambda} z$.

Similarly, we may prove the other equality $P(x,y,z) = x \circ_{\mu\lambda} (y \circ_{\mu\lambda} z)$, for any $x, y, z \in H$.

Therefore, the commutative quasi-hypergroup $(H, \circ_{\mu\lambda})$ is also associative, so it is a commutative hypergroup. ■

Generally, $(H, \circ_{\mu\lambda})$ is not a join space, as we notice in the following example.

EXAMPLE 54 On the set $H = \{a,b,c,d,x\}$, define the fuzzy set endowed with the membership functions μ and λ as follows:

$$\mu(a) = 0.10 \ \lambda(a) = 0.35$$

$$\mu(b) = 0.40 \ \lambda(b) = 0.45$$

$$\mu(c) = 0.25 \ \lambda(c) = 0.70$$

$$\mu(d) = 0.15 \ \lambda(d) = 0.30$$

$$\mu(x) = 0.08 \ \lambda(x) = 0.50$$

Then,

$$a/b = \{u \in H \mid a \in b \circ_{\mu\lambda} u\} =$$

$$= \left\{ u \in H \mid \begin{array}{l} \mu(b) \wedge \lambda(b) \wedge \mu(u) \wedge \lambda(u) \leq \mu(a) \wedge \lambda(a) \\ \mu(b) \vee \lambda(b) \vee \mu(u) \vee \lambda(u) \geq \mu(a) \vee \lambda(a) \end{array} \right\} =$$

$$= \left\{ u \in H \mid \begin{array}{l} 0.40 \wedge \mu(u) \wedge \lambda(u) \leq 0.10 \\ 0.45 \vee \mu(u) \vee \lambda(u) \geq 0.35 \end{array} \right\} = \{a, x\}.$$

Similarly, $c/d = \{c, x\}$. Therefore, $a/b \cap c/d \neq \emptyset$.

On the other hand,

$$a \circ_{\mu\lambda} d = \left\{ z \in H \mid \begin{array}{l} 0.10 \leq \mu(z) \wedge \lambda(z) \\ 0.35 \geq \mu(z) \vee \lambda(z) \end{array} \right\} = \{a, d\}$$

and

$$b \circ_{\mu\lambda} c = \left\{ z \in H \mid \begin{array}{l} 0.20 \leq \mu(z) \wedge \lambda(z) \\ 0.45 \geq \mu(z) \vee \lambda(z) \end{array} \right\} = \{b, c\},$$

whence $a \circ_{\mu\lambda} d \cap b \circ_{\mu\lambda} c = \emptyset$, which is a contradiction with the main property of a join space.

§2. In the following, we construct on H some fuzzy sets endowed with two membership functions μ and λ such that $(H, \circ_{\mu\lambda})$ is a join space. In the definition of the hyperproduct $\circ_{\mu\lambda}$ we use here, the conditions $0 \leq \mu(x) \leq 1$ and $0 \leq \lambda(x) \leq 1$, for any $x \in H$, are not requested.

We start with an auxiliar result, proved by Corsini [46].

Lemma 5.4.2. *Let $\mu : H \to [0,1]$ be a fuzzy subset of a non-empty set H. Defining on H the following hyperoperations*

$$a \overset{<}{\underset{\vee}{\circ}} b = \{x \in H \mid \mu(x) \leq \mu(a) \vee \mu(b)\}$$
$$a \overset{>}{\underset{\wedge}{\circ}} b = \{x \in H \mid \mu(x) \geq \mu(a) \wedge \mu(b)\},$$

the hypergroupoids $(H, \overset{<}{\underset{\vee}{\circ}})$ and $(H, \overset{>}{\underset{\wedge}{\circ}})$ are join spaces.

Proof. First we prove the associativity law. For any $a, b, c \in H$, we have

$$\begin{aligned}(a \overset{<}{\underset{\vee}{\circ}} b) \overset{<}{\underset{\vee}{\circ}} c &= \{x \in H \mid \exists u \in a \overset{<}{\underset{\vee}{\circ}} b : x \in u \overset{<}{\underset{\vee}{\circ}} c\} = \\ &= \{x \in H \mid \exists u \in H : \mu(u) \leq \mu(a) \vee \mu(b), \mu(x) \leq \mu(u) \vee \mu(c)\} = \\ &= \{x \in H \mid \mu(x) \leq \mu(a) \vee \mu(b) \vee \mu(c)\} = \\ &= a \overset{<}{\underset{\vee}{\circ}} (b \overset{<}{\underset{\vee}{\circ}} c)\end{aligned}$$

and similarly,

$$(a \overset{>}{\underset{\wedge}{\circ}} b) \overset{>}{\underset{\wedge}{\circ}} c = \{x \in H \mid \mu(x) \geq \mu(a) \wedge \mu(b) \wedge \mu(c)\} = a \overset{>}{\underset{\wedge}{\circ}} (b \overset{>}{\underset{\wedge}{\circ}} c).$$

Since, for any $a, b \in H$, $a, b \in a \overset{<}{\underset{\vee}{\circ}} b \cap a \overset{>}{\underset{\wedge}{\circ}} b$, it follows that $(H, \overset{<}{\underset{\vee}{\circ}})$ and $(H, \overset{>}{\underset{\wedge}{\circ}})$ are commutative hypergroups.

Set now a, b, c, d arbitrary elements in H. Without less the generality, we may suppose that $\mu(a) \leq \mu(c)$. Then, $\mu(a) \leq \mu(a) \vee \mu(d)$, whence $a \in a \overset{<}{\underset{\vee}{\circ}} d$, and $\mu(a) \leq \mu(c) \leq \mu(b) \vee \mu(c)$, whence $a \in b \overset{<}{\underset{\vee}{\circ}} c$. So $a \overset{<}{\underset{\vee}{\circ}} d \cap b \overset{<}{\underset{\vee}{\circ}} c \neq \emptyset$.

On the other hand, $\mu(c) \geq \mu(b) \wedge \mu(c)$, whence $c \in b \overset{>}{\underset{\wedge}{\circ}} c$, and $\mu(c) \geq \mu(a) \geq \mu(a) \wedge \mu(d)$, whence $c \in a \overset{>}{\underset{\wedge}{\circ}} d$. Therefore, $a \overset{>}{\underset{\wedge}{\circ}} d \cap b \overset{>}{\underset{\wedge}{\circ}} c \neq \emptyset$. Thus, we can conclude that $(H, \overset{<}{\underset{\vee}{\circ}})$ and $(H, \overset{>}{\underset{\wedge}{\circ}})$ are join spaces. ■

Proposition 5.4.3. *Let μ and λ be the membership functions of a fuzzy set defined on H. Then, $(H, \circ_{\mu\lambda})$ is a join space in each of the following situations:*

(1) *whenever $\mu = \lambda$;*
(2) *whenever $\mu \wedge \lambda =$constant or $\mu \vee \lambda =$constant;*
(3) *whenever $\mu \wedge \lambda =$constant and $\mu \vee \lambda =$constant;*
(4) *whenever $\lambda = \mu^c$ (where μ^c is the complementary of μ).*

Proof. (1) If $\mu = \lambda$, we notice that $(H, \circ_{\mu\mu})$ coincides with the hypergroup $(H, \circ_\mu)$ defined in Theorem 5.1.1, so it is a join space.

(2) If $\mu \wedge \lambda$ =constant, then, for any $x, y \in H$, we have

$$x \circ_{\mu\lambda} y = \left\{ u \in H \mid \begin{array}{c} (\mu \wedge \lambda)(x) \wedge (\mu \wedge \lambda)(y) \leq (\mu \wedge \lambda)(u) \\ (\mu \vee \lambda)(x) \vee (\mu \vee \lambda)(y) \geq (\mu \vee \lambda)(u) \end{array} \right\} =$$

$$= \{u \in H \mid (\mu \vee \lambda)(x) \vee (\mu \vee \lambda)(y) \geq (\mu \vee \lambda)(u)\} =$$

$$= x \underset{\vee}{\overset{<}{\circ}} y$$

as in Lemma 5.4.2 for the fuzzy subset $\mu \vee \lambda$.

Similarly, if $\mu \vee \lambda$ =constant, then, for any $x, y \in H$, we obtain $x \circ_{\mu\lambda} y = x \underset{\wedge}{\overset{>}{\circ}} y$ as in Lemma 5.4.2 for the fuzzy subset $\mu \wedge \lambda$. Thus, $(H, \circ_{\mu\lambda})$ is a join space.

(3) If the membership functions $\mu \wedge \lambda$ and $\mu \vee \lambda$ are constant, then it results that the hypergroup $(H, \circ_{\mu\lambda})$ is total, so it is a join space.

(4) Suppose $\lambda = \mu^c$ and denote $\mu \wedge \lambda = A$, $\mu \vee \lambda = B$. Then, the hyperproduct $\circ_{\mu\lambda}$ can be written as

$$x \circ_{\mu\lambda} y = \left\{ u \in H \mid \begin{array}{c} A(x) \wedge A(y) \leq A(u) \\ B(x) \vee B(y) \geq B(u) \end{array} \right\}.$$

But, $B = \mu \vee \lambda = \mu \vee \mu^c = (\mu^c \vee \mu)^c = A^c$ and then

$$B(x) \vee B(y) \geq B(u) \Leftrightarrow A(x) \wedge A(y) \leq A(u),$$

whence

$$x \circ_{\mu\lambda} y = \{u \in H \mid A(x) \wedge A(y) \leq A(u)\} = x \underset{\wedge}{\overset{>}{\circ}} y.$$

Therefore, by Lemma 5.4.2 it follows again that $(H, \circ_{\mu\lambda})$ is a join space. ■

§3. Next we discuss on the (intuitionistic) fuzzy subhypergroups of the hypergroup $(H, \circ_{\mu\lambda})$.

Let A be a fuzzy subhypergroup of $(H, \circ_{\mu\lambda})$; then the following axioms hold:

- $A(x) \wedge A(y) \leq \inf_{u \in x \circ_{\mu\lambda} y} A(u)$, for all $x, y \in H$.
- For all $x, a \in H$, there exists $y \in H$ such that $x \in a \circ_{\mu\lambda} y$ and $A(a) \wedge A(x) \leq A(y)$.

Since $\{x, y\} \subset x \circ_{\mu\lambda} y$, for any $x, y \in H$, the second condition is always verified for $(H, \circ_{\mu\lambda})$.

Using the notion of α-cut, Davvaz [76] proved that A is a fuzzy subhypergroup of a hypergroup $(H, \circ)$ if and only if, for any $\alpha \in (0, 1]$, $A_\alpha = \{x \in H \mid A(x) \geq \alpha\}$ is a subhypergroup of H. Therefore, for μ and λ fuzzy sets on H, since $(\mu \wedge \lambda)_\alpha = \mu_\alpha \cap \lambda_\alpha$, it follows the following implication:

If μ and λ are fuzzy subhypergroups of H, then $\mu \wedge \lambda$ is also a fuzzy subhypergroup of H.

But for the hypergroup $(H, \circ_{\mu\lambda})$ there exists a stronger result, given below.

Corollary 5.4.4. *The fuzzy subset $\mu \wedge \lambda$ is always a fuzzy subhypergroup of the hypergroup $(H, \circ_{\mu\lambda})$.*

Regarding the fuzzy subset $\mu \vee \lambda$, it is not always a fuzzy subhypergroup of $(H, \circ_{\mu\lambda})$, as one can notice from Example 54. Besides, if A is a fuzzy subhypergroup of $(H, \circ_{\mu\lambda})$, then its complement A^c could be not a fuzzy subhypergroup, as we see below.

EXAMPLE 55 Define on $H = \{1, 2, 3\}$ the membership functions:

$$\mu(1) = 0.2,\ \mu(2) = 0.6,\ \mu(3) = 0.4$$

$$\lambda(1) = 0.7,\ \lambda(2) = 0.3,\ \lambda(3) = 0.8$$

Then, the hypergroup $(H, \circ_{\mu\lambda})$ is represented by the following table:

H	1	2	3
1	1,2	1,2	H
2		2	2,3
3			3

Consider the fuzzy subset A of H defined by $A(1) = 0.25$, $A(2) = 0.6$, $A(3) = 0.7$. It is easy to notice that A is a fuzzy subhypergroup of $(H, \circ_{\mu\lambda})$. Since $A^c(1) = 0.75$, $A^c(2) = 0.4$, $A^c(3) = 0.3$, for $2 \in 1 \circ_{\mu\lambda} 1$, the relation $A^c(2) \geq A^c(1)$ is not true, so A^c is not a fuzzy subhypergroup of $(H, \circ_{\mu\lambda})$.

Recall now that an intuitionistic fuzzy subhypergroup $A = (\mu, \lambda)$ of a hypergroup $(H, \circ)$ satisfies the following axioms:

(1) $\mu(x) \wedge \mu(y) \leq \inf_{u \in x \circ y} \mu(u)$, for all $x, y \in H$.
(2) For all $x, a \in H$, there exist $y, z \in H$ such that $x \in (a \circ y) \cap (z \circ a)$ and $\mu(a) \wedge \mu(x) \leq \mu(y) \wedge \mu(z)$.
(3) $\lambda(x) \vee \lambda(y) \geq \sup_{u \in x \circ y} \lambda(u)$, for all $x, y \in H$.
(4) For all $x, a \in H$, there exist $y, z \in H$ such that $x \in (a \circ y) \cap (z \circ a)$ and $\lambda(a) \vee \lambda(x) \geq \lambda(y) \vee \lambda(z)$.

Remark that the second and the fourth condition of the above definition are fulfilled, for any intuitionistic fuzzy subset $A = (\mu, \lambda)$ of the hypergroup $(H, \circ_{\mu\lambda})$.

Proposition 5.4.5. *The intuitionistic fuzzy subset $A^{\vee}_{\wedge} = (\mu \wedge \lambda, \mu \vee \lambda)$ is an intuitionistic fuzzy subhypergroup of the hypergroup $(H, \circ_{\mu\lambda})$, whenever $A = (\mu, \lambda)$ is an intuitionistic fuzzy subset of H.*

Proof. It is straightforward, since we can write

$$x \circ_{\mu\lambda} = \left\{ u \in H \mid \begin{array}{c} (\mu \wedge \lambda)(x) \wedge (\mu \wedge \lambda)(y) \leq (\mu \wedge \lambda)(u) \\ (\mu \vee \lambda)(x) \vee (\mu \vee \lambda)(y) \geq (\mu \vee \lambda)(u) \end{array} \right\}.$$

■

Bibliography

[1] Ajmal, N., Prajapati, A.S.: Fuzzy cosets and fuzzy subpolygroup. J. Fuzzy Math. 9, 305–316 (2001)
[2] Ameri, R., Zahedi, M.M.: Hypergroup and join space induced by a fuzzy subset. Pure Math. Appl. 8, 155–168 (1997)
[3] Ameri, R., Mahjoob, R.: Spectrum of prime fuzzy hyperideals. Iran. J. Fuzzy Syst. 6(4), 61–72 (2009)
[4] Angheluţă, C., Cristea, I.: Fuzzy Grade of the Complete Hypergroups. Iran. J. Fuzzy Syst. 9(6), 43–56 (2012)
[5] Angheluţă, C., Cristea, I.: On Atanassov's Intuitionistic Fuzzy Grade of Complete Hypergroups. J. Mult.-Valued Logic Soft Comput. 20(1-2), 55–74 (2013)
[6] Anthony, J.M., Sherwood, H.: Fuzzy groups redefined. J. Math. Anal. Appl. 69, 124–130 (1979)
[7] Atanassov, K.T.: Intuitionistic Fuzzy Sets. Fuzzy Sets and Systems 20, 87–96 (1986)
[8] Atanassov, K.T.: New operations defined over the intuitionistic fuzzy sets. Fuzzy Sets and Systems 61, 137–142 (1994)
[9] Atanassov, K.T.: Intuitionistic fuzzy sets: Theory and applications. Physica-Verlag, Heilderberg (1999)
[10] Bandelt, H.J., Hedlikova, J.: Median algebras. Discrete Math 45, 1–30 (1983)
[11] Ben-Yaacov, I.: On the fine structure of the polygroup blow-up. Arch. Math. Logic 42, 649–663 (2003)
[12] Bhakat, S.K., Das, P.: On the definition of a fuzzy subgroup. Fuzzy Sets and Systems 51, 235–241 (1992)
[13] Bhakat, S.K., Das, P.: $(\in, \in \vee q)$-fuzzy subgroup. Fuzzy Sets and Systems 80, 359–368 (1996)
[14] Biswas, R.: Rosenfeld's fuzzy subgroups with interval valued membership functions. Fuzzy Sets and Systems 63, 87–90 (1994)
[15] Bonansinga, P.: Quasicanonical hypergroups (Italian). Atti Soc. Peloritana Sci. Fis. Mat. Natur. 27, 9–17 (1981)
[16] Bonansinga, P.: Weakly quasicanonical hypergroups (Italian). Atti Sem. Mat. Fis. Univ. Modena 30, 286–298 (1981)
[17] Bonansinga, P., Corsini, P.: Sugli omomorfismi di semi-ipergruppi e di ipergruppi. Boll. Un. Mat. Italy 1-B, 717–727 (1982)
[18] Borzoei, R.A., Hasankhani, A., Rezaei, H.: Some results on canonical, cyclic hypergroups and join spaces. Ital. J. Pure Appl. Math. 11, 77–87 (2002)
[19] Campaigne, H.: Partition hypergroups. Amer. J. Math. 6, 599–612 (1940)
[20] Comer, S.D.: Hyperstructures associated with character algebra and color schemes. New Frontiers in Hyperstructures, pp. 49–66. Hadronic Press (1996)

[21] Comer, S.: Multi-valued algebras and their graphical representations (Preliminary draft), Math. and Comp. Sci. Dep. the Citadel. Charleston, South Carolina 29409 (July 1986)
[22] Comer, S.D.: The Cayley representation of polygroups, Hypergroups, other multivalued structures and their applications (Italian), pp. 27–34. Univ. Studi Udine, Udine (1985)
[23] Comer, S.D.: Polygroups derived from cogroups. J. Algebra 89, 397–405 (1984)
[24] Comer, S.D.: Extension of polygroups by polygroups and their representations using colour schemes. Lecture notes in Meth, vol. 1004, pp. 91–103. Universal Algebra and Lattice Theory (1982)
[25] Corsini, P.: Hypergraphs and Hypergroups. Algebra Universalis 35, 548–555 (1996)
[26] Corsini, P.: Rough sets, fuzzy sets and join spaces, pp. 65–72. Honorary Volume dedicated to Prof. Emeritus J. Mittas, Aristotle's Univ. of Th., Fac. Engin. Math div, Thessaloniki, Greece (2000)
[27] Corsini, P.: Prolegomena of Hypergroup Theory, 2nd edn. Aviani Editore (1993)
[28] Corsini, P.: Join spaces, power sets, fuzzy sets. In: Proceedings of the Fifth Int. Congress of Algebraic Hyperstructures and Appl. Hadronic Press, Iasi (1994)
[29] Corsini, P.: (*i.p.s.*) Hypergroups of order 8. Aviani Editore (1989)
[30] Corsini, P.: Finite canonical hypergroups with partial scalar identities (Italian). Rend. Circ. Mat. Palermo 36(2), 205–219 (1987)
[31] Corsini, P.: (*i.p.s.*) ipergruppi di ordine 7, vol. 34, pp. 199–216. Atti Sem. Mat. Fis. Un. Modena, Modena (1986)
[32] Corsini, P.: Feebly canonical and 1-hypergroups. Acta Univ. Carol. Math. Phys. 24, 49–56 (1983)
[33] Corsini, P.: Recent results in the theory of hypergroups (Italian). Boll. Un. Mat. Ital. A 2(6), 133–138 (1983)
[34] Corsini, P.: Contributo alla teoria degli ipergruppi, pp. 1–22. Atti Soc. Pelor. Sc. Mat. Fis. Nat, Messina (1980)
[35] Corsini, P.: Sur les homomorphismes d'hypergroupes, vol. 52, pp. 117–140. Rend. Sem. Univ, Padova (1974)
[36] Corsini, P.: A new connection between Hypergroups and Fuzzy Sets. Southeast Asian Bull. Math. 27, 221–229 (2003)
[37] Corsini, P.: Hyperstructures associated with fuzzy sets endowed with two membership functions. J. Combin. Inf. Syst. Sci. 31(1–4), 275–282 (2006)
[38] Corsini, P., Cristea, I.: Fuzzy grade of i.p.s. hypergroups of order less than or equal to 6. PU. M. A. 14(4), 275–288 (2003)
[39] Corsini, P., Cristea, I.: Fuzzy grade of i.p.s. hypergroups of order 7. Iran. J. Fuzzy Syst. 1(2), 15–32 (2004)
[40] Corsini, P., Cristea, I.: Fuzzy sets and non complete 1-hypergroups. An. Şt. Univ. Ovidius Constanţa Ser. Mat. 13(1), 27–54 (2005)
[41] Corsini, P., Leoreanu, V.: Join Spaces Associated with Fuzzy Sets. J. Combin. Inform. Syst. Sci. 20(1-4), 293–303 (1995)
[42] Corsini, P., Leoreanu-Fotea, V.: On the grade of a sequence of fuzzy sets and join spaces determined by a hypergraph. Southeast Asian Bull. Math. 34(2), 231–242 (2010)

[43] Corsini, P., Leoreanu-Fotea, V., Iranmanesh, A.: On the sequence of hypergroups and membership functions determined by a hypergraph. J. Mult.-Valued Logic Soft Comput. 14(6), 565–577 (2008)
[44] Corsini, P., Mahjoob, R.: Multivalued functions, fuzzy subsets and join spaces. Ratio Math 20, 1–41 (2010)
[45] Corsini, P., Tofan, I.: On fuzzy hypergroups. Pure Math. Appl. 8(1), 29–37 (1997)
[46] Corsini, P.: Hyperstructures associated with ordered sets. Bull. Greek Math. Soc. 48, 7–18 (2003)
[47] Corsini, P., Leoreanu, V.: Applications of Hyperstructures Theory. Advances in Mathematics. Kluwer Academic Publisher (2003)
[48] Corsini, P., Leoreanu, V.: Hypergroups and binary relations. Algebra Universalis 43, 321–330 (2000)
[49] Corsini, P., Leoreanu, V.: About the heart of a hypergroup. Acta Univ. Carolinae 37, 17–28 (1996)
[50] Cristea, I.: Complete Hypergroups, 1-Hypergroups and Fuzzy Sets. An. Şt. Univ. Ovidius Constanţa Ser. Mat. 10(2), 25–38 (2002)
[51] Cristea, I.: A property of the connection between fuzzy sets and hypergroupoids. Ital. J. Pure Appl. Math. 21, 73–82 (2007)
[52] Cristea, I.: On the fuzzy subhypergroups of some particular complete hypergroups (I). World Appl. Sci. J. 7, 57–63 (2009)
[53] Cristea, I.: On the fuzzy subhypergroups of some particular complete hypergroups (II). In: Proc. Tenth Internat. Congress on Algebraic Hyperstructures and Applications, pp. 141–149. Czech Republic, University of Defence, Brno (2008)
[54] Cristea, I.: Hyperstructures and fuzzy sets endowed with two membership functions. Fuzzy Sets and Systems 160, 1114–1124 (2009)
[55] Cristea, I.: About the fuzzy grade of the direct product of two hypergroupoids. Iran. J. Fuzzy Syst. 7(2), 95–108 (2010)
[56] Cristea, I., Davvaz, B.: Atanassov's intuitionistic fuzzy grade of hypergroups. Inform. Sci. 180, 1506–1517 (2010)
[57] Das, P.S.: Fuzzy groups and level subgroups. J. Math. Anal. Appl. 84, 264–269 (1981)
[58] Davvaz, B.: On H_v-rings and fuzzy H_v-ideals. J. Fuzzy Math. 6(1), 33–42 (1998)
[59] Davvaz, B.: Product of fuzzy H_v-subgroups. J. Fuzzy Math. 8(1), 43–51 (2000)
[60] Davvaz, B.: Interval-valued fuzzy ideals in a hyperring. Ital. J. Pure Appl. Math. 10, 117–124 (2001)
[61] Davvaz, B.: Product of fuzzy H_v-ideals in H_v-rings. Korean J. Comp. Appl. Math. 8(3), 685–693 (2001)
[62] Davvaz, B.: Fuzzy fundamental groups. Pure Math. Appl. 14(1-2), 15–19 (2003)
[63] Davvaz, B.: T_H and S_H-interval-valued fuzzy H_v-subgroups. Indian J. Pure Appl. Math. 35(1), 61–69 (2004)
[64] Davvaz, B.: Characterizations of sub-semihypergroups by various triangular norms. Czechoslovak Math. J. 55(4), 923–932 (2005)
[65] Davvaz, B.: On connection between uncertainty algebraic hyperstructures and probability theory. International Journal of Uncertainty, Fuzziness and Knowledge-Based Systems 13(3), 337–345 (2005)

[66] Davvaz, B.: Polygroup Theory and Related Systems, World Scientific (2013)
[67] Davvaz, B.: Rough subpolygroups in a factor polygroup. J. Intell. Fuzzy Syst. 17, 613–621 (2006)
[68] Davvaz, B.: Fuzzy weak polygroups. In: Algebraic hyperstructures and applications (Alexandroupoli-Orestiada, 2002), Spanidis, Xanthi, pp. 127–135 (2003)
[69] Davvaz, B.: A brief survey of the theory of H_v-structures. In: Algebraic hyperstructures and applications (Alexandroupoli-Orestiada, 2002), Spanidis, Xanthi, pp. 39–70 (2003)
[70] Davvaz, B.: Rough polygroups. Ital. J. Pure Appl. Math. 12, 91–96 (2002)
[71] Davvaz, B.: F-approximations in polygroups. Int. Math. J. 2, 761–765 (2002)
[72] Davvaz, B.: On polygroups and weak polygroups. Southeast Asian Bull. Math. 25, 87–95 (2001)
[73] Davvaz, B.: Polygroups with hyperoperators. J. Fuzzy Math. 9, 815–823 (2001)
[74] Davvaz, B.: TL-subpolygroups of a polygroup. Pure Math. Appl. 12, 137–145 (2001)
[75] Davvaz, B.: On polygroups and permutation polygroups. Math. Balkanica (N.S.) 14, 41–58 (2000)
[76] Davvaz, B.: Fuzzy H_v-groups. Fuzzy Sets and Systems 101, 191–195 (1999)
[77] Davvaz, B.: Fuzzy and anti fuzzy subhypergroups. In: Proc. International Conference on Intelligent and Cognitive Systems, pp. 140–144 (1996)
[78] Davvaz, B.: On H_v-subgroups and anti fuzzy H_v-subgroups. Korean J. Comp. Appl. Math. 5(1), 181–190 (1998)
[79] Davvaz, B.: On H_v-groups and fuzzy homomorphisms. J. Fuzzy Math. 9(2), 271–278 (2001)
[80] Davvaz, B., Corsini, P.: Generalized fuzzy sub-hyperquasigroups of hyperquasigroups. Soft Computing 10(11), 1109–1114 (2006)
[81] Davvaz, B., Corsini, P.: On (α, β)-fuzzy H_v-ideals of H_v-rings. Iran. J. Fuzzy Syst. 5(2), 35–47 (2008)
[82] Davvaz, B., Corsini, P., Leoreanu-Fotea, V.: Fuzzy n-ary subpolygroups. Comput. Math. Appl. 57, 141–152 (2009)
[83] Davvaz, B., Leoreanu-Fotea, V.: Hyperring Theory and Applications. International Academic Pres (Hadronic Press, Inc.), USA (2007)
[84] Davvaz, B., Karimian, M.: On the γ_n-complete hypergroups and K_H hypergroups. Acta Math. Sin (Engl. Ser.) 24, 1901–1908 (2008)
[85] Davvaz, B., Zhan, J., Shum, K.P.: Generalized fuzzy polygroups endowed with interval valued membership functions. J. Intell. Fuzzy Syst. 19, 181–188 (2008)
[86] Davvaz, B., Karimian, M.: On the γ_n^*complete hypergroups. European J. Combin. 28, 86–93 (2007)
[87] Davvaz, B., Corsini, P.: Generalized fuzzy polygroups. Iran. J. Fuzzy Syst. 3, 59–75 (2006)
[88] Davvaz, B., Majumder, S.K.: Atanassov's intuitionistic fuzzy interior ideals of Γ-semigroups. UPB Scientific Bulletin, Series A: Applied Mathematics and Physics 73(3), 45–60 (2011)
[89] Davvaz, B., Vougiouklis, T.: n-ary hypergroups. Iran. J. Sci. Technol. Trans. A Sci. 30, 165–174 (2006)
[90] Davvaz, B., Iranmanesh, M.A.: Fundamentals of Group Theory (Persian). Yazd University (2005)

[91] Davvaz, B., Poursalavati, N.S.: On polygroup hyperrings and representations of polygroups. J. Korean Math. Soc. 36, 1021–1031 (1999)
[92] Davvaz, B., Hassani Sadrabadi, E., Cristea, I.: Atanassov's intuitionistic fuzzy grade of i.p.s. hypergroups of order less than or equal to 6. Iran. J. Fuzzy Syst. 9(4), 71–97 (2012)
[93] Davvaz, B., Hassani Sadrabadi, E., Cristea, I.: Atanassov's intuitionistic fuzzy grade of i.p.s. hypergroups of order 7. J. Mult.-Valued Logic Soft Comput. 20(5-6), 467–506 (2013)
[94] Davvaz, B., Hassani Sadrabadi, E., Cristea, I.: Atanassov's intuitionistic fuzzy grade of the complete hypergroups of order less than or equal to 6. Hacet. J. Math. Stat. (in printing)
[95] Davvaz, B., Hassani Sadrabadi, E., Leoreanu-Fotea, V.: Atanassov's intuitionistic fuzzy grade of a sequence of fuzzy sets and join spaces determined by a hypergraph. J. Intell. Fuzzy Syst. 23(1), 9–25 (2012)
[96] De Salvo, M.: Feebly canonical hypergroups. Graphs, designs and combinatorial geometries (Catania, 1989), J. Combin. Inform. System Sci. 15, 133–150 (1990)
[97] De Salvo, M.: K_H-hypergroups (Italian). Atti Sem. Mat. Fis. Univ. Modena 31, 112–122 (1982)
[98] De Salvo, M., Lo Faro, G.: On the n^*-complete hypergroups. Discrete Math 208/209, 177–188 (1999)
[99] Dietzman, A.P.: On the multigroups of complete conjugate sets of elements of a group, C. R (Doklady). Acad. Sci. URSS (N.S.) 49, 315–317 (1946)
[100] Dramalidis, A.: Dual H_v-rings. Rivista Mat. Pura Appl. 17, 55–62 (1996)
[101] Dresher, M., Ore, O.: Theory of Multigroups. Amer. J. Math. 60, 705–733 (1938)
[102] Dubois, D., Gottwald, S., Hajek, P., Kacprzyk, J., Prade, H.: Terminological difficulties in fuzzy set theory–The case of "Intuitionistic Fuzzy Sets". Fuzzy Sets and Systems 156, 485–491 (2005)
[103] Dudek, W.A., Davvaz, B., Jun, Y.B.: On intuitionistic fuzzy subhyperquasigroups of hyperquasigroups. Inform. Sci. 170, 251–262 (2005)
[104] Eaton, J.E.: Theory of cogroups. Duke Math. J. 6, 101–107 (1940)
[105] Falcone, G.: On finite strongly canonical hypergroups. Pure Math. Appl. 11, 571–580 (2000)
[106] Feng, Y., Jiang, Y., Leoreanu-Fotea, V.: On the grade of a sequence of fuzzy sets and join spaces determined by a hypergraph II. Afr. Mat. 24, 83–91 (2013)
[107] Freni, D.: A new characterization of the derived hypergroup via strongly regular equivalences. Comm. Algebra 30, 3977–3989 (2002)
[108] Freni, D.: Une note sur le cur d'un hypergroupe et sur la clôture transitive β^* de β (French) [A note on the core of a hypergroup and the transitive closure β^* of β]. Riv. Mat. Pura Appl. 8, 153–156 (1991)
[109] Goguen, J.A.: L-fuzzy sets. J. Math. Anal. Appl. 18, 145–174 (1967)
[110] Harrison, D.K.: Double coset and orbit spaces. Pacific J. Math. 80, 451–491 (1979)
[111] Hasankhani, A., Zahedi, M.M.: Weak, dual weak and strong FH-homomorphisms of F-polygroups. Pure Math. Appl. 10, 279–291 (1999)
[112] Hasankhani, A., Zahedi, M.M.: On F-polygroups and fuzzy sub-F-polygroups. J. Fuzzy Math. 6, 97–110 (1998)

[113] Hasankhani, A., Zahedi, M.M.: Fuzzy sub-F-polygroups. In: Hyperstructures and their applications in criptography, geometry and uncertainty treatment (Chieti, 1994), vol. 12, pp. 35–44. Ratio Math (1997)

[114] Higman, D.C.: Coherent configurations. I. Ordinary representation theory. Geom. Dedicata 4, 1–32 (1975)

[115] Hosseini, S.N., Sh Mousavi, S., Zahedi, M.M.: Category of polygroup objects. Bull. Iranian Math. Soc. 28, 67–86 (2002)

[116] Ioulidis, S.: Polygroupes et certaines de leurs propriétés (French) [Polygroups and certain of their properties]. Bull. Soc. Math. Greece (N.S.) 22, 95–104 (1981)

[117] Iranmanesh, A., Babareza, A.H.: Transposition hypergroups and complement hypergroups. Algebraic hyperstructures and applications, 41–48 (2004)

[118] Iranmanesh, A., Iradmusa, M.N.: The combinatorial and algebraic structure of the hypergroup associated to a hypergraph. J. Mult.-Valued Logic Soft Comput. 11, 127–136 (2005)

[119] Jantosciak, J.: Transposition hypergroups: noncommutative join spaces. J. Algebra 187, 97–119 (1997)

[120] Jantosciak, J., Massouros, C.G.: Strong identities and fortification in transposition hypergroups, Algebraic hyperstructures and applications. J. Discrete Math. Sci. Cryptogr. 6, 169–193 (2003)

[121] Jantosciak, J.: Homomorphisms, equivalences and reductions in hypergroups. Riv. Mat. Pura Appl. 9, 23–47 (1991)

[122] Karimian, M., Davvaz, B.: On the γ-cyclic hypergroups. Comm. Algebra 34, 4579–4589 (2006)

[123] Kehagias, A.: An example of L-fuzzy join space, Tomo LI. Rendiconti Circ. Mat. Palermo, Serie II, pp. 503–526 (2002)

[124] Kehagias, A.: Some remarks on the lattice of fuzzy intervals. Inform. Sci. 181, 1863–1873 (2011)

[125] Kehagias, A., Konstantinidou, M.: Lattice-ordered join space: an Applications-Oriented Example (2001) (unpublished manuscript)

[126] Klir, G.J., Folger, T.A.: Fuzzy sets, uncertainty, and information. Prentice-Hall International, Inc (1988)

[127] Konstantinidou, M., Serafimidis, K.: Sur les filets des hypergroupes canoniques strictement réticulés (French) [Threads of strictly lattice-ordered canonical hypergroups]. Riv. Mat. Univ. Parma 13(4), 67–72 (1987)

[128] Konguetsof, L., Vougiouklis, T., Kessoglides, M., Spartalis, S.: On cyclic hypergroups with period. Acta Univ. Carolin. Math. Phys. 28, 3–7 (1987)

[129] Koskas, M.: Groupoides, demi-hypergroupes et hypergroupes (French). J. Math. Pures Appl. 49(9), 155–192 (1970)

[130] Krasner, M.: A class of hyperrings and hyperfields. Internat. J. Math. Math. Sci. 6, 307–311 (1983)

[131] Krasner, M.: Approximation des corps values complets de caracteristique $p \neq 0$ par ceux de caracteristique 0, Colloque d Algebre Superieure (Brussels, 1956), pp. 129–206. Centres Belge Rech. Math, Louvain (1957)

[132] Kumar, I.J., Saxena, P.K., Yadav, P.: Fuzzy normal subgroups and fuzzy quotient. Fuzzy Sets and Systems 46, 121–132 (1992)

[133] Kuroki, N.: On fuzzy semigroups. Inform. Sci. 53, 203–236 (1991)

[134] Leoreanu, V.: New results on the hypergroups homomorphisms. J. Inform. Optim. Sci. 20, 287–298 (1999)

[135] Leoreanu, V.: About the simplifiable cyclic semihypergroup. Ital. J. Pure Appl. Math. 7, 69–76 (2000)
[136] Leoreanu-Fotea, V., Davvaz, B.: Fuzzy hyperrings. Fuzzy Sets and Systems 160, 2366–2378 (2009)
[137] Liu, W.J.: Fuzzy Invariant subgroups and fuzzy ideals. Fuzzy Sets and Systems 8, 133–139 (1982)
[138] Lyndon, R.C.: Relation algebras and projective geometries. Michigan Math. J. 8, 21–28 (1961)
[139] Maddux, R.: Embedding modular lattices into relation algebras. Algebra Universalis 12, 242–246 (1981)
[140] Marty, F.: Sur une généralization de la notion de groupe. In: 8th Congress Math, pp. 45–49. Scandenaves, Stockholm (1934)
[141] Mashinchi, M., Salili, S.H., Zahedi, M.M.: Lattice structure on fuzzy subgroups. Bull. Iranian Math. Soc. 18(2), 17–29 (1992)
[142] Massouros, C.G.: Quasicanonical hypergroups. In: Algebraic hyperstructures and applications (Xánthi, 1990), pp. 129–136. World Sci. Publ, Teaneck (1991)
[143] Massouros, C.G.: Canonical and join hypergroups. An. Ştiinţ. Univ. Al. I. Cuza Iaşi. Mat (N.S.) 42, 175–186 (1996)
[144] McMullen, J.R., Price, J.F.: Reversible hypergroups. Rend. Sem. Mat. Fis. Milano 47, 67–85 (1977)
[145] McMullen, J.R., Price, J.F.: Duality for finite abelian hypergroups over splitting fields. Bull. Austral. Math. Soc. 20, 57–70 (1979)
[146] Migliorato, R.: On the complete hypergroups. Riv. Mat. Pura Appl. 12, 21–31 (1994)
[147] Mittas, J.: Hypergroupes canoniques. Math. Balkanica, Beograd 2, 165–179 (1972)
[148] Mordeson, J.N., Bhutani, K.R., Rosenfeld, A.: Fuzzy Group Theory. STUDFUZZ. Springer (2005)
[149] Mousavi, S.S., Zahedi, M.M., Hosseini, S.N.: Category of topological polygroup objects. Pure Math. Appl. 14, 121–140 (2003)
[150] Nguyen, H.T., Walker, E.A.: A first course of Fuzzy Logic. CRC Press, Boca Raton (1997)
[151] Nieminen, J.: Join space graphs. J. Geom. 33, 99–103 (1988)
[152] Pu, P.-M., Liu, Y.-M.: Fuzzy topology I. J. Math. Anal. Appl. 76, 571–599 (1980)
[153] Park, J., Chung, S.C.: An algorithm to compute some H_v-groups. Korean J. Comp. & Appl. Math. 7(2), 433–453 (2000)
[154] Pelea, C.: About a category of canonical hypergroups. Ital. J. Pure Appl. Math. 7, 157–166 (2000)
[155] Prenowitz, W.: Spherical geometries and multigroups. Canadian J. Math. 2, 100–119 (1950)
[156] Prenowitz, W.: Projective geometries as multigroups. Amer. J. Math. 65, 235–256 (1943)
[157] Prenowitz, W., Jantosciak, J.: Geometries and join spaces. J. Reine Angew. Math. 257, 100–128 (1972)
[158] Rosenberg, I.G.: Hypergroups induced by paths of a direct graph. Ital. J. Pure Appl. Math. 4, 133–142 (1998)
[159] Rosenberg, I.G.: Hypergroups and join spaces determined by relations. Ital. J. Pure Appl. Math. 4, 93–101 (1998)

[160] Rosenfeld, A.: Fuzzy groups. J. Math. Anal. Appl. 35, 512–517 (1971)
[161] Roth, R.L.: Character and conjugacy class hypergroups of a finite group. Ann. Mat. Pura Appl. 105(4), 295–311 (1975)
[162] Roth, R.L.: On derived canonical hypergroups. Riv. Mat. Pura Appl. 3, 81–85 (1988)
[163] Rotman, J.J.: An introduction to the theory of groups, 4th edn. Graduate Texts in Mathematics, vol. 148. Springer, New York (1995)
[164] Schweigert, D.: Congruence relations of multialgebras. Discrete Math 53, 249–253 (1985)
[165] Schweizer, B., Sklar, A.: Statistical metric spaces. Pacific J. Math. 10, 313–334 (1960)
[166] Sen, M.K., Ameri, R., Chowdhury, G.: Fuzzy hypersemigroups. Soft Computing 12(9), 891–900 (2008)
[167] Serafimidis, K.: Sur les hypergroupes canoniques ordonnés et strictement ordonnés (French) (Ordered and strictly ordered canonical hypergroups). Rend. Mat. 6(7), 231–238 (1986)
[168] Serafimidis, K., Konstantinidou, M., Mittas, J.: Sur les hypergroupes canoniques strictement réticulés (French) (On strictly lattice-ordered canonical hypergroups). Riv. Mat. Pura Appl. 2, 21–35 (1987)
[169] Serafimidis, K., Kehagias, A., Konstantinidou, M.: The L-fuzzy Corsini join hyperoperation. Ital. J. Pure Appl. Math. 12, 83–90 (2002)
[170] Spartalis, S., Vougiouklis, T.: The fundamental relations on H_v-rings. Riv. Mat. Pura Appl. 14, 7–20 (1994)
[171] Spartalis, S.: Quotients of $P - H_v$-rings, New frontiers in hyperstructures(Molise, 1995). Ser. New Front. Adv. Math. Ist. Ric. Base, Hadronic Press, Palm Harbor (1996)
[172] Ştefănescu, M.: Some interpretations of hypergroups. Bull. Math. Soc. Sci. Math. Roumanie (N.S.) 49(97), 99–104 (2006)
[173] Ştefănescu, M., Cristea, I.: On the fuzzy grade of the hypergroups. Fuzzy Sets and Systems 159, 1097–1106 (2008)
[174] Sureau, Y.: On structure of cogroups. Discrete Math 155, 243–246 (1996)
[175] Sureau, Y.: Contribution a la theorie des hypergroupes operant transitivement sur un ensemble, These de Doctorate d'Etat, Universite de Clermont II (1980)
[176] Suzuki, M.: Group theory I, Translated from the Japanese by the author, Grundlehren der Mathematischen Wissenschaften (Fundamental Principles of Mathematical Sciences), vol. 247. Springer, Berlin (1982)
[177] Tallini, G.: On Steiner hypergroups and linear codes. In: Hypergroups, other multivalued structures and their applications, pp. 87–91. Univ. Studi Udine, Udine (1985)
[178] Tofan, I., Volf, C.: On some connections between hyperstructures and fuzzy sets. Ital. J. Pure Appl. Math. 7, 63–68 (2000)
[179] Utumi, Y.: On hypergroups of groups right cosets. Osaka Math. J. 1, 73–80 (1949)
[180] Varlet, J.C.: Remarks on distributive lattices. Bull. de l'Acad. Polonnaise des Sciences, vol. 11, pp. 1143–1147. Serie des Sciences Math. Astr. et Phys. XXIII (1975)
[181] Vougiouklis, T.: Convolutions on WASS hyperstructures. Combinatorics (Rome and Montesilvano, 1994). Discrete Math. 174, 347–355 (1997)

[182] Vougiouklis, T.: H_v-groups defined on the same set. Discrete Math 155, 259–265 (1996)
[183] Vougiouklis, T.: Some results on hyperstructures. Contemporary Math 184, 427–431 (1995)
[184] Vougiouklis, T.: A new class of hyperstructures. J. Combin. Inform. System Sci. 20, 229–235 (1995)
[185] Vougiouklis, T.: Hyperstructures and their Representations. Hadronic Press, Inc., Palm Harbor (1994)
[186] Vougiouklis, T.: Representations of hypergroups by generalized permutations. Algebra Universalis 29, 172–183 (1992)
[187] Vougiouklis, T.: The fundamental relation in hyperrings. In: The general hyperfield, Algebraic hyperstructures and applications (Xanthi, 1990), pp. 203–211. World Sci. Publishing, Teaneck (1991)
[188] Vougiouklis, T.: The very thin hypergroups and the S-construction, Combinatorics (Ravello, 1988), Mediterranean, Rende. Res. Lecture Notes Math, vol. 88(2), pp. 471–477 (1991)
[189] Vougiouklis, T.: Groups in hypergroups. Annals Discrete Math 37, 459–468 (1988)
[190] Vougiouklis, T.: Representations of hypergroups by hypermatrices. Riv. Mat. Pura Appl. 2, 7–19 (1987)
[191] Vougiouklis, T.: Fundamental relations in hyperstructures. In: Proceedings of the 2nd Panhellenic Conference in Algebra and Number Theory (Thessaloniki, 1998). Bull. Greek Math. Soc. vol. 42, pp. 113–118 (1999)
[192] Vougiouklis, T., Spartalis, S.: P-cyclic hypergroups with three characteristic elements, Combinatorics 86 (Trento, 1986), North-Holland, Amsterdam. Ann. Discrete Math, vol. 37, pp. 421–426 (1988)
[193] Vougiouklis, T.: Representation of hypergroups. Hypergroup algebra, pp. 59–73. Ipergruppi, str. mult. appl. Udine, (1985)
[194] Vougiouklis, T.: Cyclicity in a special class of hypergroups. Acta Univ. Carolin. Math. Phys. 22, 3–6 (1981)
[195] Vougiouklis, T., Spartalis, S., Kessoglides, M.: Weak hyperstructures on small sets. Ratio Math 12, 90–96 (1997)
[196] Wall, H.S.: Hypergroups. Amer. J. Math. 59, 77–98 (1937)
[197] Yamak, S., Kazancı, O., Davvaz, B.: Divisible and pure intuitionistic fuzzy subgroups and their properties. Int. J. Fuzzy Syst. 10, 298–307 (2008)
[198] Yatras, C.N.: Subhypergroups of M-polysymmetrical hypergroups. In: Yatras, C.N. (ed.) Algebraic hyperstructures and applications (Iaşi, 1993), pp. 123–132. Hadronic Press, Palm Harbor (1994)
[199] Yatras, C.N.: M-polysymmetrical hypergroups. Riv. Mat. Pura Appl. 11, 81–92 (1992)
[200] Yatras, C.N.: Homomorphisms in the theory of the M-polysymmetrical hypergroups and monogene M-polysymmetrical hypergroups. In: Proceedings of the Workshop on Global Analysis, Differential Geometry and Lie Algebras (Thessaloniki, 1995), pp. 155–165. Geom. Balkan Press, Bucharest (1997)
[201] Yuan, X., Zhang, C., Ren, Y.: Generalized fuzzy groups and many-valued implications. Fuzzy Sets and Systems 138, 205–211 (2003)
[202] Zadeh, L.A.: Fuzzy sets. Information and Control 8, 338–353 (1965)
[203] Zadeh, L.A.: The concept of a linguistic variable and its application to approximate reasoning-1. Information and Control 8, 199–249 (1975)

[204] Zahedi, M.M.: Fuzzy α-cosets of fuzzy sub-F-polygroups. In: Proceedings of the Third Seminar on Fuzzy Sets and its Applications (Zahedan, 2002), pp. 186–191. Sistan Baluchestan Univ. Dept. Math, Zahedan (2002)

[205] Zahedi, M.M., Bolurian, M., Hasankhani, A.: On polygroups and fuzzy subpolygroups. J. Fuzzy Math. 3, 1–15 (1995)

[206] Zahedi, M.M., Hasankhani, A.: Some results on F-polygroups. In: New frontiers in hyperstructures (Molise, 1995). Ser. New Front. Adv. Math. Ist. Ric. Base, pp. 219–230. Hadronic Press, Palm Harbor (1996)

[207] Zahedi, M.M., Hasankhani, A.: F-polygroups I. J. Fuzzy Math. 4, 533–548 (1996)

[208] Zahedi, M.M., Hasankhani, A.: F-polygroups II. Inform. Sci. 89, 225–243 (1996)

[209] Zahedi, M.M., Mousavi, S.S., Hosseini, S.N.: On generalized polygroup objects. Pure Math. Appl. 12, 219–234 (2001)

[210] Zahedi, M.M., Torkzadeh, L., Borzooei, R.A.: Hyper I-algebras and polygroups. Quasigroups Related Systems 11, 103–113 (2004)

[211] Zhan, J., Yin, Y., Xu, Y.: A new view of fuzzy ideals in rings. Ann. Fuzzy Math. Inform. 2(1), 99–115 (2011)

Index

$(\in, \in \vee q)$-fuzzy H_v-subgroup 101
$(\in, \in \vee q)$-fuzzy subpolygroup 61
$(\in, \in \vee q)$-*fuzzy left* and *fuzzy right cosets of* P *determined by* x *and* μ 68
$(\in, \in \vee q)$-*fuzzy normal* 65
$(\in, \in \vee q)$-fuzzy left (right) H_v-ideal 135
F-hypergroup 126
F-inverse 79
F-polygroup 79
F-subpolygroup Notice that this condition 81
FP-relation 87
F_i-hypergroup 125
H_v-group 34
H_v-homomorphism 34, 36
H_v-ideal 36
H_v-ring 35
L-fuzzy subsets 41
P-hypergroup 22
R_3-reproducibility 174
T-product 114
TL-sub-semihypergroup 121
TL-subpolygroup 57
α-cut 41
π-fuzzy subsemihypergroup 122
f-invariant 41
m-decomposition 209
Łukasiewicz and Kleene-Dienes implications 73
F-*identity* element 79
t-*level cut* of μ 108

abelian 1
addition 8
adjacent 218
anti fuzzy H_v-subgroup 95
anti fuzzy reproduction axiom 95
automorphism 35

belong to 61
binary operation 1
boundary region 76

canonical hypergroup 28
Cartesian product 3
center 11
center of a group 4
closed (invertible, ultraclosed, conjugable) 23
closed on the left (on the right) 23
commutative 1
commutative H_v-ring 36
commutative (with unit element) 29
commutative ring 9
complement 40
complete hypergroup 208
conjugable on the right 23
connected 218
core 28
cyclic group 2

degree of membership 39, 106
degree of non-membership 106
dihedral group 2
direct hyperproduct 18
direct product 3, 53
division ring 10
Double coset algebra 18

endomorphism 6
epimorphism 13
exact 77

field 10
finite group 1

full conjugation 19
fundamental equivalence relation 35, 36
fundamental group 28
fundamental relation 28, 156
fundamental ring 36
fundamental theorem of homomorphisms 7
fundamental theorem of ring homomorphism 14
fuzzifying subpolygroup 73
fuzzy H_v-homomorphism 98
fuzzy H_v-ideal 127
fuzzy H_v-subgroup of H under a t-norm 113
fuzzy H_v-subgroup with thresholds 104
fuzzy closed 94
fuzzy congruence relation 88
fuzzy grade 176
fuzzy hypercongruence 154
fuzzy hypergroupoid 146
fuzzy hyperoperation 146
fuzzy hyperring 146
fuzzy ideal 127
fuzzy implication 72
fuzzy isomorphic 98
fuzzy isomorphism 98
fuzzy left t-coset 51
fuzzy left t-coset of μ, 45
fuzzy left (right) H_v-ideal with thresholds (r, s) of 141
fuzzy max determined 129
fuzzy normal subgroup 45
fuzzy point 79
fuzzy point with support x and *value* t 61
fuzzy regular relation 154
fuzzy right t-coset 51
fuzzy right t-coset of μ 45
fuzzy rough set 76
fuzzy rough subpolygroup 77
fuzzy semihypergroup 146
fuzzy similarity relation 87
fuzzy strongly regular relation 155
fuzzy sub F-polygroup 88
fuzzy subgroup 42
fuzzy subgroup of G *generated* by μ 44
fuzzy subhypergroup 91
fuzzy subpolygroup 47
fuzzy subpolygroup with thresholds 64
fuzzy subset 39

general linear group 2
generalized de Morgan lattice 168
generated by the set X 4
good homomorphism 21, 25
grade 39
group 1
group of integers modulo n 1
group of permutations 2

heart 28
homomorphism 6, 13, 25, 80
homomorphism (good homomorphism) 33
homomorphism of fuzzy hyperrings 152
homomorphism of type 2 21
homomorphism of type 3 21
homomorphism of type 4 21
hyperedges 218
hypergraph 218
hypergroup 22
hypergroupoid 17
hypergroups with partial scalar identities 216
hyperideal 30, 32
hyperideal generated by 33
hyperoperation 17
hyperoperation extracted 123
hyperproduct 113
hyperring 33

i-v fuzzy H_v-ideal 144
i-v fuzzy subset 143
i.p.s. hypergroups 216
ideal 11
ideal generated 12
identity (neutral) element 1
image 7, 41, 143
improper subgroup 4
inclusion fuzzy homomorphism 98
inclusion homomorphism 21, 25, 34, 36
infinite group 1
integral domain 10

interval-valued L-fuzzy subset 77
interval-valued fuzzy subset 143
intuitionistic fuzzy H_v-subgroup 107
intuitionistic fuzzy grade 192
intuitionistic fuzzy sets 106
invariant 160
inverse of μ 44
inverse image 7, 41, 144
invertible on the left (on the right) 23
isomorphic 13, 36
isomorphism 6, 13, 35

Jacobson radical 16

kernel 7, 14, 21
Klein 4-ring 9
Klein 4-group 2
Krasner hyperring 28

L-fuzzy-3 H_v-group 174
L-fuzzy-3 H_v-join space 174
L-fuzzy-3 hypergroup 174
L-fuzzy-3 $- p$ hypergroup 174
L-fuzzy-3 $- p$ join space 174
left 127, 144
left (right) hyperideal 30
left congruence 4
left coset 5
left fuzzy closed with respect to 94
left fuzzy coset 160
left fuzzy reproduction axiom 91
less than or equal to 35
level rough set 77
level set 43
lower 108
lower and *upper approximations* 74, 75
lower intuitionistic fuzzy grade 190

maximal ideal 14
monoid 1
monomorphism 13
multiplication 8
multiplication by components 3
multiplicative hyperring 30

natural homomorphism 13
normal 18, 30, 48, 57, 81, 88
normal subgroup 5

or H_v-subgroup 91

p-join hyperoperation 168
path 218
permutations 2
polygroup 17
Prenowitz algebras 18
prime 15
product 44, 58
proper ideal 12

quasi-coincident with 61
quasihypergroup 22
quaternion group 2
quotient group 47
quotient ring 13
quotient rough polygroup 76

radical of an ideal 16
regular (strongly regular) 25
regular fuzzy H_v-ideal 133
regular left fuzzy reproduction axiom 133
regular on the right (on the left) 25
regular right fuzzy reproduction axiom 133
right 127, 144
right anti fuzzy reproduction axiom 95
right congruence 4
right coset 5
right fuzzy closed with respect to 94
right fuzzy coset 160
right fuzzy reproduction axiom 91
ring 8
ring of Gaussian integers 9
ring of real quaternions 9
ring with unit element 9

semigroup 1
semihypergroup 17
strong homomorphism 21
strong fuzzy H_v-subgroup 115
strong fuzzy grade 176
strong fuzzy homomorphism 98
strong homomorphism 34, 36, 80
strong left fuzzy reproduction axiom 115
strong level set 48
strong right fuzzy reproduction axiom 115
stronger than 56

strongly distributive 30
strongly regular on the right (on the left) 25
subfuzzy hyperideal 153
subfuzzy hyperring 153
subgroup 4
subhypergroup 23
subhyperring 30, 32
subnormal 160
subpolygroup 18
subring 11
subsemihypergroup 17
supersemihypergroup 17
symmetric 160
symmetric group 2

t-norm 55
total hypergroup 22
trivial ideal 12

ultraclosed on the left (on the right) 23
unitary 32
upper 108
upper intuitionistic fuzzy grade 191

very good homomorphism 25

weak homomorphism 34, 36
weak identity 30
weak normal 81
weaker than 56

zero element 9
zero hyperideal 33
zero-divisor 10
zero-invariant 85

Printed in the United States
By Bookmasters